Western North Carolina Farm & Garden Calendar

The Farming and Gardening Survival Book

Good for all eastern states in Hardiness Zones 5, 6, 7.

Can be used in most hardiness zones by adding or subtracting months depending on your climate.

**Be prepared.
You need this book.**

Know what to do every month of the year.

Each month has 5 sections:
Garden-Maintenance
Garden-Plant
Garden-Harvest
Greenhouse
Farm Animals

After all the months, there are How-To Guides:
Garden Tips
How to Save Seeds
Plant Families
Plant Health

At the end are 3 types of Index:
Plants by Common Name
Plants by How Used
General Index

Western North Carolina Farm & Garden Calendar
The Farming and Gardening Survival Book

Copyright ©2012-2013 Nancy Shirley. All rights reserved.
www.nantahala-farm.com   farm@nantahala-farm.com

Published by Nancy Shirley. Printed in the United States.
ISBN: 978-0-9890851-0-6

What is Nantahala?
    Nantahala is a Cherokee word that means "Land of the Noonday Sun". In some areas in the Nantahala National Forest in Western North Carolina the sun only reaches the bottom of deep gorges when it is directly overhead at noon.

    The Appalachian Trail and the Bartram Trail go through this forest. The Nantahala area includes the Nantahala River, Nantahala Gorge, and Lake Nantahala. They are popular for fishing, canoeing, kayaking, white water rafting, mountain biking, and hiking. The Nantahala Outdoor Center is nationally known as a popular recreation resort in the National Forest.

Permission Required to Reproduce
No part of this book may be reproduced or transmitted in any form by any means, graphic, electronic, or mechanical, including photocopying, recording, taping or by any information storage or retrieval system, without permission in writing from the publisher.

No Liability is Assumed
No liability is assumed with respect to the use of the information contained herein. Although every precaution has been taken in the preparation of this book, the publisher and author assume no responsibility for errors or omissions. Neither is any liability assumed for damages resulting from the use of information contained herein.

A Competent Professional Should Be Consulted
This publication contains the opinions and ideas of its author. It is intended to provide helpful and informative material on the subject matter covered. It is sold with the understanding that the author and publisher are not engaged in professional services in this book. If the reader requires personal assistance or advice, a competent professional should be consulted.

Consult Your Health Care Specialist
This book is not intended as a substitute for the medical advice of physicians or health care specialists. The reader should regularly consult a physician or health care specialist in matters relating to his/her health and particularly with respect to any symptoms that may require diagnosis or medical attention

The Reader is Responsible for Use of Plants
By purchasing and using this book, the reader agrees to assume responsibility for identification of the plants he or she consumes. The reader also assumes responsibility for collecting the proper parts at the proper time and preparing them in an appropriate fashion.

The Reader is Responsible When Consuming New Foods
Because of food allergies, intolerances, or other chemical peculiarities, any person may have an adverse reaction to a food that is generally known to be wholesome. By choosing to consume new foods, the reader accepts these risks and assumes responsibility for any negative consequences associated with them.

This Book Covers All Aspects of Farming and Gardening

Besides information about how to garden, it also includes farm animals, foraging, food storage, and food preparation. Plus garden tips, folklore, seed saving, seed starting, herbs, plant families, and plant health. Every plant includes its botanical name along with common names. After each description of a plant or animal, there are recommended books and where else in the Calendar that plant or animal is mentioned.

Part One: Month to Month Guidance

Each month explains what you need to do to stay on top of your food production plan. At the beginning of every month the Calendar lists an overview of what needs to be done.

Garden-Maintenance
Then there is a section each month on Garden-Maintenance. How to maintain your plants and animals so they stay in the best condition. For instance, in February you need to prune your fruit and nut trees. It goes into detail about how to properly prune each type of tree.

Garden-Plant
Each month has a section for planting. For instance, in March Garden-Plant discover ideas for predicting frost and reducing frost damage. Learn what cool weather cover crops and grains need to be planted such as clover, fava beans, fescue, flax and oats. See how to sow and care for them.

Garden-Harvest
Each month has a harvest section. There is emphasis on the right time to harvest and how to store food. For instance, in May Garden-Harvest it gives details about making hay without large machinery, how to harvest grain by hand, and how to test/dry/store grain. Learn how to thresh, winnow and dehull grain. Find out how to test for correct moisture content of grain before storing it.

Greenhouse, Hoop House or Cold Frame
Every month has a section on Greenhouse, Hoop House or Cold Frame.
For instance, in March Greenhouse learn how to start seedlings and sprouts for asparagus, broccoli, broccoli raab, cabbage, celery, chrysanthemum, ground cherry, kohlrabi, pepper, sweet potato, tomato, and tomatillo. Details on how to pre-sprout white/red/purple potatoes for greater productivity.

Farm Animals
Every month has what you need to do for your animals. Included are all the basic farm animals such as bees, cats, cattle, chickens, dogs, ducks, goats, homing pigeons, horses, pigs, rabbits, sheep and turkey. In April Farm Animals get details about how to set up nesting areas for broody hens. How to break up a broody hen. Also natural formulas for worming animals.

Part Two: "How To Guides" and Index

There are 4 Garden Guides and a 3-part index. The Guides include: Garden Tips, How to Save Seed, Plant Families, and Plant Health.

One index is alphabetical by common plant name. The second index is by how plants are used. The third index is a general index of all farm/garden topics.

Part Two: "How To Guides" and Index

Garden Tips
Appalachian Folklore
How Much is a Bushel
Companion Planting
Garden Seeders / Spreaders
How to Stratify Seeds
How to Make Potting Mix
How to Use Soil Blocks
Inoculating Seeds
Prevent Seedling Damping Off
Thinning Seedlings and Plants
How and Why to Fertilize
Rock Dusts
Soil pH
Crop Rotation and Cover Crops
Pasture
Special Forage Crops
Weeds and Soil Type
Prevent Fungal Disease
Propagation by Layering/Rooting
Propagation by Stem Cuttings
Make Your Own Rooting Hormone

How to Save Seeds
Open and Hybrid Pollination
Inbreeding / Outbreeding
Pollination Methods
Annual / Biennial / Perennial
Genetic Vigor and Inbreeding
How Difficult Seed is to Save
Harvesting / Preparing Seed
Storing Seed for Planting
Years Seeds Remain Viable

Plant Families
Care, Diseases, Pests and Seed Saving

Families Included:
Amaranthaceae, Amaryllidaceae, Brassicaceae/Cruciferae, Chenopodiaceae, Compositae, Convolvulaceae, Cucurbitaceae, Gramineae/Poaceae, Labiatae, Leguminosae/Fabaceae, Liliaceae, Malvaceae, Polygonaceae, Portulacaceae, Solanaceae, Tetragoniaceae, and Umbelliferae/Apiaceae.

Ways to Improve Plant Health
Nutrient Deficiency Symptoms
Disease Symptoms
Fungi Disease
Bacterial Disease
Viral Disease
Insects & Nematodes
Insect Damage Symptoms
Greenhouse Pests and Disease
Products to Control Insects
Large Animal Pests
Sprays and Dusts
Make Traps for Flying Insects
Make Traps for Crawling Insects
Make Traps for Slugs and Snails
Row Covers & Crop Protectors
Make Organic Insect/Pest Sprays
Organic Fungicide Sprays

3 Types of Index

1. Alphabetical Index by Common Plant Name
Fruits, Grain, Herbs, Nuts, Seeds and Vegetables

2. Plants Indexed by How Used
Fruit- Annual/Biennial
Fruits & Nuts- Perennial
Grain, Grass, Seeds, Cover Crops
Herbs/Flowers- Annual/Biennial
Herbs/Flowers- Perennial
Leafy Greens- Annual/Biennial
Leafy Greens- Perennial
Root Crops
Tomato and Related
Vegetables- Annual/Biennial
Vegetables- Perennial

3. General Index of All Farm and Garden Topics

WESTERN NORTH CAROLINA FARM & GARDEN CALENDAR
Farming & Gardening Survival Book for Hard Times

Good for Tennessee, Georgia, South Carolina, Virginia and most eastern states. Includes all of North Carolina. USDA Hardiness Zones 5-7   Appalachian Mountains   www.nantahala-farm.com   By Nancy Shirley.

To every thing there is a season, and a time to every purpose under the heaven:
A time to be born, and a time to die; a time to plant, and a time to pluck up that which is planted.

Do you wonder what you should do each month to keep your garden and farm productive? This month by month guide shows you when to plant, harvest and maintain your garden, greenhouse and farm. It can be used in most hardiness zones by adding or subtracting months.

Each month is divided into 5 sections: Garden-Maintenance, Garden-Plant, Garden-Harvest, Greenhouse, and Farm Animals. This means a plant is medicinal according to folklore or old-time medicine. (For all medical needs, contact your doctor. This book is for entertainment only.)

At the end of this manual are how to guides: **Garden Tips, Plant Families, How to Save Seed, and Plant Health.** Plus an **index of fruits and vegetables** and a regular index.

Learn how to sow, grow and harvest:

Cool Season Crops: Beet, broccoli, brussels sprouts, cabbage, carrot, cauliflower, celery, chard, collard, kale, kohlrabi, lettuce, mustard, parsley, parsnip, pea, potato, rutabaga, spinach, turnip and more.

Warm Season Crops: Bean, corn, cucumber, lima bean, melon, pumpkin, squash and more.

Hot Season Crops: Eggplant, pepper, okra, sweet potato, tomato, watermelon and more.

Main crops such as potatoes and corn can be planted every year. Other crops can be planted every other year or every few years, and then dry, can or freeze enough to last several years. Or set up farmers markets to trade/sell/buy fruits, vegetables, seeds, plants, eggs and animals.

Many vegetables especially leafy greens withstand frost very well and can be harvested fresh for 8 months outside and all year in a greenhouse or cold frame.

This works best with **perennials** because they are the first greens to appear in spring and the last to go dormant in fall. Grow perennials such as chives, comfrey, dandelion, lambs quarters, mint, oregano, plantain, purslane, rosemary, sage, salad burnet, sorrel, stinging nettle, thyme, violets and winter savory.

Annual leafy greens that can be harvested during cold weather are arugula, claytonia, collard greens, creasy greens, kale, leeks, mache, mizuna, minutina, mustard greens and radicchio.

Root crops good for storing over the winter are: beets, carrots, celeriac, garlic, horseradish, jerusalem artichokes, kohlrabi, onions, parsnips, parsleyroot, potatoes, rutabaga, salsify, sweet potatoes, turnips, winter radish and others.

Above ground vegetables, nuts/seeds and grain good for storing over the winter are: acorns, amaranth, beans, buckwheat, cabbage, corn, flax, hazelnuts (filberts), hickory nuts, oats, peas, pumpkin, rye, sunflower, walnuts, wheat, winter squash and others.

For fruit over the winter, grow trees and bushes such as apples, blackberry, blueberry, elderberry, grapes, peach, pear, plum and raspberry. Most store for several months in a cool location. All can be dried, frozen or canned.

  **Crops Grown for Maximum Calories & High Yield Per Acre**

Food that has high calories per pound and a high yield include **burdock, garlic, jerusalem artichokes, potatoes, leeks, parsnips, salsify and sweet potatoes.** Potatoes are the #1 survival food. For survival farming at least 30% of your land should grow these high calorie root crops.

Beans, onions, peanuts, rutabaga, soybeans and turnips with tops are not as productive but are still very valuable. It may be possible to have 2 harvests of onions per year.

Root crops that are not as high calorie per area are beets, carrots, mangel beets, and radishes but are still very valuable. It is good to have a diversified diet.

If you have a lot of land, then **fruit trees/bushes, nut trees, sunflower seeds, winter squash, corn and other grains** are good. Grains are best if they are easy to thresh and winnow.

Read the book "How to Grow More Vegetables and Fruits, Nuts, Berries, Grains and Other Crops Than You Ever Thought Possible on Less Land Than You Can Imagine" by John Jeavons.)

 **Foods that Produce Carbon as a By-product for Other Farm Needs**

Carbon production in farming is low nitrogen by-products such as **fiber, straw, hay, dry stalks, and dry leaves** that can be used in compost, as cover crop, green manure (to be dug into soil), animal fodder, bedding for animals, mulch, and other uses. They include by-products from grains such as **amaranth, barley, cereal rye, corn, oats, quinoa, sorghum, triticale and wheat. Fava beans** are grown to maturity for the dry bean and for dry biomass (carbon). **Sunflowers and filberts (hazelnuts)** can also be raised for these purposes.

 **Old fashioned storage** such as drying, canning and root cellars are described for vegetables and fruits. **Seed saving** is included to save money on seed costs and to preserve heirloom and endangered varieties.

Some **foraging** (looking for food or provisions) is included. (For **good DVDs about foraging in Appalachia**, watch "Backyard Remedies" sold at The Whole Store in Murphy, North Carolina, www.wholestorenaturalfoods.com.)

 **Acres Needed to Grow Food:** FAO (Food and Agriculture Organization of the United Nations) states that the minimum amount of agricultural land needed for sustainable food production with a diversified diet (including meat) such as that eaten by North Americans and Western Europeans is half a hectare per person. (A hectare of land is about 2.47 acres so that is about **1.2 acres per person**.) This assumes adequate water supplies and decent growing land.

The absolute minimum land needed to support one person on a **restricted, vegetarian diet is 1/4 acre**. There can be no post-harvest waste, and farmers must know exactly when/how to plant, fertilize, irrigate and maintain crops.

Length of growing season, what is planted, richness of soil, weather and other factors strongly influence how much food can be grown. In some instances if properly managed, having animals can increase the productivity of land. **Permaculture** methods of growing increase the long-term sustainability of land. Properly managed orchards are an example of permaculture that leads to good sustainability and productivity.

Read "Gaia's Garden: A Guide To Home-Scale Permaculture" by Toby Hemenway and "Edible Forest Gardens" by Dave Jacke and Eric Toensmeier. For **life and farming wisdom in Appalachia**, read the Foxfire books. Read "Gardening When It Counts" by Steve Solomon and "The Resilient Gardener" by Carol Deppe.)

 The eastern 2/5 of **North Carolina** is a **coastal plain and tidewater.** Moving west, the next 2/5, about 200 miles wide, is a **piedmont plateau.** In the west, the land slopes upward from gentle to rugged rolling hills to the **southern Appalachian Mountains** containing the Blue Ridge and Great Smoky Mountains.

Preface p. 2

JANUARY
January 1 daylight is 9 hours 51 minutes. January 31 daylight is 10 hours 27 minutes.

Usually does not go below 0 degrees. Snow usually melts in a few days/weeks. Some warm days with highs in the 50s.

Garden- Maintenance

Log trees for **firewood** in the dormant season from **late autumn (October) to late winter (February).** (For details about **logging and splitting firewood,** see October Garden-Maintenance.)

USDA Hardiness Zones *Zone is lower at higher elevations.*

A **Hardiness Zone** is a geographic area where specific types of plants are capable of growing all year, as defined by weather (climate) conditions. This includes its ability to withstand the minimum temperatures of the zone. For example, a plant that is "hardy to zone 6" means it can **withstand a minimum temperature** of 0 to minus 10 degrees.
   Perennials in that zone are able to withstand those temperatures. **Annuals** that can not withstand those temperatures are planted only during warmer parts of the year.

Garden- Harvest

 Harvest after frost and through most of winter outside: **collard greens, creasy greens, kale, kohlrabi, leeks, sunchokes (jerusalem artichokes), turnips and others.**

Greenhouse, Hoop House or Cold Frame (January)

Harvest greenhouse crops but only when plants are **not frozen.** If you pick when frozen, they will be droopy when they thaw. Little or no watering needed in greenhouse or cold frame because plant growth is slow.

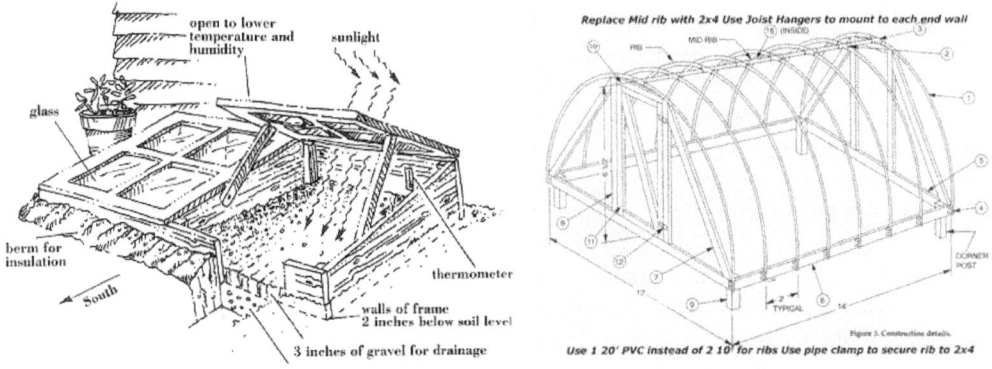

*The first image is a **cold frame**. The second image is a **hoop house**. It is covered with clear plastic.*
   A cold frame is an unheated, mini-greenhouse put in a sunny location, preferably **facing south**. It needs good drainage. It can be stand alone or built next to an existing building.

It is used to keep frost-hardy plants alive all **winter,** keep frost-sensitive plants alive longer in the **fall,** and start frost-sensitive plants earlier in the **spring**. It can be used to harden off seedlings in the spring. **Hardening off** is when seedlings from the house or greenhouse are put in a temporary location to get used to colder temperatures, more sun and more wind. (For more details about hardening off, see "Garden Tips" in back of this manual.)

Open on warm, sunny days so plants do not overheat. Close it at night and/or when the temperature is colder.

A cold frame is **easy to build.** The simplest design is just an old window put on top of some cinder blocks. Or the base can be built with wood. Make it 3-4 feet wide so you can reach the plants in the back.

Farm Animals (January)

 **Animal Bedding:** Every month add dolomite lime, rock phosphate, gypsum, and/or rock dusts (a powder available at quarries) to all manure in stalls, under cages, in animal pens, and in chicken coop. This **reduces odor** and is an easy way to add extra nutrients to the garden when move manure outside. Manure is high in nitrogen and phosphoric acid. (For more about **fertilizers**, see "Garden Tips" in the back of manual.)

Dolomite lime is a natural liming material which supplies 21% calcium and 12% magnesium.

Gypsum (a sulphate mineral high in calcium and sulphur) changes the structure of ammonia (nitrogen) so it is odorless and not lost to the air. **It reduces the odor in a barn or coop.**

Rock dusts such as rock phosphate make the manure less prone to smell and is good for the garden.

Rock Phosphate is a natural phosphate that has minor minerals and phosphorus.

Add **leaves, hay, straw and/or sawdust.**

(📖 Read "Soil Fertility and Animal Health, Albrecht Papers" by William A. Albrecht and Charles Walters.)

 Chicken, turkey and duck egg production is up a little. Peak egg production is **April and May**. Good egg production is **February through July.** Lowest egg production is **September through December.**

(For **chicken/duck/turkey breeds**, see February Farm Animals. For **incubators and brooding chicks,** see March Farm Animals. For **broody hens** see April Farm Animals. For **molting**, see September Farm Animals.)

 Feed animals kelp, alfalfa meal and other supplements. (For details see November Farm Animals.)

 **Worm** cattle, goats, horses, pigs, sheep, dogs, cats and other animals. Worm homing pigeons, turkeys, ducks and chickens with **garlic in water.** Put 5-6 cloves in a blender with several cups water. After blending, add to 3-4 gallons of water. Give to birds as only source of water for 24 hours.

(For **natural wormer recipes**, see April Farm Animals. For **mites and lice**, see March Farm Animals.)

 ## Honey Beekeeping in January

A **beekeeper** collects honey and other hive products (beeswax, propolis, pollen, royal jelly). Bees pollinate crops. **A beehive is a colony of bees that has 3 types of bees:** a queen bee who is usually the only breeding female; about 30,000-50,000 female worker bees; and male drones ranging from thousands in the spring to very few in winter.

In **January and February** the queen is surrounded by thousands of workers to stay warm. She may lay a few eggs. There is **little activity except on a warm day** (45-50 degrees or warmer) when workers take cleansing flights.

One beehive can consume up to 25 pounds of **stored honey** this month. Make sure they have at least 15 pounds (6 frames of capped honey in a shallow super, or 2-3 frames in a deep super). If not, add sugar or honey.

If there is snow, make sure the entrance to the hive is clear for proper ventilation.

Order bees and any equipment you will need later. Repair and maintain **bee boxes and frames.**

(For **harvesting honey**, see September Farm Animals.) (📖 Read "Beekeeping For Dummies" by Howland Blackiston.)

 # FEBRUARY
February 1 daylight is 10 hours 29 minutes. February 28 daylight is 11 hours 25 min.

First of 3 "hunger" months because food stored from the summer and fall may be getting low.
And hunting is difficult in harsh weather. Farmers Almanac says the February full moon is called **Hunger Moon**. Weather similar to January. End of February some rain, some snow.
Groundhog Day is February 2. If it is cloudy when it comes out of its burrow, then spring comes early. If it is sunny and it sees its shadow, cold winter weather continues for 6 more weeks. It is popular folklore from Germany.

 ## Garden- Maintenance  (February)

 Log trees for **firewood** in the dormant season from **late autumn (October) to late winter (February)**. (For details about **logging and splitting firewood,** see October Garden-Maintenance.)

 ### Apply Dormant Oil Spray to Fruit and Nut Trees (optional) in February or March

Control insects such as aphids, mites, pear psylla and scale insects by spraying **dormant or delayed dormant oil** on trees before leaves emerge or buds open. Applications of a dormant Superior or Supreme horticultural oil with an insecticide (malathion, endosulfan, or permethrin) controls these insects if applied during late winter or early spring.

Only apply if it is needed. You may only need the dormant oil but not the insecticide. Contact your County Extension office for more information.

Restrictions: Do not apply oil if a heavy freeze is expected since damage to the tree could occur. Oil sprays must have enough time to dry before freezing weather, usually at least 10 to 12 hours. **Do not use sulfur such as lime-sulfur or Bordeaux mix within 3-4 weeks of any oil spray** especially when temperatures are above 80 degrees.

Prune trees before spraying. Never spray insecticides when trees are in bloom. It kills the bees that pollinate the blossoms. Always read and follow label directions when using any pesticide. The amount of spray is based on the size of the tree. Sprays should be applied to thoroughly wet the branches until they drip.

Spray apple and pear trees with a mixture of dormant oil plus insecticide (permethrin or endosulfan) to control scale, aphids and mites. Mix 5 tablespoons Superior or Supreme dormant oil with an insecticide (according to label rates) in each gallon of spray.

You can do the dormant oil spray before or after the fungicide spray. Though you may want to do the dormant oil spray first (a month before the fungicide spray).

(For **more about oil sprays and insecticides**, see "Plant Health" in the back of this manual.)

 ### Apply Dormant Fungicide Sprays to Fruit and Nut Trees in February or March

Apply fungicides such as copper and sulfur in dormant season. Use as last resort when nothing else has worked. **Do not use within 3-4 weeks of dormant oil sprays.** Sulfur is usually not good for apricot trees. Do not apply once the blossoms have opened because it kills bees and other beneficial insects who pollinate the flowers.

Only apply if it is needed. Contact your County Extension office for more information.

(See **Fungi Disease** in "Plant Health" in back of this manual. For **Bordeaux Mix,** see How to Make Organic Fungicide Sprays and Solutions in "Plant Health" in back of this manual. For **Sulfur, Lime-Sulfur and Bordeaux Mix**, see Sprays and Dusts in "Plant Health".)

   ### Bramble Fruit Propagation in February:

Red Raspberry Propagation:
Can be propagated from suckers, root cuttings and by dividing roots. They produce small plants (suckers) near the mother plant that grow up from the roots. When dormant, divide mother plant with a sharp spade and replant.

Black and Purple Raspberry Propagation:
When dormant, divide the tip-rooted plant with a sharp spade and replant. (For **tip rooting raspber-**

ries, see July Garden-Maintenance. For **general tip rooting information,** see "Garden Tips" section.)
(See April Garden-Plant. For **pruning and more propagating brambles,** see October Garden-Maintenance.)

Prune Fruit / Berry / Nut Trees and Bushes when dormant:
Apple, apricot, Asian pear, blueberry, currant/gooseberry, elderberry, fig, grape, hardy kiwi, peach/nectarine, pear, plum, rosa rugosa, walnut. (Image below on right is fruit spur.)

(📖 Read "The Pruning Book" by Lee Reich.) (See April Garden-Plant. See below for details for each tree/bush.)

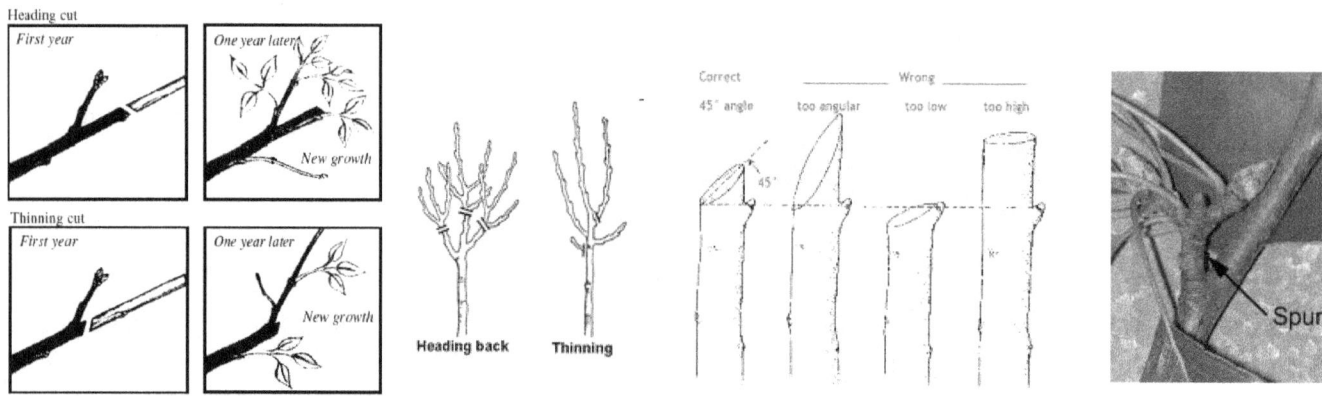

Heading Cuts and Thinning Cuts
There are 2 types of pruning cuts: (See first image above on left.)
Heading cuts stimulate growth of buds closest to the wound. The direction in which the top remaining bud is pointing determines the direction of new growth. Heading cuts stimulate rapid regrowth from buds below the cut. These vigorous shoots make shrubs bushier, but not smaller.
Thinning cuts remove branches at branch collar. They reduce density without stimulating regrowth.

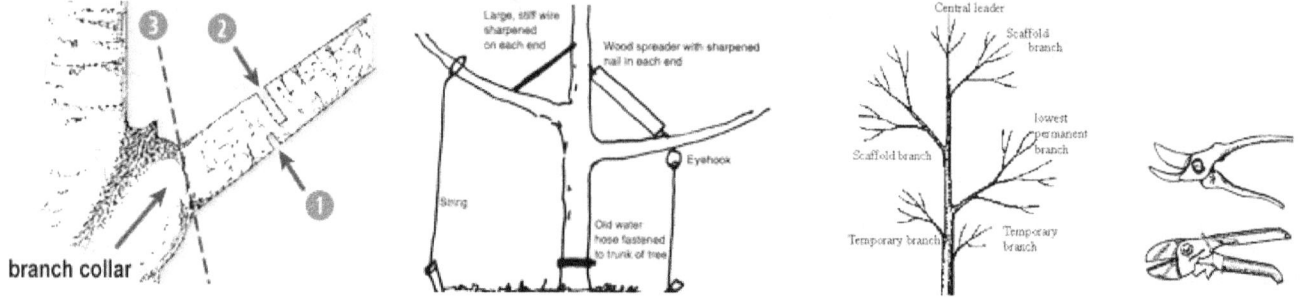

Above image on left. A basic pruning cut is 3 steps:
1. To prevent tearing, make first undercut about 10-18 inches from limb's attachment (depends on branch size).
2. To lessen weight of branch, make a second cut from the top, directly above or a few inches further out on limb.
*3. To make the last cut, remove the stub by cutting back to the **branch collar** or branch bark ridge.*
Pruning cuts should be made just outside the branch collar. The branch collar should not be damaged or removed.
*Avoid cutting into branch bark ridge (**branch collar**) where new tissue promotes wound closure.*
(Bark ridge forms in branch crotch and partially around stem from growth of stem and branch against one another.)
Above image second from left. Several ways to **spread scaffold branches** so they grow in proper direction. One way is to use an old garden hose wrapped around the branch with a rope inside it that is tied to a stake.
Above image third from left is parts of a tree. From top to bottom: temporary branch, lowest permanent branch, scaffold branch, and central leader.

 **Caring for Tools:** Keep tools **sharp**. This includes hoes, pruning shears, shovels and other tools.

Wash the blade with soap, then dry. Buff off rust with medium-coarse steel wool. Place the beveled edge of the blade facing you. Put a medium diamond hand file next to the blade at the **same angle as the bevel**. Draw file along blade towards tip. It usually takes 10-20 strokes. Then use a fine diamond hand file a few times on both sides.

Disinfect pruning tools in a bleach solution (1 part bleach to 9 parts water) after each cut. Hold the shears in the solution for at least 2 seconds. This is important if tree has disease so do not spread to other trees.

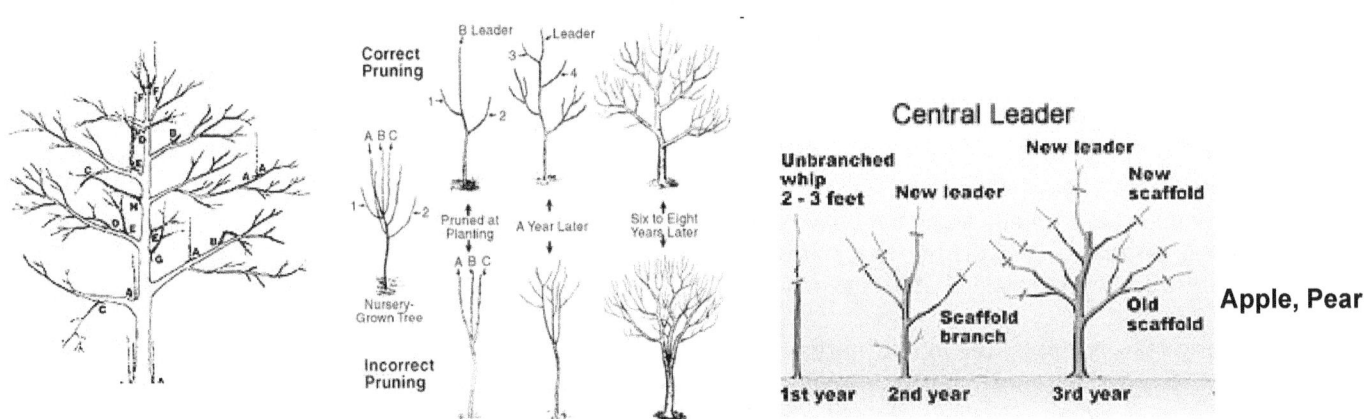

Above images are for pruning apple, asian pear, and pear trees (non-stone fruit trees).
Above image on left is where to prune: A. Suckers, B. Stubs or broken branches, C. Downward-growing branches, D. Rubbing or criss-crossing branches, E. Upward growing interior branches, F. Competing leaders, G. Narrow crotches, H. Whorls.

Apple Pruning in February: (See April Garden-Plant. For **thinning fruit**, see June Garden-Maintenance. For **more pruning,** see July Garden-Maintenance. See August/September Garden-Harvest with **harvest details in August.**) (Read "The Apple Grower: A Guide for the Organic Orchardist" by Michael Phillips.)

Apples produce fruit on spurs formed on branches 1 year old or more. Do not cut fruit spurs (spurs are 1 to 3 inches long, thick, stubby looking).

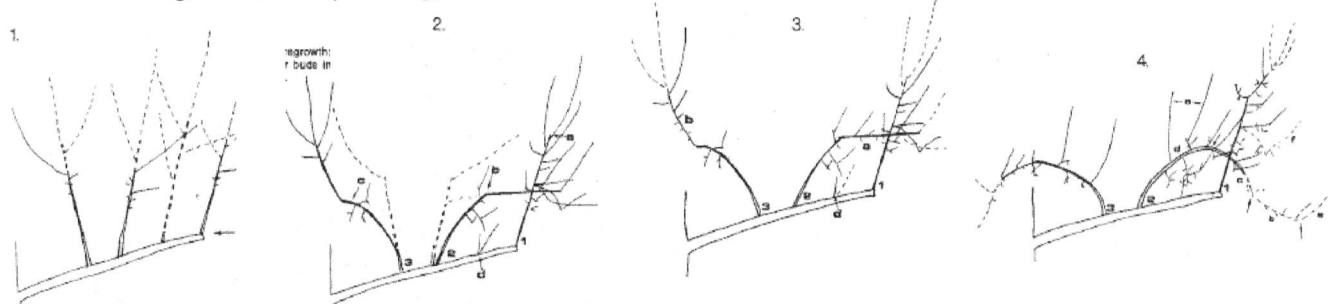

Above 4 images on left: Four steps in pruning a scaffold limb.
 1. Dashed lines are second season of growth. Undashed lines are first season of growth that may develop flower buds in second season. Remove excess shoots entirely and prune remaining ones to outward-growing side branches.
 2. Growth is 2 years old and branched. Dashed lines show wood to be removed.
 3. Dashed lines show what needs to be pruned. Cut back to a single upright shoot. This reduces growth of shoots below and encourages fruiting wood close to the secondary scaffold branch.
 4. The same limb 4 years after the first pruning. **a-** last year's growth **b-** 2-year-old parts **c-** 3-year-old which fruited previous summer **d-** 4-year-old parts. Remove upright shoots at e and elsewhere. Parts a, b and c are removed because they are far from the secondary scaffold branch. Branch 1 has been headed (cut) every year. Remove fruiting wood from it.

Pruning young apple tree: Lengths are for standard trees, reduce for dwarf trees. Training to a central leader (main trunk) produces a tree with a pyramid shape. (See "Heading Cuts and Thinning Cuts" above.)

If newly planted tree is a **whip** (it has no branches and looks like a long stick), then cut the trunk at height of about 32 inches. This stimulates branches to grow along the trunk, and the topmost bud will become the central leader.

For a new tree that already has **side branches**, cut back trunk to 32 inches. Cut off any branches along the trunk between the ground and 24 inches high. Cut back any remaining side branches to 2-4 inches, leaving no more than 2 buds on each branch stub.

First Summer (July): Make sure top shoot becomes the leader (trunk). Pinch back all other shoots.

First Winter or Early Spring (February): If there has been a lot of new growth, choose 3 to 5 branches for the first set of scaffold branches. (A scaffold branch is one of the main limbs radiating from the trunk of a tree.) These branches should spiral around the trunk with about 4 inches vertical distance between each branch. Cut off the other side branches and any vertical branches that may compete with the leader. Prune back the main leader shoot, but keep it as the highest part of the tree to maintain pyramid shape.

Second Summer (July): Make sure the top shoot is growing vertically, cut off any competing shoots.

Second Winter or Early Spring (February): Select another set of scaffold branches 2 to 3 feet higher than the first set. If the tree did not grow enough the second year, do this the third winter.

Thereafter: Keep doing the above until you have 3-4 sets of scaffold branches. Keep shape by pruning out watersprouts (excessive growth branches), suckers and crossing or diseased branches. Keep lower branches longer than upper ones to maintain shape. On young trees cut back 1/2 to 1/3 of branch that is growing excessively.

Pruning mature apple tree in February:

Every year remove broken/diseased branches, crossing limbs, weak stems, branches growing inward to center, branches growing vertically or straight down. Prune enough new growth to allow light to filter into the interior when the tree has leafed out so fruit can ripen. Shorten any branches that are too long.

If too many apples have formed and crowded each other, **thin fruit spurs** to only a few per branch.

Pruning old, neglected apple tree in February:

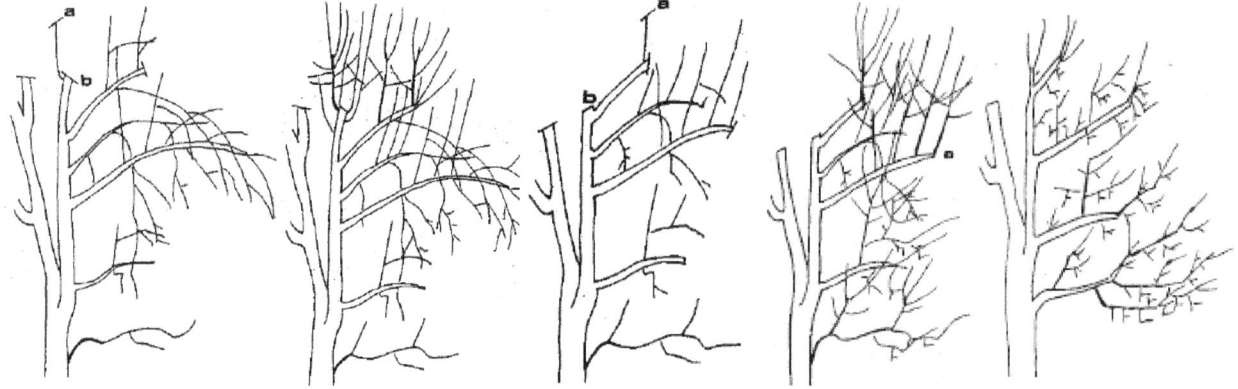

Lengths and heights are for 24 foot tall apple tree. Images are half the tree. (Top= cut off. Leader= main trunk.)

First Year of Restoration: First image on left: **a-** Top the leader at 18 feet tall to reduce suckering. **b-** Top the main scaffold limb at 16 feet. **c-** In upper third of tree, remove and/or shorten horizontal and hanging wood. Remove dead wood.

Second Year of Restoration: Second image: Limbs left above permanent limbs suppress growth on limbs below them. Long shoots grew from near the cuts and on the upper sides of horizontal limbs that were cut last year. Remove them. You should see some new growth in the lower limbs; do not cut them.

Second Year of Restoration: Third image. **a-** Top (cut off) highest branches at 17 feet. **b-** Top main scaffold limb at 14 feet. Thin out shoots on upper limbs. Keep all live wood on lower limbs so they replace fruiting wood cut from top.

Third Year of Restoration: Fourth image. Remove more than 50% of new growth in top limbs. Redirect lower limbs upward by thinning to leave an upright shoot at the end of the branch. Lower limbs receive enough light and begin to grow. Remove only a little wood from lower limbs.

Fourth Year of Restoration: Fifth image. The tree is 17 feet tall with an 18 foot spread. It has a round, cone shape with well developed lower limbs and a lot of young fruiting wood.

All Years: An old, neglected tree requires extensive corrective pruning. The main goal is to open up the interior to allow good light penetration. In most cases it is best to spread the **corrective pruning over 2-3 years.** When severe pruning is done in winter, trees should **not be fertilized that spring.**

First remove all upright, vigorous growing shoots at their base that are shading the interior. As with young apple trees, select 3-5 lower scaffold branches with good crotch angles and spaced around the tree. Limbs with poor angles and excess scaffold limbs, should be removed at their base.

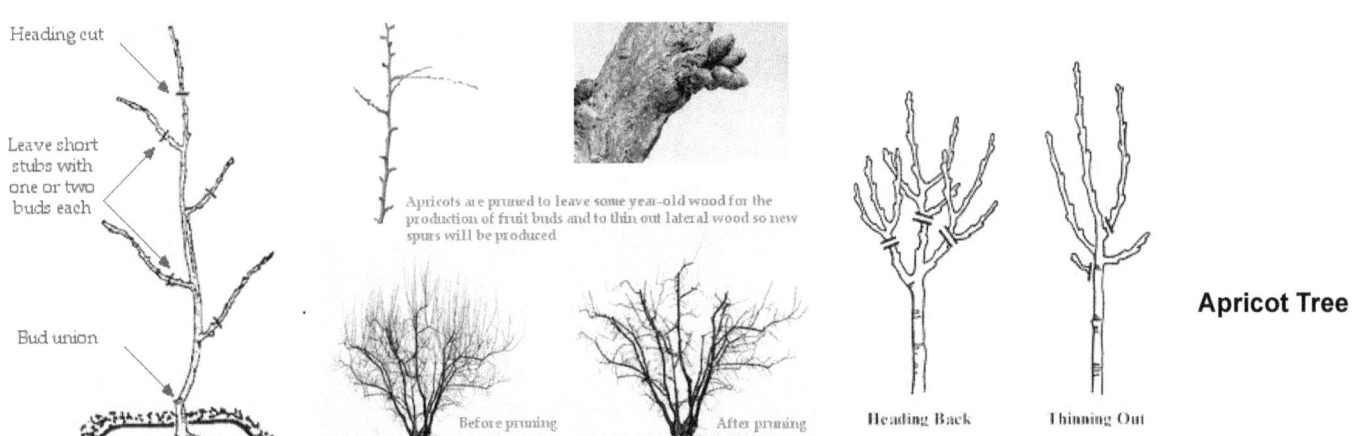

First image above is first year pruning of *apricot tree*. *Cut 24-32 inches high.*
Second image: Prune to leave some year-old wood for production of fruit buds and thin out laterals so new spurs form.

Apricot Pruning in February: (See April Garden-Plant. For **thinning** see June Garden-Maintenance. See August Garden-Harvest. See **peach pruning** below for more information about **stone fruit trees**.)

Prune out 2-year or older branches that begin to bend down. Cut them back entirely to an upright growing branch (use a thinning cut that is close to a bud). Do not head them. Remove shoots from center of tree, and cut out interfering limbs and dead/diseased wood. (See "Heading Cuts and Thinning Cuts" above at beginning of fruit tree section.)

Apricots bear almost all fruit on 1-year-old wood. Fat, full buds are fruit buds, and smaller buds are leaf buds. In most varieties the fruit buds are on the tip section of branch.

Asian Pear Pruning in February: (See April Garden-Plant, August Garden-Harvest.)

Prune similar to apple trees. (See details above.) Pruning should encourage several limbs with wide angle branches off main scaffold limbs. **All fruit are borne on fruit spurs on 2- to 6-year-old wood.** Older wood and spurs give smaller fruit than those on 2- to 4-year-old wood.

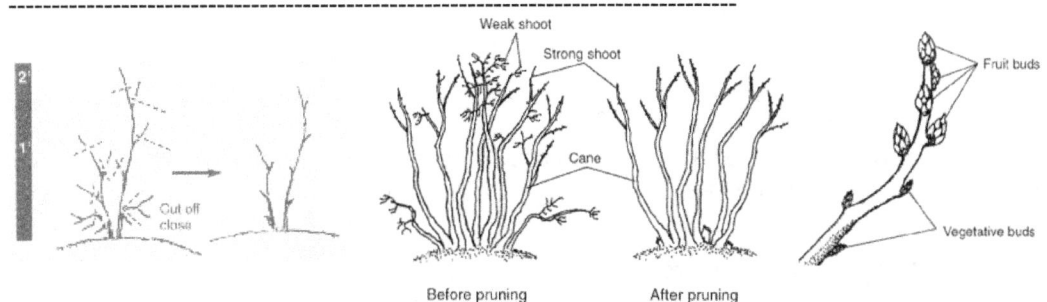

Left image above: blueberry pruning at planting.   Middle image: dormant pruning.   Right image: types of buds.

Blueberry Pruning in February: (For **planting** see April Garden-Plant. For **propagation** see March/December Garden-Maintenance. See July Garden-Harvest.)

Should be pruned every year but the first 3-4 years not much pruning is needed. **Buds for next year's fruit form in late summer through early fall on second year wood.** The goal is to renew plant by pruning old wood (canes) and forcing new canes to grow from base of plant. Prune fast growing, tall branches to keep bush at good height. Cut out thin, short shoots.

On **older blueberries**, cut branches with few flower buds. Flower buds are easily visible on 1-year-old branches. Sometimes severe pruning is needed and branches are cut to 2-3 feet above ground. However, some types do not readily grow new canes from the ground so these should not be cut as severely. (You can tell the type by how many new canes are growing.) Can cut half of canes back to 2-3 feet or taller. Then next year cut other half.

The safest way is to cut side branches short instead. Can prune off some of height if bush is tall.
Branches 3-6 years old are the most productive. At 6-10 years old, blueberries are at their best production. There should be about 15 canes. Prune when dormant in late winter or early spring.

Pruning Black Currants

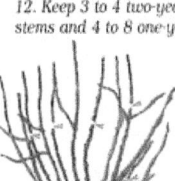

In the spring of the first year: following planting, prune back all shoots to 1 or 2 buds.

Third spring: cut out weak, damaged or diseased shoots as well as shoots in excess of 10 to 12. Keep 3 to 4 two-year-old stems and 4 to 8 one-year-old stems.

Second spring: following the severe spring pruning, new vigorous shoots have grown from the base of plants during the last season. Select 4 to 7 vigorous stems and remove

Fourth spring and subsequent years: prune out to ground level damaged, weak or diseased shoots. Remove all three-year-old stems, thin out to 4 the two-year-old stems and to 4 to 8 for the one-year-old stems.

Pruning **currants & gooseberries**.

 Currants and Gooseberries (ribes) Pruning in February: (See April Garden-Plant. See June Garden-Harvest.)

Pruning late winter or early spring (February):

Ribes species produce fruit at the base of 1-year-old wood. **Fruiting is strongest on spurs of 2- and 3-year-old wood.** A cane that grows from the base of plant will produce fruit for 3 years.

Remove any branches that lie on the ground and any branches that are diseased/broken. After the **first year**, remove all but 6-8 of the most vigorous shoots.

At the end of the **second year**, leave the 4 or 5 best one-year-old shoots and up to 3 or 4 two-year-old canes.

At the end of the **third year**, prune so approximately 3 or 4 canes of each age remain.

By the **fourth year**, the oldest set of canes should be removed and new canes allowed to grow.

A **mature plant** should have about 8 bearing canes, with younger canes eventually replacing the oldest. Remove all canes at ground level that are older than 3 years, leave 4 inch stub.

Fruit ripens in June or July. Plants get 3-6 feet tall.

 **Elderberry Pruning in February:** (For **propagation** see March/July Garden-Maintenance. See April Garden-Plant, August Garden-Harvest.)

Elderberries send up many new canes each year. The canes usually reach full height (8-10 feet) in one season and develop side branches in the second. Flowers and fruit develop on the tips of the current season's growth, often on new canes but especially on side branches.

Second-year elderberry canes with good lateral (side) development are the most fruitful. In the third or fourth year, older wood tends to lose vigor and become weak.

Pruning late winter or early spring: Remove all dead, broken or weak canes, plus all canes more than 3 years old. Leave an equal number of one, two and three-year-old canes.

 Fig Pruning in February: (See April Garden-Plant. For **propagating** see March/July Garden-Maintenance. See August Garden-Harvest.)

Prune when dormant in early spring or late fall after all figs are harvested. Remove branches in middle of tree to let in more sun. Prune branches that cross over one another and that grow in instead of out. Trim off thin, weak and spindly branches. Prune every year for the first 3 years. After that, prune to maintain shape.

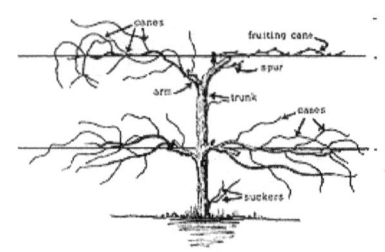

At left, mature **grape** plant. At right, mature plant pruned.

Above images are 4-arm cordon trellis. The bottom wire is 2 1/2 to 3 feet from the ground. The second wire is 5 to 6 feet from the ground. The wires are held up with 4 inch x 4 inch wooden posts that are 8 feet tall. Place posts about 8 feet apart with concrete base preferred. Use 1 1/4 inch staples to attach galvanized #9 wire between the posts.

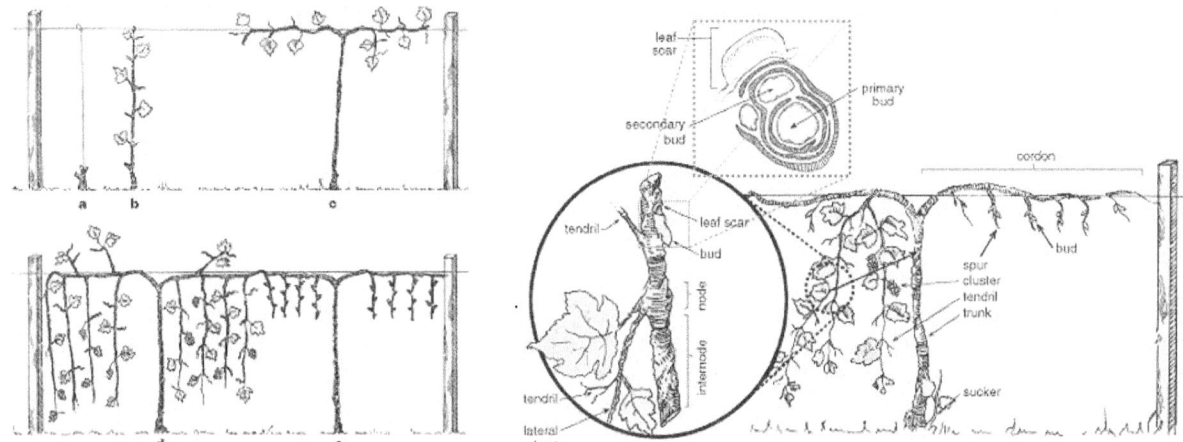

Grape Trellis System and Training: *(Above image on left is 2-arm cordon trellis.)*

a. *After planting*, cut off all shoots on vine except for two. Tie a string from base of plant to top of wire.

b. About a month later, train the strongest growing shoot up the string, and cut off other shoot. When the vine gets to the top of the string **(first or second year)**, pinch off about 1 inch of terminal (end) growth to force vine to branch.

c. Train the 2 lateral (side) branches and tie to wire in opposite directions. Remove all flower clusters in **year two**.

d. During the **third growing season**, the curtain (cane growth from lateral-growing cordons) is produced. Position growth downward and allow only one cluster of fruit to set this year.

e. In the next dormant season **(late February or early March)**, prune lateral (side) growth on cordons, leave a cane (spur) every 8-12 inches. Cut back each spur, leave only 4 buds at base of cane. Remove all other prunings from plant. In later years, choose new spurs to renew curtain.

Grape (Vining) Pruning for February: (See April Garden-Plant. For **propagation** see March/December Garden-Maintenance. See September Garden-Harvest.)

Read "The Grape Grower: A Guide to Organic Viticulture" by Lon Rombough.)

Vining grapes such as Eastern Concord or Niagara do well with 2- or 4-arm Cordon system of trellising. They produce fruiting shoots that grow downward. **All grapes bear fruit on branches from previous summer (biennial fruit production).** Do not wrap grape vines around wire instead tie to wire.

Prune **late winter or early spring**. Mature plant may need up to 90% wood pruned off. Vines grow too many leaves and not enough grapes if given too much nitrogen.

Four-arm Cordon Trellis System:

For new plant, in **spring** select one good branch and cut off all others. Do not allow any lateral (side) growth until plant reaches first wire of trellis 3 feet off of ground. Once 2 lateral shoots reach this height, train one to go in one direction on wire, the other in other direction. Then let shoots grow from these 2 branches to the next wire 2 1/2 to 3 feet up. Then fruiting spurs will grow from these lateral shoots. Cut spurs so spaced 6-12 inches apart, leaving current fruit spur and bud spur for next year. These spurs are cut short.

Fruit is produced on first 3-5 buds on canes grown previous summer. Only allow 17-20 bunches of grapes per branch. During **summer** canes can grow as they wish.

Then in **winter** select best 5 or so lateral (side) shoots from same branches as before to replace previous shoots. Leave 3-4 buds per foot on cordon (horizontal main trunk). Cut shoots 3 inches above top wire. Cut off old shoots leaving 2 buds on shoot.

Old, unattended grape vine:

Remove 1/2 to 3/4 of vine the first year when dormant. The first **summer** select several strong canes and tie to trellis. Then next **dormant season**, prune away all but those canes.

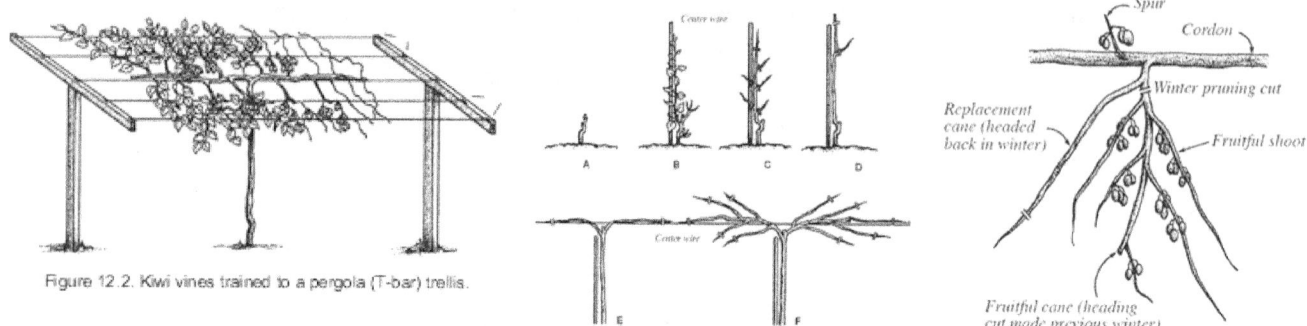

Figure 12.2. Kiwi vines trained to a pergola (T-bar) trellis.

Above images are all hardy kiwi. The middle image is the first 2 years of training kiwi vine:
  A. *Prune to 2 buds at planting.*
  B. *Train 1 shoot as trunk, remove all others **(growing season, year 1)**.*
  C. *Cut trunk at top (head back) as shoot growth at terminal (end) loses vigor **(growing season, year 1)**.*
  D. *Remove lateral shoots. Let trunk grow beyond wire, then cut off below top wire **(growing season, year 1)**.*
  E. *Choose 2 shoots to form cordons (lateral trunks). Cut to 1/4 inch diameter when dormant **(grow, year 1)**.*
  F. *Shoot growth, **year 2**. Pruning cuts in **dormant season of year 2** also are shown by //.*
The image on right: A fruiting branch *of hardy kiwi. Fruit are produced on shoots growing from last year's growth. Winter pruning cuts are shown by //.*

 Hardy Kiwi Pruning in February: (See April Garden-Plant. For **more pruning** see June Garden-Maintenance. For **propagation** see July/October Garden-Maintenance. See September Garden-Harvest.)

(Read "The Backyard Berry Book: A Hands-On Guide to Growing Berries, Brambles, and Vine Fruit in the Home Garden" by Stella B. Otto.)

Winter Pruning (February): Remove canes that bore fruit the last 2 summers. **New fruiting canes develop at the base of previous year's growth.** Keep the best 1-year-old side canes that have not borne fruit. Trim back to 8-12 buds. Fruiting canes should be 8-12 inches apart on cordon (horizontal main trunk). They bear fruit for 2 years.

Every 2-3 years up to 60% of side branches are pruned. Fruit is only produced on shoots of current season growth from 1-year-old wood.

Summer Pruning: In late summer branches can be trimmed off where wood is 1/4 inch diameter.

Can **propagate** from hardwood cuttings in late winter or softwood cuttings in July. (For general details about herbaceous, softwood, semi-hardwood and hardwood cuttings, see "Garden Tips" in back of manual.)

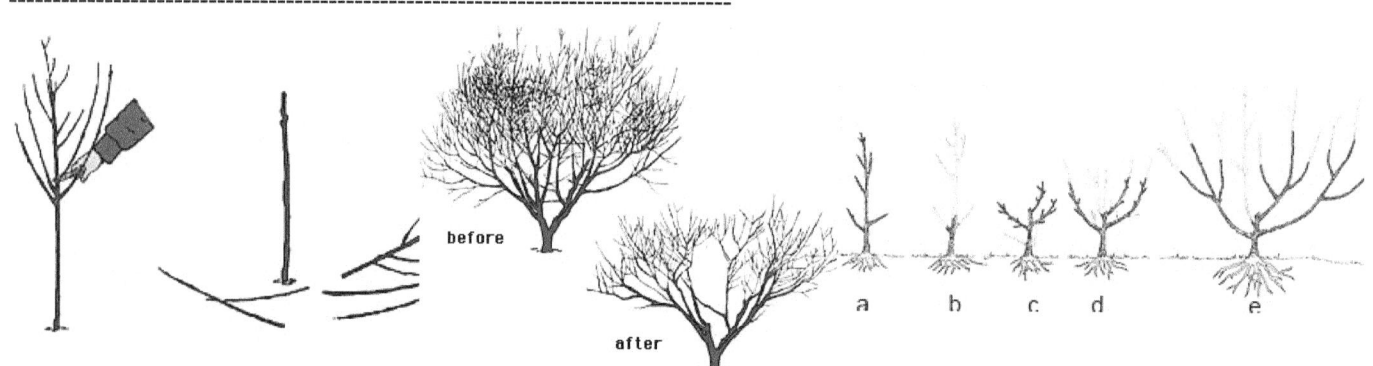

Pruning for stone fruit trees such as peach, nectarine or plum.
  First image above on left: *At planting cut tree about 2 feet tall. Remove all side branches regardless of tree size.*
  Second image above: *On bearing trees trim sub-scaffold and other branches to keep a good tree height for picking fruit. (Scaffold branches are main limbs off of trunk.) Fruit set on 1-year-old shoots that regrow each year. Thin out crowded shoots so all receive good sunlight. Remove low branches that are in the way when moving around tree.*
  Third image: a. *Ideally 3 or 4 branches should be evenly spaced around trunk at 26-30 inches above ground.*
    b. *If no good scaffold branches, cut trunk back to 30" above ground and remove lateral branches.*
    c. *Select 3 or 4 that are evenly spaced around the trunk during the **first growing season**.*
    d. *In **February or March of the second growing season**, cut back scaffold branches and remove suckers or watersprouts growing vertically. Suckers are shoots that grow at the base of a tree. Watersprouts are thin*

branches that usually grow straight up in the air. They do not usually bear fruit.
  *e. In **February or March of the third growing season**, cut back scaffold branches. Remove suckers growing vertically and shoots growing toward the center of tree. Also cut watersprouts.*

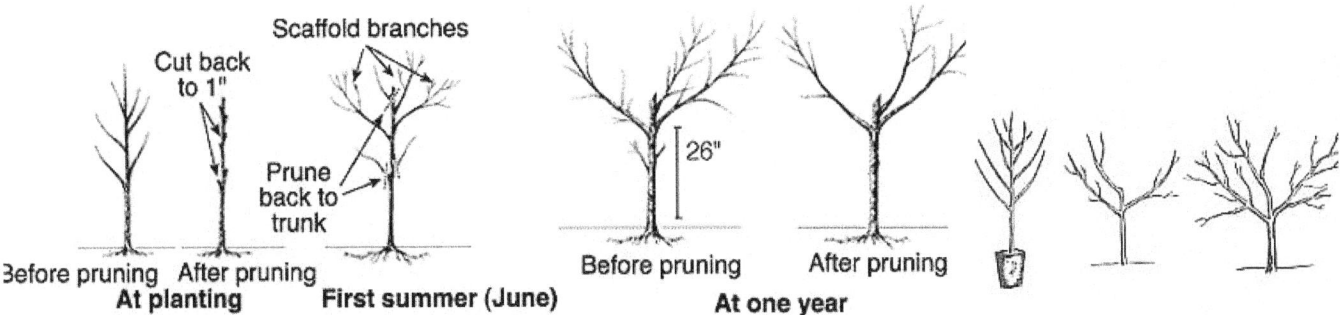

 **Peach / Nectarine Pruning in February:** (See April Garden-Plant. See August Garden-Harvest.)

Trees do well with up to 60% heavy pruning into funnel shape. **Fruit grows on twigs and branches that grew the previous summer.** Only allow blossoms to set fruit in middle 1/3 of these branches. Cut off top 1/3 of these branches. Cut off all branches that bore fruit last summer. **When pruning always favor new branch growth.**

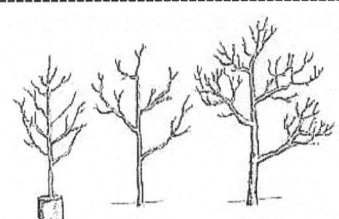

Pruning for apple and pear trees (not stone fruit).

 Pear Pruning in February: (See April Garden-Plant. See August Garden-Harvest.)

Prune similar to apple trees. (See above for details.) **Pear trees produce fruit on spurs (with knobby ends) on older wood. These same spurs produce fruit every year.** On a new tree (whip) the leader (main trunk) should be cut fairly short such as 24-36 inches tall. Cut 1/3 off of 1-year-old branches.

 Plum Pruning in February: (See April Garden-Plant. See August Garden-Harvest.)

(See **peach pruning** above for more information about stone fruit trees.)

Plum trees produce fruit on spurs on branches 2 years or older. They bear fruit on their tips. Prune tree to funnel shape. Until the tree is 10 years old, cut back long, fast growing branches to 2/3 of length. Cut out branches completely that do not add to shape of tree.

Little pruning is needed after basic shape is established. When the tree is 10 years old or more, new growth must be cut out almost completely.

Wild roses such as **Rosa Rugosa pruning** can go 3-5 years with almost no pruning other than removing old blooms. However, all dead/damaged branches and canes should be removed each year. Occasionally, an old cane can be removed at the base to promote new canes. (See April Garden-Plant, September Garden-Harvest.)

 Walnut Pruning: (See April Garden-Plant. See September Garden-Harvest.)

January to early March pruning: On a young tree do not let top grow too high. Select a dominant leader to be main trunk, then remove all additional leader stems. Choose strongest branches growing from the central leader to form the tree's scaffold branches. (See "Heading Cuts and Thinning Cuts" above at beginning of fruit/nut tree pruning.)

After 3 years, the central leader and scaffold branches should be developed, giving the tree its form. To maintain size and form, prune heavily every 3-5 years: shorten scaffold branches, and remove uppermost growing tips, any crossing/weak side shoots and dead/weak wood. Walnut trees do not need a lot of pruning.

🌱 Garden- Plant 🌱 (February)

The ground is frozen. Only seed planting is in greenhouse or other protected area.

Shiitake Mushrooms

Shiitake mushrooms or black forest mushrooms (Lentinus edodes) are the second most popular eating mushroom in the world. They are high in protein, and vitamins A, B, B12, C, and D. They are medicinal. It is anti-viral.

These mushrooms are a wood-decay fungus that **grows on hardwood logs such as oak**. They need shade, warm temperatures, and a moist environment. Western North Carolina is a perfect place for them to grow.

Tree Cutting

Logs need to be inoculated with Shiitake spawn (vegetative stage of mushroom mycelium). Mycelium is a mass of interwoven filaments or threads. The best trees for production in North Carolina are white oak, red oak, chestnut oak, and sweetgum.

Trees should be cut in the dormant season (November to early March) with late winter (February) being best. Cut before buds start coming out (budbreak). Shiitake requires high carbohydrates to grow, and carbohydrates are highest when trees are dormant. Also the bark needs to remain on the logs. If sap has started to flow, bark is more likely to be damaged or fall off.

Log Cutting

After trees are cut they should sit for 10 days with the branches left on. Then they can be cut into logs. Logs can be a wide variety of lengths and widths but the best is around 3-5 feet long, and 3-8 inches in diameter. Handle logs gently so bark does not come off.

Trees cut in late fall or early winter (November to January) should be inoculated within 3 months. Trees cut in late winter (February, early March) should be inoculated within 1 month. In western North Carolina it is best to inoculate before April. In the rest of North Carolina or in warmer areas, it is best to inoculate before March.

Store cut logs so they remain moist but are not waterlogged. If the weather is unusually wet, the logs may need to be covered with plastic but there should still be good air flow. In dry weather burlap, muslin or similar material may need to be put on the logs.

Inoculation or Adding Spawn to Logs

Logs are inoculated with **Shiitake mycelium called spawn that is usually mixed with sawdust**. It is a fuzzy white growth similar to bread mold. Spawn is sold at mushroom supply companies. Store spawn in a refrigerator. Let sit at room temperature for a few days before inoculating.

Use a 12 mm (.47 inch) drill bit to drill 1 inch deep holes into logs. Special bits with a stop are sold by mushroom supply companies. (See second image above.) Use a high-speed, heavy-duty drill with an electric cord. A battery drill drains too quickly. Even better is an angle grinder with the correct drill bit.

Drill holes 3-6 inches apart in rows 3-4 inches apart all around the log. The closer the holes are, the faster the log becomes impregnated with mycelium. However, spawn costs are higher.

The spawn/sawdust mixture is then pushed into each hole. It is easier if you use a **sawdust plunger or inoculator** available at mushroom supply companies. (See third image above.) Press the plunger into the spawn 4 times, then place over hole. Press down 4 times to release spawn.

Cover each hole with hot **beeswax or cheesewax** using a brush or dauber. The log ends can also be sealed to keep in moisture. Heat wax in an old pot. One pound seals about 10 logs.

Mark one end of each log with a metal or plastic tag with date and type of spawn.

Spawn Run or Incubation Period

The inoculated logs are stacked in a warm, shady area that has good air circulation but is not windy. Around 80-90% shade is best. They need rain. If the weather is dry, soak the logs every few weeks for 8-12 hours.

Logs need to be stacked so there is air between the logs. Lincoln log (crib, criss-cross), X pattern, or lean-to stacking methods are good. Criss-cross stacks are about 4 feet tall. Bottom logs are not inoculated. Lean-to (slope) stacks are best if land is hilly. (See 3 images above.)

About 200 logs will fit on 1/4 acre. It takes 6 to 18 months for mushrooms to develop.

Log Moisture

The best log moisture to grow Shiitake mushrooms is between 35-55%. If moisture goes below 25%, the mycelium and mushrooms die. When moisture goes below 40%, or there has been no rain for 2-3 weeks, the logs should be watered. A sprinkler system can be used for 6-12 hours once a week. Or better is to **immerse the logs in cold water**. Put the logs in a nearby stream. Or put in a trash can or livestock tub/tank filled with water for 24 hours.

Fruiting or Mushroom Formation

It is possible to get a few mushrooms sometime in **September-November.** Though most develop after a little more than one year, usually in **April-May**. Thinner logs with more spawn produce sooner. After 1 year logs can be restacked to provide more space for mushrooms.

Mushrooms form when the entire log has been colonized with the fungus/mycelium. The first sign of this is the ends of the logs are white and fuzzy, and sometimes brown. The bark feels spongy.

Mushrooms form naturally for 3-6 years, once every spring and fall.

Mushroom formation can be forced to harvest 3-6 times a year. These logs produce for 2-3 years.

Forcing is best done after 1 year. Soak logs in cold water starting in spring for 24 hours to encourage fruiting. Mushrooms usually appear within 1 week of soaking.

During fruiting in low humidity, sprinkle water on the logs 2-3 times a day. Protect from heavy rain by placing plastic over the logs. The plastic should not touch the logs.

Logs can be soaked every 4-8 weeks, up to 3 times a year. Do not soak in the winter. A cord of wood of about 125 logs yields about 500 pounds or 4 pounds per log of mushrooms.

Diseases and Pests

In North Carolina the most common disease fungi are Trichoderma (green mold), Hypoxylon (black mold), and Coriolus versicolor (turkey tail). Green mold forms if logs are too wet. Black mold forms if logs are too dry or too old. Turkey tail fungus forms on bruised logs with missing bark and on the ends of logs. Use fresh logs and seal the ends.

Slugs and snails like to eat mushrooms. When first stacking logs, put gravel on the ground. Periodically sprinkle lime and/or wood ash on the ground. Remove leaves and debris away from the logs.

Harvest

Mushrooms need to be **harvested daily during fruiting**. Pick while there is still a small curl at the end of the cap. This occurs about 5-7 days after the mushroom first appears. Twist or cut at base of stem. Handle gently. Consult an expert before eating any mushrooms so you are sure they are Shiitake. Consult your doctor and/or health care specialist before eating mushrooms.

Store between 32-36 degrees at 85% humidity in a plastic or paper container with holes. They last about 2 weeks under these conditions. They can be dried at around 120 degrees.

Let your logs rest for 4-8 weeks before starting forced fruiting again. Keep log moisture between 30-40%.

Grow Your Own Spawn

Mushrooms reproduce through spores. Creating your own mushroom spawn requires pure, high quality mushroom strains, sophisticated equipment, and sterile conditions to do it right. It is better to buy from a company that specializes in it. But if you want to try it, there is a pdf online called "Shiitake Spawn Preparation Chiefly with Sawdust". It is very detailed.

The basic idea is to have a very **sterile environment** with sterile equipment. Even the air needs to be sterile so a HEPA air filter is needed. You need an autoclave (a strong container using high pressures and temperatures) or pressure cooker. Equipment needs to be boiled in a pot for 30 minutes, or less time in a pressure cooker. Handle with

sterile gloves.
   Put a small piece of mushroom from the inner cap on agar in a **petri dish**. Agar is a gelatinous substance made from seaweed. Keep moist. Keep temperatures between 65-75 degrees. You can use an incubator. The mycelium grows and fills the dish in about 2 weeks. It will be long and threadlike.
   Sterilize water with grain such as wheat, rye, corn or millet. Put in a pressure cooker or autoclave for 52 minutes at 15 psi. After the grain is sterilized and cooled, put mycelium on top of the grain in jars. Keep sterile. Shake the jar occasionally. Let grow for 2-4 weeks. You can then take it and create about 10 more jars of mycelium with grain that will grow again.
   Mix with **sterilized hardwood sawdust** such as oak. Refrigerate it if you will not be using it right away. It only lasts a month or two at most.
   There are many different methods, and different species of mushrooms require varying conditions to grow.
   (For more information contact the North Carolina Mushroom Growers Association at www.ncmushrooms.org.
   Read "Growing Gourmet and Medicinal Mushrooms" by Paul Stamets and "Mycelium Running: How Mushrooms Can Help Save the World" by Paul Stamets. For mushroom supplies: www.fieldforest.net.)

Garden- Harvest (February)

 **Harvest** after frost and through most of winter outside: collard greens, creasy greens, kale, kohlrabi, leeks, sunchokes (jerusalem artichokes), turnips and others.

Greenhouse, House, Hoop House or Cold Frame (February)

Hardy Greens: Arugula, claytonia, mache, minutina, mizuna, mustard greens, and radicchio.
Vegetables: Globe artichoke, and onion/leek.

 Only in late winter and early spring in the **greenhouse:** When growing seedlings for transplanting outside in late spring or early summer, use heat such as portable kerosene **heater** when temperatures drop unusually low. This is only to prevent seedlings from being killed, not for growing tender annuals all year. This is a low cost way to lengthen growing season.
   Instead or in addition row covers, **frost blankets** or sheets can be put over plants if there is a cold snap or at night. It is better if the cloth does not touch the plants so use wire or poles to hold up the cloth just above the plant.

 **Hardy vegetables** are good for spring and fall gardens, or for winter greenhouse/cold frame. All taste best when they mature in cool weather. They remain good for weeks (sometimes months) after the first hard frost. **Harvest** when not frozen or else they will be droopy when they thaw.
   Extra hardy vegetables can take very hard frosts (usually 0 to 15 degrees). They include arugula (15 degrees), claytonia (0 degrees), mache (0 degrees), minutina (15 degrees), and mizuna (15 degrees),.
   Hardy vegetables can take hard frosts (usually 25 to 28 degrees). They include broccoli, brussels sprouts, cabbage, collards (down to 20 degrees), English peas, kale (down to 20 degrees), kohlrabi, mustard greens, parsley, radish, spinach (down to 20 degrees), and turnip.
   Semi-hardy vegetables can take light frosts (usually 29 to 32 degrees). They include beets, carrot, cauliflower, celery, Chinese cabbage, endive, Irish potatoes, lettuce, radicchio, rutabaga, salsify, and Swiss chard.

Sow in greenhouse frost hardy greens in February. The plants below are all **very frost hardy salad greens** perfect for the winter greenhouse, hoophouse or cold frame. **Harvest only when plants are not frozen.** If you harvest when frozen, then they will be droopy when they thaw. Little or no watering needed.
   Read the book "Oriental Vegetables: Complete Guide for the Gardening Cook" by Joy Larkcom. Also

read "The Winter Harvest Handbook" by Eliot Coleman.)

 **Arugula** (brassicaceae family; Eruca sativa or Eruca vesicaria subspecies sativa) is a fast growing, **annual in the mustard family**. Also known as roquette, garden rocket, rucola, and rugola.

Cover seeds with 1/8 to 1/4 inch of soil. Sow 1 inch apart. Seeds sprout in 3-10 days even in cold weather. Germinates best at 40-55 degrees. Thin to 6 inches apart in rows 12-18 inches apart. Can eat thinnings.

Succession plant every 2-3 weeks. Grows 6-12 inches tall before flowering, then with flowers it is 24-36 inches tall.

Care: Likes soil pH of 6.0-6.8. Likes full sun but will tolerate shade. **Frost hardy to 15 degrees.** Heat causes plants to form flowers (bolt). Does not cross pollinate with other members of the mustard family. Does self seed.

Harvest 3-5 weeks (21-35 days) after sow. Matures in 45-60 days. Flowers and leaves are edible. Eat raw or cooked. (See September Greenhouse but details are here.)

 **Claytonia** (portulacaceae family; Claytonia parvifolia, miners lettuce, winter purslane) is a leafy green **annual.** Seeds like soil between 50 and 55 degrees. Sow 1/4 inch deep but not deeper. Thin 8 inches apart. Grows to 1 foot tall. Matures in 40 days. Succession plant every 3 weeks.

Care: Grows quickly. Likes soil pH 6.5-7.0. Likes full sun but will tolerate partial shade. **Frost hardy to 0 degrees.**

Harvest: Flowers and leaves are edible. Tastes good even when plant flowers. Leaves are bitter in hot weather.

 Mache (corn salad) (valerianaceae family; Valerianella eriocarpa) is an **annual hardy to 0 degrees.** Sow 3-4 seeds per inch, 1/4 inch deep, with rows 10 inches apart. Germinates in 10-20 days. Thin to 4-6 inches apart. Takes 45-60 days to mature. Grows 2 inches tall. **Grow like spinach.** Plants bolt (form flowers) in summer.

Care: Likes soil pH of 6-7. Likes full sun but will tolerate partial shade.

Eat leaves raw or cooked. (See September Greenhouse but details are here.)

Minutina (plantaginaceae family; Plantago coronopus; buckshorn plantain, staghorn, or erba stella) is an **annual**, oriental salad green with a nutty flavor and crunchy texture. It was grown in Colonial America.

Sow 30 seeds/foot. Sow 1/4 inch deep. Thin to 6-8 inches apart. Matures in 50-60 days. Grows 10 inches tall.
Care: Frost hardy to 15 degrees. It likes cold, rainy weather and moist soil. Does well in full sun. Self seeds.
Uses: Best eaten when 5 inches tall (or less) when it is tender and before it flowers. Flowers and leaves are edible. Eat raw or cooked. Used as a medicinal for fevers.

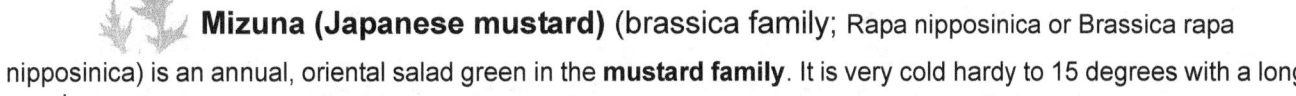 **Mizuna (Japanese mustard)** (brassica family; Rapa nipposinica or Brassica rapa nipposinica) is an annual, oriental salad green in the **mustard family**. It is very cold hardy to 15 degrees with a long growing season.

Sow 1/4 inch deep, 2 inches apart. Germinates in 5-15 days. When 1 inch tall, thin to 6-10 inches apart. Grows to 14-16 inches tall. Matures in 6-8 weeks but can start eating in 3 weeks. Goes to seed in summer.
Care: Likes full sun but tolerates some shade. Likes moist soil. Likes soil pH of 5.5.
Uses: Flowers and leaves are edible. Tender and good eating in a salad. Can eat cooked.
(See October Greenhouse but details are here. For mustard greens, see below.)

 **Mustard greens** (brassicaceae family; Brassica juncea) also known as mustard spinach, leaf mustard and white mustard. **Sow late winter (February) in greenhouse. Or sow early spring (March or early April) outside 3 weeks before last frost date. Also sow outside late summer (August), 8-10 weeks before first frost date.** Better planted in late summer.

It is easy to grow. It is a cool season **annual. Can take light frost down to the mid 20s.** Grown as food and as **green manure / cover crop**.

Sow 1/4-1/3 inch deep, spaced 3 inches apart. Germinates in 7-10 days. Thin seedlings to 5-9 inches apart in rows 1 foot apart. Can eat thinnings. Harvest in 30-40 days. Matures in 45-50 days.

With flower stalks they can be 3 feet tall. The flowers are small and bright yellow.

Care: Likes full sun. Likes shade if the weather is hot. Needs moderate amount of fertilizer. Likes soil pH of 5.5-7.0. Keep evenly watered. Water-stressed (not enough water) plants tend to go to seed earlier.

Varieties: The leaves of mustard greens have either a crumpled or flat texture and may have edges that are either toothed, scalloped, frilled or lacy. Colors range from light to dark green and can also be red.

Western varieties include **curly-leaf or common mustards** which have frilly roundish leaves, and **mustard spinach** which has large smooth dark green leaves.

Asian varieties include **mizuna**, a Japanese green (see above).

Using: Mature **leaves** are 15-24 inches long but leaves are better eaten when smaller when the plant is 5 inches tall and leaves are 4 inches long. Use small leaves for salads. Large, older leaves may have to be simmered an hour.

The greens contain oxalic acid (also in spinach) which should be eaten in moderation. The oxalic acid is reduced by cooking. Do not cook in aluminium or iron pots since leaves will turn black.

The **leaves, seeds and stem are edible**. High in vitamins A, K, C and E, and good source of folic acid, calcium, selenium, chromium, iron and zinc.

The brown seeds are used to make **Dijon mustard**. To harvest the seeds, leave the stalks on the plants until the seed pod has dried. Can use whole or grind the seeds.

(See March/August Garden-Plant but most details are in this section. For **cover crops** see March Garden-Plant.)

 **Radicchio** (asteraceae family; Cichorium intybus; Italian chicory) is a member of the **chicory family. It is a biennial.** Leaves are deep red with white veins. **Can take light frost down to 30 degrees. Sow 2 weeks before last frost in the spring, or 8 weeks before the first frost in the fall (August or September).**

Cover seeds with 1/8 inch of soil. Space seeds 1-2 inches apart in rows 12 inches apart. Germinates in 7-14 days. When 1 inch tall, thin to 8 inches apart in rows 2 feet apart. The head matures in 80-90 days. It is 4-12 inches tall. Succession plant every 2 weeks.

Care: Likes full sun except some shade is good in summer. Likes rich soil. Likes pH 6.0-6.5. Keep well watered.

Using: Harvest heads when 3-6 inches across. Cut off heads at base. Once it goes to seed, the leaves are too bitter to eat. Light frost makes the leaves less bitter. It has a slightly bitter taste. Can eat raw or cooked.

(See August/September Garden-Plant but most details are in this section.)

Sow these vegetables in greenhouse in February:

 Globe artichokes (compositae or thistle family; Cynara scolymus) are a **biennial or perennial related to chicory, dandelion, lettuce and salsify**. It produces its harvest (the flower bud) the second year. However, you can **fool nature** and do it in 1 year. You grow the young plants in warm, then cool temperatures. The best variety for this system is Imperial Star or Grande Beurre. The variety Green Globe is not good to use, except Cal Green Globe is OK.

 For more information read the book "Four-Season Harvest" by Eliot Coleman.)

Matures in 90-100 days (13-15 weeks or 3-4 months) but flower buds are produced the second year. So with this system, buds are ready in **August or September.**

Mid February through early March soak globe artichoke seeds in water for 2 days. Then **sow seeds in house** at 70-80 degrees in sunny location. Sow seeds 5 inches apart, 1 inch deep. Sow more than you need because some plants will not be true to type. Remove plants that are small or albino (white). Germination takes 10-20 days.

Six weeks later (**early April), move to greenhouse.** Temperature should not go below 25 degrees. But it should be as cool as possible. To fool the plants they need at least 10-12 days at less than 50 degrees.

Transplant outside at 12 weeks (**mid May or early June**). Plant 24 inches apart in 30 inch wide bed. Plants can be up to 3 feet across. Slugs like artichokes. (For how to make slug and snail traps, see "Garden Tips" at end of manual.)

Harvest in **August or September.** Immature flower bud is eaten. In a good season you get 8-12 heads per plant.

(**To move from house to greenhouse,** see April Greenhouse. **To move from greenhouse to outside,** see June Garden-Plant. See August/September Garden-Harvest with **harvest details in August.**)

 Sow **onion and long-season leek** seeds in February greenhouse.

Sow **onion** (amaryllidaceae family; Allium cepa or Allium fistulosum). Sow in greenhouse in February and **transplant outside in April**. Also can directly **sow seed outside in April or 2-3 weeks before first frost date in spring. Or sow outside 8-10 weeks before first frost in fall.**

The seed does not store more than 1-2 years. Sow 1/2 inch deep in rows 10-12 inches apart. Can be sown thickly, and then eat thinned green onions. Thin to 2-4 inches apart. Matures in 92-115 days depending on variety.

Most onions are **biennial** with flowers forming the second year. **Onions are frost hardy.** Onions grow slowly.

Care: Likes soil pH of 5.5-7.0. They will grow in most soils. Likes well drained soil. Likes full sun.

When seedling tops fall over, cut plants to 3 inches tall. Can eat tops.

Onions can be purple, yellow or white. Some produce bulbs and some do not (scallions, green onions). They vary in hotness or pungency based on how much **sulfur** they contain. Vidalia are low in sulfur. If you add sulfur to your soil, your onions will be more pungent or strong tasting.

Two types of onions based on day length:

Long-day onions are best for those north of the 36th parallel. (North Carolina is a little bit south of the 36th parallel.) They need about 14-16 hours of daylight. They include Sweet Spanish Onions and Walla Walla onions.

Short-day onions are best for those south of the 36th parallel. They need about 12-14 hours of daylight. They include Red Burgundy, Texas Grano and Yellow Granex.

There are several varieties of onions:

Common slicing onion such as white and yellow (Spanish). **Purple** onions with a mild taste. **Scallions** or green onions with little or no bulb. **Pearl** are small pickling onions. **Shallots** have a mild taste and small bulb. **Leeks** are like scallions but mild. **Vidalia** onion is based on where grown in Georgia.

(For **fertilizing** see April Garden-Maintenance. For **perennial onions** see July Garden-Harvest. For **planting perennial onions** see September Garden-Plant. For **planting seeds and transplanting,** see April Garden-Plant but all details are in this section. For **fungal diseases and onion pests**, see Amaryllidaceae Family in "Plant Families".)

Sow **leeks** (amaryllidaceae family; Allium ampeloprasum) in greenhouse **8-10 weeks before last frost in spring.**

Sow 1/4 inch deep. Transplant outside in **May** at 4-8 inches apart in rows 24 inches apart.

Care: Likes soil pH of 6.2-7.0. Likes full sun and well drained soil. Tolerates rainy weather. Few diseases or pests.

They are a **biennial member of the allium genus** that includes onions, garlic, shallots and chives. (See "Plant Families" in back of manual.) They do not form much of a bulb. To have seed, mulch first year plants heavily in winter or bring into root cellar, then grow again the next year when flowers form.

Uses: The entire plant is edible. Young leeks are good raw. Cook older leeks. Long-season (winter) leeks store better than short-season (autumn) leeks.

Varieties:

1. Autumn leeks (short season) are less hardy with shorter days to maturity. They can be **sown very early in spring in a greenhouse, and then transplanted outside around the first spring frost date.** They should be mature by autumn. Their leaves are usually light green. **Some mature as early as 75-85 days.**

King Richard is a popular autumn leek with 12 inch tall white stalks and more than 2 inches in diameter. It is a fast grower, maturing in 75 days.

2. Winter leeks (long season) are hardy enough to stay outside during the winter (**hardy to 8-15 degrees**). They can be sown outside through early summer. They are harvested from when they reach pencil size until late winter or early spring (if warm winter). Mulch in winter. The leaves are dark blue-green. **Most mature in 105-170 days.**

Saint Victor is a hardy, productive variety that has deep blue-green foliage with a purplish tinge. **Alaska, Giant Musselburgh, Broad London, and Carina** are also good winter varieties.

(For **leeks** see May Garden-Plant but most details are in this section. See September/October Garden-Harvest with **harvest details in September.** For **chives** see March Greenhouse.)

 # Farm Animals   (February)

 **Open paddocks** previously planted with **brassica forage crops** such as kale, kohlrabi, mangel beet, rutabaga and turnip. Let animals graze for part of the day only. Sow July-September for grazing September to March. (See September/November Farm Animals. See **Fall and Winter Pasture** in "Garden Tips".)

Chicken, turkey and duck egg production is up with longer days. Pullets (young female chickens) that hatched mid-summer start laying eggs.

Peak egg production is April and May. Good egg production is **February through July.** Lowest egg production is **September through December.** (The longest day is in June. Molting is in September and shortest day is in December.)

You can start **incubating eggs** in an incubator **mid-February or March** depending upon what housing you have for the chicks (or ducklings or turkey poults) when they hatch. Chicken gestation (time hen sits on eggs) is 21 days. Turkey gestation is 25-30 days. Ducks are 28 days. Baby birds need temperatures of 90-95 degrees after they hatch.

(For **broody hens**, see April Farm Animals. For **incubators and brooding,** see March Farm Animals.) Read "Pastured Poultry Profits" by Joel Salatin and "Storey's Guide to Raising Chickens" by Gail Damerow.)

Line Breeding vs Inbreeding Animals

Inbreeding is the mating of 2 animals who are closely related to each other. The opposite, **outcrossing,** is when the 2 parents are completely unrelated. Or sometimes more realistically no relatives within 4-5 generations. Outcrosses have the most genetic diversity.

Line Breeding (linebreeding) is inbreeding between 2 animals that are not very closely related to each other. The difference between linebreeding and inbreeding varies depending on the species.

A parent breeding with its offspring, or two siblings (brother-sister) mating is always considered **inbreeding**. In humans it is incest. You usually do not want this kind of mating. If you do this kind of inbreeding, you must outcross the next generation.

Uncle-niece, aunt-nephew, grandfather-granddaughter, grandmother-grandson, half-sibling matings, and first cousin matings are considered inbreeding by some people and **linebreeding** by others. In particular half-sibling matings are more towards inbreeding.

Line breeding is used to increase a particular trait in animals on a more consistent basis. It may not always work out well because you are increasing both good and bad genetics (disease). Keep careful records of your breeding and the outcomes. You have to cull (kill) animals that do not have desirable traits.

Hybrid Vigor occurs when two completely unrelated animals (of the same species) breed. Their offspring are usually bigger and healthier than the offspring of closely related cousins. However, this only applies to the direct offspring of this match, and not to the offspring of the offspring of the hybrid vigor mating.

 **Chickens** (Gallus gallus domesticus) have been domesticated for thousands of years. A hen is a female chicken. A rooster is a male chicken. A chick is a baby chicken.

Chickens need a **coop** or small building where they roost at night that keeps them safe from predators such as coyotes or raccoons. During the day they like to roam around a **pasture** (free range) looking for greens and bugs to eat.

The better their diet, the more nutritious their eggs. Feed hens **chicken laying feed** which has extra calcium for their eggs. They like some **scratch** thrown on the ground which is crushed corn and other grains such as wheat or milo (sorghum grain). They need granite **grit** (small pieces) to help them digest food in their gizzard (part of stomach).

The **number of eggs a hen lays** depends on breed and care they are given. They lay between 150-325 eggs a year. An average is about 250 eggs a year. Commercial breeds lay more eggs per year than heritage breeds.

The **productive life** for egg laying for a hen is 2-3 years. They lay eggs their entire life but lay less each year. They can live to be 10-15 years old or older. After 3 years most people replace the hens (unless she is your friend!).

There are **hundreds of different breeds** with 60 popular in the United States.

Most large chicken breeds (standard size) have a **bantam** (bantie or banty) counterpart sometimes called a miniature. They are 1/5 to 1/4 the size of the standard breed, but exhibit all of the standard breed's characteristics.

Some popular breeds:

Dominique chickens are a heritage breed meaning they have been around for hundreds of years. They are endangered. Dominiques have a rose comb versus the Barred Plymouth Rock's single comb. Dominiques are a cold-hardy dual-purpose (meat and eggs) bird. Hens are caring mothers. A hen can lay 230 eggs per year. A cock weighs about 7 pounds, a hen weighs about 5 to 5.5 pounds. Nantahala Farm (www.nantahala-farm.com) sells them.

Rhode Island Red is one of the most popular chicken breeds. It has rusty feathers. It readily lays eggs and provides good quality meat. It is a robust breed. A hen can lay 300 eggs per year. A cock usually weighs 7-9 pounds, a hen weighs 6-7 pounds.

Leghorn chickens are very good egg layers and a popular breed. Its temperament is erratic and noisy. It may suffer from frostbite if raised in very cold conditions. A hen can lay 300 eggs per year. A cock weighs 7-8 pounds, a hen weighs 5-6 pounds.

Jersey Giant cocks can weigh over 13 pounds, making it an ideal choice for a good meat producing chicken. It is very healthy and can adapt well. It is black and white. A hen can lay 150-200 eggs per year.

Plymouth Rock was a very popular homestead breed and remains popular today. It comes in different colors but the most common ones are white, bur, barred and silver penciled. It is a friendly bird with a calm nature. A hen can lay 200-220 eggs per year. A cock weighs 9-10 pounds, a hen weighs 7-8 pounds.

(For **broody hens**, see April Farm Animals. For **incubators and brooding yourself**, see March Farm Animals. For **molting**, see September Farm Animals. For **mites and lice**, see March Farm Animals.)

Cattle (subfamily Bovinae, genus Bos, Bos primigenius) can breed any month of the year. Gestation (pregnancy) is 9 months. Most farmers breed **early spring**, so calves are born **late winter**. A cow is female. A bull is an intact adult male. A steer is a castrated adult male. A baby is a calf. Cows have 4 nipples/teats (goats/sheep have 2).

Cattle are **ruminants** like sheep, goats and deer. (A ruminant is a hoofed, horned mammal of the suborder Ruminantia that has a stomach divided into four compartments. It chews a cud.) They eat grass and weeds. They need hay in the winter. Cattle live to be about 20 years old.

Care is similar to goats and sheep. One cow needs anywhere from 2-6 acres of pasture with the amount depending on the size of the cow, type of land, how it is managed, and how much rain there is. Cattle need about 2 pounds of hay per 100 pounds of body weight every day. A cow, bull or steer drinks up to 20 gallons of water a day. (For **how to milk, clabbering milk, and other animal care,** see goats below. Also see sheep below.)

They need their hooves trimmed at least twice a year to prevent hoof rot and lameness.

Oxen are usually castrated adult male cattle 4 years or older that are used as **draft animals**. Though females and uncastrated males are sometimes used. Oxen are trained like draft horses. A **steer** is also a castrated male but is a younger animal that is not trained. They are used for plowing, transport, moving logs, threshing grain by trampling, and powering machines that grind grain or supply water.

Common Breeds:

Ayrshire: They are orange to dark brown, with or without white colored legs. They are medium-sized with long horns. Well adapted to rocky farms and harsh winters. Average output is 17,000 pounds (~2000 gallons) of milk per 305 day-cycle/year, with 3.9% butterfat, 3.3% total protein. Cows weigh 1,000-1,300 pounds. Bulls weigh 1,850-2,000 pounds. They are active and challenging. They are used as oxen.

Brown Swiss: Light brown in color. Hardy animals tolerant of harsh climate. They are calm and easy going. They produce large quantities of milk, close behind Holsteins. Average output is 21,000 pounds (~2450 gallons) of milk per cycle with 4.0% butterfat, 3.5% total protein. Cows weigh 1,400-1,500 pounds. Bulls weigh 2,200-2,400 pounds. They are used as oxen.

Guernsey: Fawn and white in color. Hardy, calm and docile. They are small, about three-fifths the size of a Holstein, but produce up to 14,700 pounds (~1700 gallons) milk with 4.5% butterfat, 3.5% total protein each cycle. A cow weighs about 1,100-1200 pounds. Bulls weigh 1,700-1,800 pounds. Not usually used as oxen.

Holstein-Friesian: Large black and white cow bred to make large quantities of milk from grass. Production of large volumes of milk relative to other cows. Average milk output up to 28,000 pounds (~3260 gallons) per cycle, with 2.5-3.6% butterfat, 3.2% total protein. Cows weigh 1,000-1,500 pounds. Bulls weigh 2,300-2,500 pounds. Has an agreeable disposition and is used as oxen.

Jersey: Comes in all shades of brown. Docile and inquisitive. Like Guernsey, they are small, but produce relatively large amounts of milk, on average 16,000 pounds (~1860 gallons) per cycle, with a high butterfat content of 4.9%, total protein 3.7%. Cows weigh 1,000 pounds. Bulls 1,500-1,600 pounds. Not usually used as oxen.

Milking Shorthorn: They are all-red, red with white markings, all-white, or red roan (hair is mixed with a different color). A medium size breed known for easy calving. A somewhat docile animal. Multi-purpose cattle that provided early settlers with not only milk, but meat and pulling power. A typical cow produces 15,400 pounds (~1880 gallons) of milk per cycle with a butterfat content of 3.8%, total protein 3.3%. A cow weighs about 1,400 pounds. A bull weighs 2,100-2,300 pounds. Easy to train as oxen and considered one of the best for it.

(See **worming and rotational grazing** in April Farm Animals. Read "The Family Cow" by Dirk Van Loon and "Oxen: A Teamster's Guide to Raising, Training, Driving & Showing" by Drew Conroy.)

 **Dogs** (Canis lupus familiaris) and **Cats** (Felis catus) breed all year.

Dogs: Small to medium dogs breed starting at 6-12 months. Larger dogs usually come into heat (ready to mate) for the first time at 12-24 months. Dogs have two heat cycles that last about a month each. Each cycle is about six months apart. Gestation (pregnancy) is 63-65 days. Litter size is 1-17 puppies with the average being about 6.

Some veterinarians like to give **core vaccinations** (except for rabies) at 6 weeks of age. However, puppies are usually protected by mother's milk at that age so the shot is effective in only 25% of them. Most veterinarians wait until 7-9 weeks for the first shot.

Cats: They breed all year but most births are **March through September.** They can get pregnant at 4-5 months old. Heat cycles last from several days to two weeks or longer. This repeats every 2-3 weeks. Litter size is 1-10 kittens with the average being about 4-5.

 **Donkey or Ass** (Equus africanus asinus) is part of the Equidae or horse family. It has been domesticated for at least 5000 years. **They are used as draft or pack animals.**

They are also **good at protecting goats, sheep, llamas, cattle and other animals from predators** such as coyotes. They have a herd instinct so stay with the rest of the animals all the time. They chase predators and attack them by rearing up and hitting them with their front feet.

A male donkey or ass is called a jack. A female is called a jenny or jennet. A young donkey is a foal. Jack donkeys bred with a female horse produces a mule. Donkeys can live to be 30 or 50 years old. The height at the withers (highest part of the back at base of the neck) ranges from 31-63 inches. They weigh from 180 to 1,060 pounds.

A jenny is usually pregnant for 12 months but varies from 11 to 14 months. They usually give birth to one foal.

Donkeys have **a reputation for being stubborn.** If they are afraid of something, it is difficult to force them to go near it or do what you want. They are not as emotionally connected to people as horses are. However, when you have earned their trust, they are willing and pleasant to work with.

They are hardy and not picky about their food. They are not ruminants. They are grazers. They eat grain, grass, weeds and hay. Their hooves need to be trimmed 3-4 times a year. They are originally from warm, dry climates so need more protection in winter than other animals. Their coat does not resist water as well as horses.

Read "The Donkey Companion: Selecting, Training, Breeding, Enjoying & Caring for Donkeys" by Sue Weaver.)

 **Ducks:** Domestic ducks (Anatidae family of birds) do not have to have a pond though they love having one to swim in. You can give them a childrens wading pool. **Better in rainy, wet environments than chickens.** They love rain. They are messier than chickens because they like to play in water. Most domestic breeds are unable to fly. Care is similar to chickens. A female duck is a hen. A male duck is a drake. A baby duck is a duckling.

Ducklings and goslings (baby goose) imprint on the first creature they see that is bigger than them. They can imprint to other ducklings. (It needs to happen in the first 30 hours.) They believe that creature whether a mother duck or a human is their species. They imprint on the species and at about 1 week old can recognize an individual of that species. They also recognize voices. If they imprint on you, they will follow you around when they are adults.

A duck eats more feed than a chicken since it is bigger but their eggs are 20-35% larger. **Ducks lay 250-325 eggs a year.** Chickens lay about the same number of eggs a year. Many bakers prefer duck eggs to chicken eggs because baked goods are softer, fluffier, and better rising.

There are two main **egg producing breeds:** Indian Runner (3-5 pounds) and Khaki Campbell (4 pounds).

The main **meat producing breeds** are Aylesbury (10-11 pounds), Muscovy (7 pounds female, 15 pounds male), Pekin (very fast growing, 8-11 pounds), and Rouen (9-12 pounds). Pekin is the most commonly eaten duck.

Dual purpose breeds that do fairly well as both layer and meat birds include Ancona (6-7 pounds), Magpie (5-6 pounds), Orpington or Buff Duck (7-8 pounds), Saxony (7-8 pounds), and Swedish (6-8 pounds). Nantahala Farm (www.nantahala-farm.com) sells Ancona ducks.

(📖 Read "Storey's Guide to Raising Ducks: Breeds, Care, Health" by Dave Holderread.)

 **Goats** (Capra aegagrus hircus) are about 1/6 the size of a cow. There are 3 types of goats: **dairy, meat and fiber**. Goats are **ruminants** (have 4 stomachs, even-toed hooves, and chew their cud). **They are browsers not grazers.** They like to eat grass, weeds, and tips of bushes and trees. They eat hay. They love grain but should be fed it in moderation. They need goat minerals and mineral salt to be available all the time. Sodium bicarbonate (baking soda) is helpful to control stomach pH and prevent bloat.

A female is a **doe or nanny**. A male is a **buck or billy**. Castrated males are wethers. They usually live about 10-12 years but can live to be 20. Breeding of does should stop at about 9-10 years old.

They need shelter from rain, snow and wind. They are curious and intelligent. They do not like being alone since they are herd animals.

One acre of pasture/browse will support about 5 full-size goats depending on pasture/browse quality and rain. However, the more land they have, the fewer problems with internal parasites because they prefer eating plants away from the ground. If the pasture is overgrazed, they are more likely to eat plants near the ground and then contact worms. Contact your veterinarian and/or County Extension office. (See **worming** in April Farm Animals. It includes **pasture management** to reduce worm infections.)

1. Common dairy goat breeds:

Alpine are medium to large goats with upright ears. They are hardy, adaptable and come in a variety of colors and patterns. Their milk has an average butterfat of about 3.5%. They are seasonal breeders meaning they only breed during certain months of the year. Does (females) weigh about 125 pounds.

LaMancha goat is a medium-large breed with a short, glossy coat of any color or pattern. They have 1-2 inch ears or small/absent ears. Does produce 1,050 to 3,510 pounds of milk per year (3.9% butterfat). They are seasonal breeders. Bucks weigh about 150 pounds. Does weigh about 130 pounds.

Nubian goats are a medium-large breed that have short, glossy, fine coats of any color or pattern. They have long ears. They produce over 3,000 pounds of milk/year (4.8% butterfat) and are meatier than other dairy breeds. They usually are seasonal breeders but sometimes breed all year. Bucks weigh 175 pounds. Does weigh 135 pounds.

Nigerian Dwarf is a miniature dairy goat with upright ears. They are bred to have the length of body and structure, in proportion, of a full size dairy goat. They come in a variety of colors and patterns. They produce a lot of milk for their size (3-4 four pounds per day). Their milk has an average butterfat of about 6.1%. They breed all year. They weigh about 75 pounds.

Oberhasli are sometimes called Swiss Alpine. They are a medium sized breed. Their coloring is reddish brown with distinctive black markings. Does may sometimes be black. Their milk has an average butterfat of about 3.6%. They are seasonal breeders. Does weigh about 120 pounds and bucks weigh about 150 pounds.

Saanens are white and large with upright ears. They are heavy milk producers but their milk is lower in butterfat than some of the other dairy breeds. Their milk has an average butterfat of about 3.5%. They are seasonal breeders. Does weigh 150 pounds or more. Bucks weigh over 200 pounds.

Toggenburg goats are medium size with upright ears. They have a light tan coat that turns chocolate color with a lot of copper in their diet. They have distinctive white markings. Their milk has an average butterfat of about 3.3%. They were originally bred for cheese production. They are seasonal breeders. Does weigh at least 120 pounds. Nantahala Farm (www.nantahala-farm.com) sells Toggenburg goats.

2. Common meat goats breeds: include those below plus **Kiko** and **Savanna** goats.

Boer: Large, well-muscled, classic meat goats. Their short coats may be red, black or spotted, though most are white with red or brown heads. Boers have long, pendulous (hanging) ears. They breed most of the year. Bucks weigh between 240-300 pounds. Does weigh 200-220 pounds.

Fainting or Myotonic goats (Fainters or Tennessee Fainting goats) are small to medium in size. Their coats can be long or short. They come in a variety of colors including black, tan, red, brown, and white. They breed all year. They weigh from 60-175 pounds.

Pygmy goats are small, about the size of Nigerian Dwarf. Most are raised as pets but they are meat goats. They are short, stocky, thickset and heavy boned. They can be any color. They breed all year. They weigh 40-70 pounds.

3. A common fiber goat is Angora. Their mohair grows 6 inches long. A single goat produces 6 1/2 to 7 1/2 inches a year. They have creamy white hair, and produce 12 or more pounds of mohair per year. The mohair is made into silk-like fabric or yarn. It is durable and warm. Does weigh 70 to 110 pounds. Bucks weigh 125 to 200 pounds.

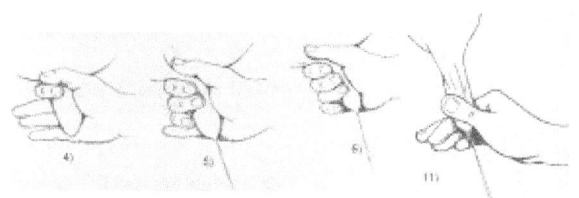  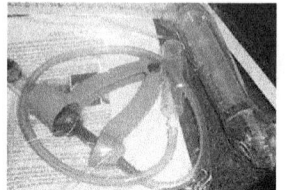 

Care for All Goats

1. Milk Production: Goat gestation (pregnancy) is 5 months. Seasonal breeders usually give birth January through April. They usually give birth to 2 kids though 1, 3 or 4 are also possible. Most births are very successful. Goats have 2 nipples.

To improve milk production feed alfalfa hay, alfalfa meal/pellets, oats, barley, beet pulp, 16-18% protein feed but not in excess. Feed live yeast cultures such as Diamond V (this is not brewers yeast) that help prevent acidosis by raising rumen (stomach) pH. Brewers yeast is good as a general supplement though. Herbs such as dill, fennel, and fenugreek increase milk. (For **beet pulp** see Fall and Winter Pasture in "Garden Tips" in back of manual. For **more about hay,** see June Garden-Harvest.)

Milk 3-4 times a day for the first 30 days for maximum production during that cycle. But 2 times a day is enough. If you do not want to bottle feed the kids, you can let them nurse full time the first week or so. After that at night separate kids from mothers. Milk the mothers in the morning and let the kids stay with the mothers all day. This way you milk once a day. When you sell the kid(s), you milk twice a day. Most kids can be sold at 2 months old.

Cows are easier to milk by hand (see first image above) than goats because they are so much bigger. If you do not want to milk goats with your fingers, you can use a **manual milker**. The best is one with a hand pump similar to that used to inject animals with. There is no needle however. Food-grade tubing is attached to the back of the pump. A 35 or 60 cc (ml) syringe is attached to the other end of the tubing (see second and third images above). If you have small goats or for some standard goats, a 35 cc syringe is better. Some larger goats or those with large nipples may need the 60 cc syringe. You pump the milk into a 1/2 gallon plastic or glass container such as a mason jar. You need a tiny brush to clean inside the tubing occasionally.

Clabbering Milk: Clabber is produced by letting raw, unpasteurized milk (goat, cow or other milk) **turn sour (curdle)**. Over time the milk thickens or curdles into a yogurt-like substance with a strong, sour taste. Pasteurized and homogenized milk will not clabber. You do not have to have a **starter** such as kefir grains or buttermilk culture though they can be used to produce different tasting sour milk.

If you do not use a starter, a **lacto-acid producing beneficial bacteria** found naturally in the environment causes the souring. This is similar to Lactobacillus acidophilus and other beneficial bacteria that helps people digest their food. This is also found in naturally fermented products such as sauerkraut. Clabbered milk is very healthy for you.

Leave the milk on the counter in a lightly sealed jar (preferably glass) for **2-3 days in a warm spot** (room temperature such as 70 degrees or warmer) until **solids (curds)** appear. The curds are like cottage cheese and float at the top. You can strain it out and let the liquid drain out of it. The **liquid is whey** and can be used in cooking or fed to animals. It is very good for them. The longer you let it sit, the more sour it gets. The warmer the temperature, the faster it sours. To stop the souring, put it in the refrigerator. You can drink or cook with it prior to total separation of curds and whey if you want a **buttermilk** type product that is not as sour.

On very rare occasions it will not sour properly and you will have to throw it away. You will learn how it is supposed to taste and smell when done properly. Find an expert to help you with this before you drink it.

2. Disbudding and Scurs: If you want goats with no horns, then kids should be disbudded (removal of horn buds) when they are about 3 days old for males and 4-8 days old for females. Do not disbud the first day. Check the size of the horn bud first. Disbud when there is a good nub. It is best to use an electric horn disbudder.

Males are more likely than females to get scurs (partial horn growth) when they are adults. If this happens, you can trim them every few months. An **OB saw** (see last image above) from a farm supply store or a wire saw is the easiest way to do this. Only cut about an inch or so at a time. How much you can cut depends upon whether the horn bleeds or not. Contact your veterinarian and/or County Extension office.

3. Hoof Trimming: Goat hooves need to be trimmed every 6-8 weeks. Otherwise, they may get hoof rot or lameness. It is easiest if you secure them in a milk stand. Use a hoof trimmer that is similar to garden shears. Trim a little at a time. The bottom of the hoof should be fairly flat when you are done. (See image of hoof trimmer at beginning of the goat section.)

4. Lice, Mites or Mange:
A natural method is to give 1 teaspoon yellow dusting sulphur per goat per day in their feed until cured.

Or use the chemical CyLence Pour-On Insecticide (1% Clyfluthrin) at 1 ml per 25 pounds for goats orally. Do not pour on goat skin. (Cattle is used at 8 ml per 400 pounds. You can pour it on their skin.) Treat again after 3 weeks and maybe again in another 3 weeks. You probably will not need it after that for a while. Also good for horn and face flies. Contact your veterinarian and/or County Extension office.

Or use "Ivermectin 1% Injectable" orally (not injected) at about 1 ml per 25-50 pounds for goats. (Cattle dosage is 1 ml per 110 pounds. Swine dosage is 1 ml per 75 pounds.) Also good for gastrointestinal roundworms, lungworms, Sarcoptic mange and Psoroptic mange. Contact your veterinarian and/or County Extension office.

5. Vaccinations are given such as CD&T every year. Kids are given it at 1-3 weeks old. Then again in 1 month. It is best to vaccinate does 1 month prior to kidding. Other vaccines may also be given. It is good to worm then too.

Not everyone vaccinates their animals. You may not need to if your herd is healthy. Consult your veterinarian and/or County Extension office. (See sheep a few pages from here for scrapie, urinary stones, and pregnancy toxemia information.)

6. Poisoning. Contact your veterinarian promptly. If this is not possible, you can try this formula but do so at your own risk. Animals react differently to various treatments. Poisoning symptoms include frothing at the mouth, vomiting, lack of coordination, crying, shaking, rapid breathing/pulse, paralysis, convulsions and possible death.

If your goat has eaten something bad but does not show symptoms yet, give 2 tablespoons or 2 crushed tablets of Pepto-Bismol mixed with 1/8 cup (for 100 pound goat) or 1/4 cup (for 200 pound goat) mineral or food-grade oil such as olive oil. The goat will have diarrhea the next day.

Or mix a rounded teaspoon of activated charcoal or a rounded teaspoon of Bentonite Clay with 1/2 teaspoon of vitamin and 1/3 cup of water. Use an oral drench syringe to give to the goat. (See image of syringe at beginning of the goat section.) Repeat every 1-6 hours. Do it every hour if the goat is very sick.

If your goat is showing symptoms and is vomiting, get your goat to a veterinarian right away. If a vet is not available, wait until vomiting stops. Then do one of the above formulas. Watch the animal carefully.

(For **breeding season,** see September Farm Animals but most details are in this section. For **worming,** see April Farm Animals.) Read "Nourishing Traditions" by Sally Falon and "Natural Goat Care" by Pat Coleby.)

Horses (Equus ferus caballus) usually breed in the **spring**. Gestation (pregnancy) is 11 months so the foal is born the following spring. One foal is born per birth. A female horse is a mare. A young male horse less than 4 years old is a colt. A male horse that has not been castrated is a stallion. An adult male horse that has been castrated is a gelding.

Horses live 25-33 years. They need to be regularly groomed.

Hooves need to be maintained by a farrier (a craftsman who trims and shoes horses' hooves). They should be trimmed every 6-8 weeks. Hooves should be frequently cleaned with a hoof pick to remove stones and dirt. Keep feet clean and dry to prevent hoof rot and lameness.

They need shelter from rain, wind and snow. They need horse minerals and mineral salt available all the time. They do not like being alone since they are herd animals.

A horse or pony needs 1.5-2.5% of its body weight in food per day. The average riding horse weighs 1000 pounds. Horses are **more sensitive than goats/sheep about hay they eat**. They need a good pasture. They prefer grass over weeds (they are not ruminants like goats/sheep). A horse needs 10-12 gallons of water a day. (For more about **hay types,** see June Garden-Harvest.)

Types of Equine Breeds:
    Cold-bloods- Larger, gentle horses for working or hauling.
    Hot-bloods- Swift, fast horses used for racing and speed.
    Warm-bloods- Great breed for equestrian sports and competitions.

The two most popular horse breeds in the United States are the Arabian and American Quarter Horse.

American Paint is a breed and not a color. It is known for its color and stock breed style. They have several color patterns and coat colors.

American Quarter Horse is one of the most popular horse breeds in both the United States and the world. It was developed for speed at quarter mile sprints.

American Saddlebred originated in Kentucky and Tennessee. They are easy riding horses. It is largely used for show purposes and is known for its controlled gaits, high head carriage and distinctive tail set.

Appaloosa is named after the Palouse River country of the northwest United States. These colorful horses are known for their beauty, endurance and surefootedness.

Arabian is a light breed. Hot blooded horses such as the Arabian can be traced back to the Arabian, the Barb and The Turkmene.

Missouri Fox Trotter was developed in southern Missouri as an easy riding horse that moves long distances at a speed of 5-8 miles per hour. The breed has roots in plantation horses of the south, Arabian and Morgan.

Morgan is believed to have been of Thoroughbred and Arabian breeding. The horse is popular for riding and driving and is known to be an excellent all around horse. (Driving is hitching a horse to a wagon, carriage, cart or other horse-drawn vehicle with a harness.)

Tennessee Walking Horse was bred for riding, driving and farm work. The horse naturally overstrides, which creates a running walk gait. The breed was developed from the Narragansett and Canadian Pacers, as well as the Saddlebred, Standardbred, Thoroughbred and Morgan.

Thoroughbred race horse was developed for speed at intermediate distances. No other breed can match it at racing distances.

Draft or work horses are large horses bred for hard, heavy tasks. They are strong, patient, and have a calm temperament. They were indispensable to pre-industrial farmers. They include breeds such as Abtenauer, American Cream, Belgian Draft, Brabant, Breton, Clydesdale, Dutch Draft, East Bulgarian, Fjord, Friesian, Galiceno, Groningen, Haflinger, Italian Heavy Draft, Jutland, Konik, Lithuanian Heavy Draft, Murgese, Nonius, Noriker, North Swedish, Orlov Trotter, Percheron, Pleven, Shire, Spotted Drafts and Suffolk Punch. **Belgians** are by far the most popular breed of draft horse in the United States.

(See **worming and rotational grazing** in April Farm Animals. Read "Complete Horse Care Manual" by Colin Vogel and "Draft Horses: An Owner's Manual" by Beth A. Valentine. The "Small Farmers Journal" quarterly magazine has articles on using draft animals in farming. Check out the "Draft Horse Journal" quarterly magazine.)

Pigs / Hogs / Swine (Sus domesticus) can breed all year. Gestation is 4 months (112 days). Sows have about 12 piglets per litter. A male pig is a boar. A young, just weaned pig is a shoat or shot. A female pig that has never given birth is a gilt.

Pigs are the most intelligent animal on the farm. They eat a wide variety of foods and love acorns. On homesteads they were fed scraps from the garden and kitchen. Most domestic pigs live 10-15 years.

You can buy a piglet in the spring, raise it all summer, and then kill it late fall for food all winter. Ideal butchering weight for **hogs** is about 240-260 pounds with 300 pounds if you like a lot of fat. Of course, this depends on the breed of pig. But this is the most efficient weight per amount of feed given. (For **butchering,** see November Farm Animals.)

They need shelter from rain, snow, sun and wind. If the weather is hot, they need mud to cool off in. They need a very strong fence if kept in a small area. Or they can be pasture raised on a large amount of land where they will be easier on the fencing. Some say the meat tastes better when pasture raised.

American Guinea Hogs were the most popular pig in the southeastern United States in the 1800s. It is an all black, lard pig that grows to 250-300 pounds. They are endangered. They are a great homestead pig.

Common breeds of pig include:

American Landrace are large white pigs with an elongated body with a less-pronounced back arch than other breeds. The sows are prolific breeders that produce large amounts of milk. It is commonly used for bacon.

Berkshire are black with white spots. They have short, perky ears and white stockings on their feet. They are medium-sized, quite hardy, and often used as the male in crossbreeding. They grow quickly and efficiently and produce large litters.

Chester White pigs are used as meat producers and for crossbreeding. These medium-sized pigs have white skin with some black spots, a straight back similar to that of Landrace pigs and floppy ears. They produce in a wide range of settings, making them popular for both large and small farms.

Duroc has the highest feed-to-meat conversion ratio of any United States pig breed. They have red skin and brown, black or red hair. They are relatively lean and muscular compared to other pigs, grow to a large size, have short, floppy ears and a short snout.

Hampshire has a distinctive black coat with a white stripe around the forequarters. Hampshires are large pigs noted for their hardiness, high-quality meat and foraging ability.

Poland China are prolific in reproduction. These black pigs have white socks, snout and tail, and a very sturdy frame. They produce 16-17 piglet litters regularly, but have a tendency to overlay large litters. This pig is extremely hardy and deals well with varying conditions.

Spotted, or Spots, pigs are descended from Poland China hogs, but have a lighter frame and distinctive black and white spotting. They are efficient feeders, converting food to meat quickly. They transmit these qualities to their offspring, even when crossbred. Spotted female pigs are noted for their gentleness and hardiness.

Yorkshire or American Yorkshire are large and long, similar to American Landrace, but smaller. It has pink skin, white hair and upright ears. Sows make excellent mothers. This breed is a major ham and bacon producer.

(📖 Read "Storey's Guide to Raising Pigs" and "Dirt Hog" by Kelly Klober.)

 **Rabbits / Bunnies** (European rabbit or Oryctolagus cuniculus) breed all year. Gestation (pregnancy) is one month. They are productive breeders for 4-6 years. The female is a doe. The male is a buck. A baby rabbit is a kitten or kit. Small rabbit breeds such as dwarfs live 10-12 years. Larger to medium sized breeds live 8-10 years.

They need a cage that is protected from rain, sun, snow and wind. They like **grass hay** such as timothy. Give as much as they want. They also like alfalfa hay (a legume) but only feed a little at a time. They eat rabbits pellets that contain a lot of alfalfa. They like dark leaf lettuces, collard greens, comfrey, turnip greens and carrot tops.

Rabbits sometimes get **ear mites.** A remedy I got from a local old timer is to mix 3 parts food-grade oil such as olive oil with 1 part horse liniment (Absorbine Veterinary Liniment). Liniment is a blend of herbal extracts and essential oils including calendula, echinacea and wormwood. Put some in each ear and rub the ear a little. Do this again in a few days to a week. Ask your veterinarian and/or County Extension office before trying it. This formula may be useful in other animals.

Common breeds of **meat rabbits** are New Zealand White and Californian. They have high growth rates, dressed weights, and ratios of meat to bone.

New Zealand rabbits are black, red or white. They weigh 10-13 pounds when mature, and are market-ready fryers (4-5 pounds) by 8 weeks. Average litter size is 8-10. It is well-proportioned with a full, well-muscled body.

Californian rabbits are white with black ears, nose, feet and tail. When mature they weigh 9-10 pounds. The average litter size is 6-8. The body is plump but fine-boned. Breeders cross New Zealand and Californian for their vigorous hybrid offspring. (📖 Read "How To Raise Rabbits" by Samantha and Daniel Johnson.)

Sheep (Ovis aries) usually breed in the **fall** like most goats do. Gestation is 5 months so lambs are born in **spring.** A female sheep is a ewe. They have 2 nipples/teats like goats. An intact (not castrated) male is a ram. A castrated male is a wether.

Sheep live about 8-13 years, sometimes 20 years. They are raised for their wool and meat. Lamb is from a sheep less than 1 year old. Mutton is from an adult sheep. Sheep are more docile than goats.

They are ruminants like goats and deer. Their care is similar to goats (see above). Contact your veterinarian. One acre will support about 6-7 sheep depending upon the quality of the pasture/browse and the amount of rain.

Sheep are sheared or clipped (hair is cut off) for their woolen fleece. Wool fabrics have greater bulk than other textiles, and retain air, making them good for insulation. Wool is spun into yarn, and then woven into fabric or knitted.

They are also sheared to make it easier for a lamb to nurse. It is good to trim hooves the same time you shear. Feet need to be trimmed about every 2 months to prevent foot rot and lameness.

Vaccinations are given such as CD&T every year. Lambs are given it a 1-2 months old. Then again in 1 month. It is best to vaccinate ewes 1 month prior to lambing. Other vaccines may also be given. It is good to worm then too. Consult your veterinarian and/or County Extension office.

Scrapie is an infectious, fatal, degenerative disease found in goats and sheep. It is more common in sheep than goats. It eats the brain similar to Mad Cow Disease. It is spread through the placenta. Do not let your animal eat the placenta. Symptoms include tremors, scratching, weakness, weight loss, and poor coordination. It is rare in North Carolina. The USDA and APHIS have scrapie eradication programs that involve certification.

Sheep may have **urinary stones** or crystals (urolithiasis). It is more common in castrated males. It is caused by too much phosphates in the feed. Prevent problems by only giving feed that is meant for that particular species.

Pregnancy toxemia and ketosis can occur in the last 4 weeks of pregnancy in sheep, goats, horses and donkeys. Babies require a lot of energy (carbohydrates) towards the end of pregnancy. If the mother is not getting enough high quality food, her body converts fat to glucose (sugar). This happens more if the mother is overweight, stressed, lacks exercise, or has too many worms.

Symptoms include swelling in the legs, stiffness, not much activity, depression, and low appetite. The breath smells like acetone (nail polish remover). Feed the ewe more corn, alfalfa hay, and/or soybeans. Additional supplements (choose one) include Propylene Glycol, Nutri-Drench, Dextrose, or Magic (1 part molasses, 2 parts Kayro, 1 part food-grade oil). Get the ewe exercising. Alfalfa hay is also good when ewes are milking. Contact your veterinarian.

Sheep are categorized as being best for providing a certain product such as **wool, meat, milk, hide,** or a combination (dual-purpose).

Breeds:

Fine-wooled breeds include American Cormo, American Merino, Booroola Merino, Debouillet, Delaine Merino, and Rambouillet. Their wool has great crimp and density. It is used for textiles.

Medium-wooled breeds (bred for meat) include Cheviot, Dorset, Hampshire, Montadale, North Country Cheviot, Oxford, Shropshire, Southdown, Suffolk, and Tunis. They are high wool production breeds.

Long-wooled breeds (bred for mutton and wool) include Border Leicester, Coopworth, Cotswold, Lincoln, Perendale, Romney and Wensleydale. They are the largest sheep (225-350 pound rams and 175-275 pound ewes).

Fur or hair breeds do not grow wool. They are raised for meat and hides. They include American Blackbelly, Barbados Blackbelly, California Reds, Dorper, Katahdin, Romanov, Royal White Sheep, and St Croix.

Dual purpose breeds include American Miniature Brecknock, Columbia, Corriedale, East Friesian, Finnsheep, Panama, Polypay and Targhee.

(For **breeding season,** see September Farm Animals.) (Read "Natural Sheep Care" by Pat Coleby.)

Turkeys (Meleagris gallopavo), were raised on small family farms in the past **for meat and pest control.** They eat lots of bugs, mosquitoes, ticks and flies. Turkey care is similar to chickens and ducks. Baby turkeys are called poults. A Tom is a male turkey. A hen is a female turkey. Most domestic turkeys can not fly.

The **eggs** are edible but they are not as productive as chickens. A **turkey lays about 100 eggs a year** compared to chickens laying 250-300 eggs a year.

A **heritage turkey** is a variety of domestic turkey that has historic characteristics that are no longer present in the majority of turkeys raised for eating (commercial). Heritage breeds can mate naturally; commercial breeds can not.

Heritage turkeys have a longer life-span and a **slower rate of growth than commercial turkeys**. Commercial breeds are slaughtered at 14-18 weeks, while heritage breeds reach a marketable weight in 28 weeks. This gives them time to develop a strong skeleton and healthy organs before building muscle mass. Auburn, Buff, Black, Bourbon Red, Narragansett, Royal Palm, Slate, Standard Bronze, and Midget White are heritage breeds.

Turkeys are **native to the United States**. The original breed was the wild Bronze turkey from which domestic varieties have descended:

Beltsville Small White is a small, white-feathered heritage breed slightly bigger and broader than Midget White.

Black, Spanish Black, or Norfolk Black has very dark plumage with a green sheen.

Bourbon Red is a smaller non-commercial breed with dark reddish feathers with white markings.

Broad Breasted White (Large White) is used mostly for commercial meat production. Commercial turkey's breast is too large for natural mating so they are artificially inseminated.

Broad Breasted Bronze (Large Bronze) is not as popular as it once was because the black pin feathers are noticeable on the dressed bird.

Chocolate is a somewhat rare heritage breed with markings similar to a Black Spanish, but light brown instead of black in color.

Midget White is a smaller, white-feathered heritage breed that looks like a miniature of the commercial Broad Breasted White turkey.

Narragansett is a popular heritage breed named after Narraganset Bay in Rhode Island. Its color pattern has black, gray, tan and white. The pattern is similar to Broad Breasted Bronze. They have a calm disposition, good maternal abilities, early maturation, good egg production, and excellent meat quality.

Slate or Blue Slate turkeys are a very rare breed with beautiful gray-blue feathers.

Standard Bronze looks like the Broad Breasted except it is single breasted and can naturally breed.

(Read "Storey's Guide to Raising Turkeys: Breeds, Care, Health" by Leonard S. Mercia.)

MARCH
March 1 daylight is 11 hours 27 minutes. March 31 daylight is 12 hours 33 minutes.

March 20 or 21 is **Vernal Equinox, night and day are about the same length.**
March comes in like a lion and goes out like a lamb.
Second of 3 "hunger" months. Sometimes rains, sometimes snows. Mud season.
Early March: Frogs are active in ponds breeding. The ground is frozen except for the first inch.
Perennials such as chives, comfrey, day lily and mint start to come up. Greenhouse plants growing fast.
Late March: The ground is not frozen anymore. High temperatures usually in the 50s. Pasture has a hint of green.
More perennials start to come up such as oregano, sorrel, strawberry, thyme, and parsley.

Garden- Maintenance

Divide Roots of Mature Perennials:

Every 3 years **divide chives** (amaryllidaceae family; Allium sativum or Allium schoenoprasum) clump when leaves are 3-4 inches tall. Dig up whole clump and divide into groups of 5-10 bulbs. Plant 6 inches apart. Can also divide in **fall.** (See May Garden-Plant.)

Every few years **divide the roots of well developed perennial plants** such as anise hyssop, catnip, echinacea, lemon balm, mint, oregano, sage, salad burnet, sorrel, St Johns wort, tarragon, thyme, and valerian.

Remove deep **mulch** at spring thaw or when daffodil leaves are a few inches out of ground. Remove from asparagus, comfrey, garlic, onions, rhubarb, strawberries and other **over-wintered perennials.**

Most plants should be **fertilized in spring** to encourage new growth. (For **fertilizing,** see April Garden-Maintenance. For details about **natural fertilizers and soil health,** see "Garden Tips" in the back of this manual.)

Fertilize **asparagus** well. Asparagus has 3 stages of growth: **harvest, fern and dormancy.** Each stage requires different fertilization. Asparagus responds well to nitrogen in the fern stage. During the other 2 stages, the nutrients are stored in the crown. Phosphorus is needed then.
(For details about **planting seeds and crowns,** see April Garden-Plant. For **salt and production,** see July Garden-Maintenance. See April Garden-Harvest.)

Fertilize fruit/nut trees and bushes. Fertilizing is best done **just before leaves start to develop.** Do not fertilize in **summer** since this stimulates late summer or fall growth that is susceptible to winter injury.
Nitrogen encourages leafy growth but too much nitrogen discourages bloom and fruit production. **Phosphorus** encourages root/flower production. **Potassium** encourages root development. It is best to have your soil tested. It is good to add rock dusts to your soil every year. Rock dusts can be added at any time. (For **How and Why to Fertilize and Rock Dusts,** see "Garden Tips" section in back of this manual.)
Planting a **perennial cover crop** of grasses, legumes and herbs is a great way to have long-term fertility in your orchard. Cover crops provide nitrogen, attract beneficial insects and recycle nutrients. Comfrey is a good cover crop in orchards. (For **cover crops,** see "Garden Tips" section in back of this manual. For **comfrey,** see March Garden-Plant.)

Fertilize grapes before the buds swell in **spring.** If there is too much vine growth and not enough grapes, apply less nitrogen. If fruit set has happened, you can apply some fertilizer in **early to mid June.**

Fertilize strawberries in late June after main harvest. If fertilize now, fruit may be soft. (For **fertilizing and more,** see June Garden-Maintenance.)

Apply Dormant Oil or Fungicide Spray to Fruit and Nut Trees in March or February
Prune trees first. (For **details about spraying**, see February Garden-Maintenance. For **more about oil sprays and insecticides**, see "Plant Health" in the back of this manual.)

Propagate Fruit and Nut Trees/Bushes:

 To **propagate blueberries step 2:** Take **cuttings** out of refrigerator and place in container filled with well-soaked sphagnum peat and coarse sand (half of each). Stick bottom of cutting into mixture with only top bud above ground. Rooting hormone is not needed. Keep moist. Should leaf in 6 weeks. Transplant in **June** to nursery area. Watch for 1 year. Then transplant to permanent site.
(For **step 1** see December Garden-Maintenance. For **step 3** see June Garden-Plant. For **pruning** see February Garden-Maintenance. See April Garden-Plant. For information about potting mixes and herbaceous/softwood/semi-hardwood/hardwood cuttings, see "Garden Tips" section in back of this manual.)

 Propagate **elderberry** from hardwood cuttings, softwood cuttings, root cuttings, or suckers.
Hardwood cuttings 12 inches long with 3 or more nodes (where leaves grow) are taken from 1-year-old canes in **fall or early spring** when plants are dormant. They are planted in spring (step 2) with top 2 buds above the soil and 1 stem node in the soil. Put about 6 inches deep in soil. Plant each cutting 1 foot apart. No rooting hormone is needed.
(For **pruning** see February/July Garden Maintenance. See April Garden-Plant and August Garden-Harvest. For **propagation with hardwood step 1**, see October Garden-Maintenance. For **propagation with semi-hardwood**, see July Garden-Maintenance. For general details about herbaceous, softwood, semi-hardwood and hardwood cuttings, see "Garden Tips" in back of this manual.)

 To **propagate grapes step 2:**
Take **cuttings** out of refrigerator and place bottom in moist, well drained soil with 2 nodes (where leaves grow) in soil. Will root in 4-8 weeks. Transplant to nursery area. Watch for 1 year. Then transplant to permanent site. Can also propagate by burying part of cane in dirt next to plant (**tip layering**).
(For **step 1** see December Garden-Maintenance. For **pruning** see February Garden-Maintenance. See April Garden-Plant. For general details about tip layering and herbaceous/softwood/semi-hardwood/hardwood cuttings, see "Garden Tips" in back of this manual.)

 Garden- Plant (March)
Cool weather cover crops: Clover, fava beans, fescue, flax and oats.
Vegetables and Fruit: Anise hyssop, beets, burdock, carrots, collard greens, cilantro/coriander, comfrey, creasy greens, dandelion, dill, garlic, horseradish, kale, lettuce, mustard greens, peas, rhubarb, sea kale, spinach, strawberry and turnip.

Root vegetables were the hidden treasure of medieval peasants because plundering armies could destroy the village, burn the barn, and steal the livestock, but it was hard for them to dig out all the potatoes, turnips and parsnips.
(For a list of **cold hardy vegetables and minimum cold temperatures**, see February Greenhouse.)

 Good gardening books: (more are listed in relevant sections)
"The New Seed Starters Handbook" by Nancy Bubel.
"The Complete Book of Vegetables, Herbs and Fruit" by Biggs, McVicar and Flowerdew.
"Specialty and Minor Crops Handbook" by the University of California
"Manual of Minor Vegetables" by James Stephens
"The Gardener's A-Z Guide to Growing Organic" by Tanya Denckla.

"Perennial Vegetables" by Eric Toensmeier.
"Gaia's Garden: A Guide To Home-Scale Permaculture" by Toby Hemenway.
"American Horticultural Society Plant Propagation" by Alan Toogood
"The Organic Gardener's Handbook of Natural Insect & Disease Control" by Ellis and Bradley.
"Insect, Disease and Weed ID Guide" by Cebenko and Martin
More about organic gardening: see http://attra.ncat.org, National Sustainable Agriculture Information Service.
An excellent monthly magazine is **Acres USA.** It covers commercial-scale organic, sustainable farming, and natural animal care. They sell **eco-agriculture books** at their web site: http://www.acresusa.com.

 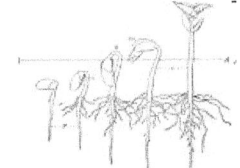

Seed Sources:

Bountiful Gardens (www.bountifulgardens.org) has heirloom, open-pollinated seed.
Fedco Garden Supplies (fedcoseeds.com) has seeds, tubers, bulbs and trees. Good prices.
Peaceful Valley Farm & Garden (www.groworganic.com) has seeds, plants and garden supplies.
Seeds of Change (www.seedsofchange.com) has certified organic seeds.
Seed Savers Exchange (www.seedsavers.org) has open-pollinated seed. Buy/trade seeds with members.
Nantahala Farm & Garden (www.nantahala-farm.com). That's who wrote this book! Some heirloom seeds.

Country Living TV: RFD-TV (www.rfdtv.com) is a cable channel that highlights rural living. Agriculture professionals and hobbyists cover farm shows, vintage farm equipment, commodities, and farm basics. There are many shows on horses, rural lifestyle, farm auctions, 4-H, Future Farmers of America, arts/crafts, cooking, and country music.

Rural TV (www.myruraltv.com) is a cable channel with shows about livestock, crops, farm machinery, farm business, country lifestyle, healthy living, country cooking, farming news, and weather.

Schools that Teach Country Living:

Foxfire Fund, Mountain City, Georgia at www.foxfire.org and 706-746-5828. They publish the Foxfire books and offer courses about rural living in southern Appalachia.

John C Campbell Folk School, Brasstown, NC, www.folkschool.org, 828-837-2775. They teach basketry, blacksmithing, book arts, clay, cooking, dance, drawing, dyeing, folklore, gardening, homesteading, glass, jewelry, knitting, leather, metalwork, music, nature studies, needlework, painting, printmaking, quilting, rugs, sewing, soap making, spinning/weaving, storytelling, woodcarving, woodturning, and writing.

County Cooperative Extension Service:

Every state has a County Extension office at its land-grant university, and a network of local or regional offices. They are staffed by experts who provide useful, practical, and research-based information to agricultural producers, small business owners, youth, consumers, and others in **rural communities**.

In **North Carolina** the associated universities are North Carolina State University and North Carolina A&T State University. The North Carolina County Extension web site is: http://www.ces.ncsu.edu/.

Programs

1. 4-H Youth Development- It cultivates important life skills in youth through various programs including agriculture and animal husbandry.

2. Agriculture- Research and educational programs help people improve marketing strategies and management skills. They help farmers and ranchers increase productivity through resource management, controlling crop pests, soil testing, livestock production practices, and other areas.

3. Leadership Development- They train people to be able to teach programs in gardening, health, safety, family/consumer issues, and 4-H youth development.

4. Natural Resources- They teach landowners and homeowners how to use natural resources wisely and protect the environment with educational programs in water quality, timber management, composting, lawn waste management, and recycling.

5. Family and Consumer Sciences- They help families by teaching nutrition, food preparation skills, child care, family communication, and health care.

6. Community and Economic Development- They help local governments investigate and create viable options for economic and community development.

See "Garden Tips" in the back of this manual for:
General Garden Tips
  Appalachian Folklore, How Much is a Bushel, Companion Planting
  Garden Seeders and Broadcast Spreaders

Seeds and Seedlings
  How to Stratify Seeds, How to Make Potting Mix, How to Make and Use Soil Blocks for Seedlings
  Inoculating Seeds, How to Prevent Damping Off of Seedlings, Thinning Seedlings and Plants
  Hardening Seedlings Before Planting Outside

Soil and Crops
  How and Why to Fertilize, Rock Dusts, Soil pH
  Crop Rotation and Cover Crops, Pasture, Special Forage Crops
  Weeds and What They Say about Soil Type, Prevent Fungal Diseases

Propagation
  Propagation by Tip Layering/Rooting, Propagation by 4 Types of Stem Cuttings
  Make Your Own Rooting Hormone

 ### Predicting Frost and Reducing Frost Damage

Folklore: "Clear moon, frost soon." "The first frost of autumn will occur exactly six months after the first thunderstorm in the spring." "When katydids or long-horned grasshoppers or dog-day cicadas start singing, you can expect first frost in about 6 weeks (in the fall)."

What is Frost: Frost (white or hoarfrost) happens when air temperatures go below 32 degrees. Ice crystals form on buds/flowers/leaves and injure or kill tender plants or parts of plants. **Clear, calm weather and falling afternoon temperatures, especially if the wind is out of the northwest, are perfect conditions for frost.** It usually indicates the approach of a large mass of polar air and a hard freeze.

If the temperature is cool, but there are clouds, it will not be as cold. During the day, the sun heats the earth. After the sun sets, this heat radiates upward, lowering temperatures at the ground. However, if there are clouds at night, they trap the heat and keep the temperatures warmer near the ground.

Wind: If there is no wind, the coldest air settles to the ground. The temperature at the ground may be freezing, even though 5 feet higher it may not be. Wind stops the cold air from settling to the ground. However, if the wind is below freezing, frost is very damaging.

Air and Soil Moisture: When moisture condenses out of humid air, it releases heat that may be enough to prevent frost damage to plants so wet weather is good if frost is approaching.

The more humid the air, the higher the dew point. **Dew point** is the air temperature (varies according to air pressure/humidity) below which water droplets begin to condense into dew. The more dew, the less frost damage.

Cold winter air is usually very dry. Winter winds remove water from plants faster than their roots can absorb it. And when the ground freezes, the plants can no longer absorb any water. This makes them more prone to frost damage.

Location: Cold air is heavier than warm air, and sinks to the lowest areas. The best location for a garden is on a gentle, south-facing slope that is heated by late-afternoon sun and protected from north winds. Buildings/trees or a body of water also protect plants from frost.

Soil: Different types of soil hold different amounts of moisture. Loose fertile soil releases more moisture into the surrounding air than sandy or poor soil. Good soil protects plants from frost.

Type of Plant: Young plants or parts of plants with new growth are the most prone to frost damage. Frost tolerance is higher in plants with maroon or bronze leaves, because they absorb and retain heat. Downy- or hairy-leaved plants also retain heat. Compact plants and plants close to each other are more protected from frost.

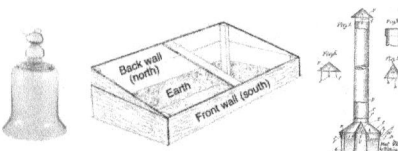 ### Frost Protection Methods

1. Water is an insulator. Plants with enough water have full cells that resist freezing. Moist soil stays warmer than dry soil. In the **fall** make sure all of your perennials are well watered before the ground freezes.

If a freeze is coming, then 1-2 days before, **water plants deeply** if the ground is dry. Water early in the day if possible. Water when air temperatures are 40 degrees or higher. Try to water only the soil, not leaves. If snow is on the ground, you do not need to water. Do not overwater since this increases soil heaving when water freezes.

Commercial fruit and vegetable growers leave **water sprinklers on all night** to wet the plants. When the water freezes, it releases heat, protecting the plants. They are covered with ice. It works if the temperatures are not too cold. Sprinklers need to run all night as long as temperatures are below freezing. It has to be done just right to work.

Or put **tanks of water** underneath fruit or nut trees. As the water freezes, heat is released.

2. Cover plants with sheets, row covers (Reemay), straw, hay, leaves or other insulating material. It is better not to cover with plastic since the plants can not breathe. (If you do use it, remove as soon as the sun comes up so plants do not overheat especially if temperatures go above 50 degrees.) Drape sheets or Reemay loosely over plants and hold them down with rocks or row cover pins. It is better if the cloth does not touch the plants but either way is OK. Cloth can be held up with wire or rope. Young trees may need cloth or other protection around the bottom of the trunk. (You can even use soil.) Remove when the temperatures warm up.

Trees withstand cold better if the ground does not have weeds on it. The earth absorbs heat in the day and releases it at night. The weeds interfere with this.

3. Cover plant with individual covering such as a milk jug, glass bowl, or other cloche. (See first image above.) One brand is Wall O' Water that has cells that are filled with water. It surrounds the plant. Remove when the weather is warm/sunny or else plants will overheat.

4. A house, greenhouse or cold frame protects plants. (See second image above.) Move potted plants inside when frost is coming. (For cold frames, see January Garden-Maintenance.)

5. Smudge pots (choofa or orchard heaters) protect fruit and nut trees from frost. (See last image above.) A smudge pot is a large pot with a chimney that is filled with oil, kerosene, diesel fuel or sump oil. It is put between the trees. The smoke and heat surround the trees reducing the chance of frost damage.

6. Fertilizing weeks before frost: Do not apply **nitrogen** late summer or early fall. Apply nitrogen late winter or early spring. Nitrogen encourages growth of new leaves which are more frost sensitive than older growth.

It is good to apply **potassium or kelp** late summer or early fall because they help plants withstand frost better. It creates thicker cell walls.

7. Anti-transpirant sprays such as Frost Away, Frostguard, Frost Shield and Wilt-Pruf help plants withstand temperatures 1-8 degrees colder than they normally would. They reduce the cooling effect of wind (wind chill).

Sow **cover crops and grains** (cool weather, spring) in March or September/October such as clover, fava, fescue, flax and oats. (See **Garden Seeders and Broadcast Spreaders** in "Garden Tips" in back of manual.)

Some cover crops are also grown for **grain** such as barley, oats, rye and wheat.

Cool weather cover crops include Austrian winter pea, barley, clover, daikon radish, fava bean, fescue, field peas, flax, fenugreek, garbanzo bean, mustard, oats, annual ryegrass, rape/canola, rye, triticale, vetch, and wheat. Sow cool season crops in early spring when cool, moist soil favor rapid growth. Cool weather cover crops should be at full height before hot weather (85 degrees or more).

(For **mustard, peas, radish (winter/daikon), and rape/canola (under kale),** see March Garden-Plant.

For **beans and fenugreek,** see May Garden-Plant.

For **vetch and ryegrass,** see September Garden-Plant.

For **barley, rye and wheat (triticale),** see October Garden-Plant.

For **clover, fava bean, fescue, flax and oats,** see below.

For **how to make hay and harvest/store grain,** see May Garden-Harvest. For **cool weather winter cover crops,** see September Garden-Plant. For **warm weather summer cover crops,** see June Garden-Plant. For more about **cool and warm weather cover crops,** see "Garden Tips" in the back of this manual.)

 Sow **clover** or trefoil (fabaceae family; Trifolium genus with over 300 species) by broadcasting. Then rake in to cover 1/4 inch. Clover is a fast growing, **cool-season** plant that adds **nitrogen** to the soil. Grows well with ryegrass. (For ryegrass see September Garden-Plant.) Beneficial insects love clover. Prefers full sun. Member of the **legume or pea family.**

Food Uses: The blossoms and seeds are edible. Young leaves are edible raw or cooked. All can be made into tea.

Popular Varieties:

1. Crimson clover (fabaceae family; Trifolium incarnatum) has dark red flowers. A **winter-hardy annual** that grows 18 inches tall. Hardy to 10 degrees. It is **sown early spring or in the fall (September-October)** for forage (feed for animals) and as a cover crop. It does not multiply by runners the way Red Clover does.

Sow 1/4-1/2 inch deep. Sow 1/2 to 2 pounds per 1000 square feet, or 20-40 pounds per acre. Likes soil pH 6.5-7.0. Tolerates shade. Reseeds itself. Does not like water logged soil. Fixes 80-125 pounds of nitrogen per acre.

As **hay** it is high in nutrition when harvested in the pre-bloom stage. **Blooms in June** on 18 inch plants.

2. Red Clover (fabaceae family; Trifolium pratense) is a popular clover with red flowers. It is a **short-lived perennial (or biennial)** that grows 15-36 inches tall. Used for **hay and silage**. (Silage is green fodder compacted and stored in airtight conditions, usually in a silo, without being dried, used as animal feed in the winter.) Fast growing. Good for breaking up clay soil. Usually **sown late summer or early fall (September/October)** as a cover crop but can be planted most of the year.

Sow 1/2 pound per 1000 square feet, or 20 pounds per acre. Likes soil pH 6.0 or higher. Grows in acid soil better than most clovers. Does not like hot summers. Fixes 200 pounds of nitrogen per acre.

3. White Clover (fabaceae family; Trifolium repens) is a **long-lived perennial** good for pasture and also used for **hay and silage**. Low growing with height of 8 inches. It has shallow roots. **Sown as a spring or fall cover crop**.

Sow 1/4 pound per 1000 square feet, or 6–10 pounds per acre. Likes soil pH 6.0-7.0. Needs good drainage. Does not tolerate drought. Flowers are white or pinkish. Fixes up to 170 pounds of nitrogen per acre.

Types of white clover: Ladino clovers grow more than twice as large as small white clover.

White Dutch clover grows to 6-10 inches tall so is classified as small or intermediate.

Sow 1/2 to 1 pound per 1000 square feet, or 15-30 pounds per acre.

(For **cool weather winter cover crops**, see September Garden-Plant.)

Sow **fava (faba) beans** (leguminosae family; Vicia faba, broad bean, bell bean) in March **when plant peas.** Grows best in temperatures 60-65 degrees, but will grow when as low as 40 and as warm as 75 degrees. **An annual that is hardy to 15 degrees. Sow 4-6 weeks before the last spring frost.**

Sow 1-2 inches deep. If **growing for their seed/bean**, plant 6 inches apart in rows 1-2 feet apart. If **growing as a cover crop,** plant closer together. Plant 3-5 pounds per 1000 square feet.

Thin plants to 8-10 inches apart. Forms a small shrub about 2-5 feet tall. Matures in 75-95 days.

Care: Likes full sun. Likes soil pH of 6.0-6.8. Likes well drained soil. Like moderately fertile soil. Fixes 150 pounds per acre of nitrogen. Tolerates frost.

Has black and white flowers that attract beneficial insects.

Pods: Pinch back top of fava bean plant when first pods begin to form to have a greater and more uniform yield. They do not form beans in warm weather. Pods are 6 to 8 inches long and have 4 to 6 flat, oval seeds that are white, yellow, green or pinkish-red.

Harvest: Pick beans for fresh use like snap beans when seeds are about the size of a pea. They are also grown to maturity and used as shelled beans.

Uses: Makes an excellent **cover crop and green manure**. Beans are **eaten** but a few people of Mediterranean descent may be allergic to them.

(See June Garden-Harvest but there are no details.)

Sow **fescue** (poaceae/gramineae/grass family; festuca genus, 300 species). Sow **February through April.** But best sown **September through November**. Best soil temperature is 50-65 degrees. Best not planted in summer. Some are annuals and some are perennials.

For a **pasture** sow 15-30 pounds per acre. Cover with 1/8-1/4 inch of soil. Seeds need to be in good contact with moist soil. Will tolerate poor soil but does better in good soil. **Fertilize in spring.**

Likes full sun but will tolerate partial shade. Drought tolerant with deep root system. Best if not mown lower than 4 inches. Somewhat dormant in very hot weather. Not many diseases or pests.

Fescue is a tufted, **cool season grass**. Related to ryegrass. Fescue **varieties** range from small grasses 4 inches tall with very fine thread-like leaves to tall grasses up to 6 feet tall with broad leaves.

There are 2 sub-species:

1. Tall Fescue (Festuca arundinacea) is a broader leaved, bunching grasses.

It is good for **North Carolina** including the mountains. It stays green and may even grow a little during the winter. It forms a strong sod tolerant of animal trampling damage, which is common on wet pasture during the winter. Winter grazing has minimal influence on yield or quality the following season. Likes soil pH 6.5 to 8.0. A perennial.

Meadow fescue (Festuca pratensis) is a tall, tufted grass similar to Tall Fescue. It grows 47 inches tall and flowers from June until August. Likes soil pH of 6.7-6.9. A perennial.

2. Fine Fescue (Festuca rubra) is a finer-leaved shorter fescue. Fine fescue grass species include creeping red, hard fescue, chewings fescue and sheep fescue. They are more shade and cold tolerant.

Creeping red fescue is a fine leaf fescue that has narrow, deep green blades. Prefers shadier and cooler areas than other fescues. Likes soil pH 5.0 to 6.5. A perennial.

For **fescue hay** fertilize in February, then again in spring after first cutting. Fertilize again after last cutting in fall.

(For **hay** see June Garden-Harvest. For **fescue** see September Garden-Plant but details are in this section. For **fertilizers**, see "Garden Tips" in back of this manual.)

Sow **flax** (linaceae family; Linum usitatissum) for **seeds or fiber**. Different varieties are grown for each. Fiber flax is taller than seed flax. Linseed oil comes from flax seed. Some varieties are annual, others perennial.

Either type of flax can be harvested early for fiber when the seeds are immature. Or either can be harvested late when seeds are mature, and used for seed and fiber.

Also grown as **cool season cover crop**. Temperatures below 28 degrees kills flax seedlings. Temperatures below 15 degrees can kill older flax plants. Grow similar to spring oats. (For **oats** see below.) **Sow after oats and before corn.** (Corn is sown in May.) **Sow about 2 to 4 weeks before the last spring frost.** In areas with mild winters such as Georgia or Louisiana it can be sown in the fall.

Sow 1-2 pounds per 1,000 square foot. Sow 1/2 to 1 inch deep. Germinates in 14-21 days. Thin seedlings to 1 foot apart. It is an **annual** that grows 1-4 feet tall. Matures in 12-14 weeks (84-98 days, 3 to 3 1/2 months) so that is **June or July.**

Care: Likes clay soil. Likes well drained, rich soil. Likes soil pH 6.0-6.5. Likes full sun. Prefers temperatures less than 90 degrees. Very good cold tolerance between seedling and bloom stages. Self pollinates. May get fungus diseases. Insects like it.

Uses: Has beautiful pale blue flowers where edible seed is produced in bolls or pods each containing 4-10 seeds. Harvest seeds when approximately 90% of seeds are brown. High in fiber and omega-3 fatty acids.

(See July Garden-Harvest.)

Sow **oats (common)** (poaceae/gramineae/grass family; Avena sativa) or **oats (hulless)** (Avena nuda) for **eating or spring cover crop** as soon as soil can be worked. A **cool season annual** that does well in cool, moist climates. It needs 6-10 weeks of cool growing temperatures and can tolerate light frost. **Sow about 4-6 weeks before the last spring frost.**

For **winter cover crop, sow in late summer or early fall** about 5-10 days before triticale and winter wheat are planted. **Sow at least 6 weeks before fall killing frost. Can sow 10 weeks before fall frost.** Oats get winter-killed when temperatures drop below 10 degrees.

For eating, **white oats** are grown in northern areas of the United States. They tolerate cool, moist weather better than red oats. **Red oats** are grown in warmer areas with mild winters.

Sow 1 pound for 350-500 square feet or 2-3 bushels (70-90 pounds) per acre. Sow about 2 seeds per square inch. Sow by broadcasting, then rake in to cover 1/2 to 1 inch with soil. If not covered, birds will eat them and germination will be less. Germinates in 7-20 days. Usually germinates quickly. Seedlings have high vigor and germinate well.

Fast growing annual. Grows 2-4 feet tall. Matures in 90-110 days (3-4 months). Harvest in **mid to late summer.** Each oat plant produces several seed heads from which oats are gathered.

Care: Likes full sun. Likes moderately rich soil. Likes same type of soil as potatoes. Likes soil pH 6.0-7.5. Likes moist but well drained soil. Not drought tolerant. Needs more water than most grains. Does poorly in hot, dry summer. In humid weather there may be problems with fungus.

Uses: Oats are fed to all types of livestock. They are 12-22% protein. Oat straw is more nutritious and easier to digest than wheat straw.

(For **oats cover crop** see September Garden-Plant but details are here. See July Garden-Harvest. For **map of when winter wheat is planted**, see October Garden-Plant. For **general grain harvesting**, see May Garden-Harvest.)

Sow Cold Hardy Crops Outside in March:

In March or April **plant/transplant fruit/nut trees and bushes**. (See April Garden-Plant.)

Stratify **anise hyssop** seeds for 2 months. (For how to stratify, see "Garden Tips" in back of book.)

Early sowing of **beets** (chenopodiaceae family; Beta vulgaris). Sow before beans but after peas and lettuce about **2-3 weeks before last frost in spring**. Soil needs to be around 60 degrees for good germination. Light frost is OK. Can succession sow every 3-6 weeks until temperatures reach 80 degrees or to **late June**. Can also sow **6 to 8 weeks before the first average frost in the fall** which is late August.

Presoak seeds 12 hours. Make sure seeds have good contact with soil. They have poor germination so sow thickly. Each seed is actually 2-6 potential plants. Sow seeds 1/2 to 1 inch deep about 1 inch apart.

Thin table beet seedlings to 3-4 inches apart in rows 1-2 feet apart. Plants should not be crowded. They do not transplant well. Roots mature in 55-80 days (8-11 weeks).

Care: Prepare the soil so it is not hard and beets will grow larger. Likes full sun but can grow in partial shade. Likes well drained soil. Likes soil pH of 6.0-6.8. Likes moderate nitrogen. Keep well watered so roots do not get tough. No serious pests or diseases. A **biennial**.

Uses: Can eat thinned plants. Leaves and roots are edible both raw and cooked.

Varieties:

1. Sugar beet roots are a creamy white color and can weigh 6-15 pounds. Grown for making sugar.

To make sugar: Place tiny beet pieces in a pan and cover with water. Boil the water and cook on medium heat for about 1 hour until beet juice is extracted. Remove the beet pulp. Then keep boiling the beet juice until 1/3 of it remains. Let the juice cool. Strain to remove the sugar crystals. Boil the remaining juice it has completely crystallized.

2. Garden or table beets are grown for their sweet flavor and small size. They are orange, pink, red, yellow, white or striped. Most are globe shaped. They grow with much of the root above the soil.

3. Mangel-wurzel fodder beets are grown for **livestock** but can also be eaten by people though they are fibrous and not as sweet. (Kohlrabi and rutabagas are also good for livestock especially over winter.) They grow up to 60 pounds. They are red, white or yellow. They grow up to 2 feet long with half or most of it above ground.

Mangel beets should be spaced at least 7 inches apart. Matures in 55-84 days. The roots store well, 6 months or longer. For winter storage, plant in **August.**

Mangels handle heat and drought better than other root crops, but they take longer to grow. Mangels produce more sugar per acre than sugar beets. Mangels are a good source for sugar and alcohol products.

(For **early beets** see July Garden-Harvest. For **late beets** see August Garden-Plant but details are here. **For details see October Garden-Harvest.** See September Greenhouse. For **forage crops,** see Pasture in "Garden Tips".)

 Sow **burdock (common)** (asteraceae or compositae family; Arctium minus-common or Arctium lappa-Japanese) seeds in **early spring**.

For big roots that are easy to dig up, make sure the soil is not compacted. Cover seeds with 1/2 to 3/4 inches of soil. Germinates in 4-7 days. Thin to 3 inches apart in rows 2 feet apart. Matures in 100 or more days (**June or July**).

It is a **biennial thistle** in the **aster/daisy sunflower family** that grows as a rosette of large leaves the first year. The second year it produces a 5 foot tall, erect, bushy flowering stem. Some varieties grow to 9 feet tall. The flowers turn into **burrs** that stick readily to fur and clothing. If you are not saving seed, remove the flower stems.

Care: Easy to grow. Likes full sun but will tolerate shade. Likes well drained soil that is somewhat rich. Prune off a few leaves if you are growing it just for the root.

Uses: You can also **forage** for burdock. The edible root usually grows 1-2 feet long but can grow 4 feet long. It is peeled and cooked. Best eaten when 1 year old. When it is over 2 years old, the root becomes too bitter to eat.

Young leaves and stems/stalks are edible raw or cooked. Stalks are peeled.

(For **eating leaves**, see March Garden-Harvest. For **digging up roots**, see June Garden-Harvest.)

Sow early **carrots** (umbelliferae family; Daucus carota). **Sow 2-3 weeks before last frost in spring.** Ready to harvest in 2 to 2 1/2 months (**May or June**). Sow storage/overwinter carrots in **June**.

(For **early carrots** see June Garden-Harvest. See August/December Greenhouse. For **late carrots and sowing/growing details**, see June Garden-Plant. See September/October Garden-Harvest with **harvest details in September.**)

 Sow **cilantro** (umbelliferae family; Coriandrum sativum) in March. The seeds are called **coriander**. A member of the **parsley family**. (See "Plant Families" in back of manual.)

An **annual** best grown in cool weather. It has stemmy, fern-like leaves.

Sow seeds 1/4 to 1/2 inch deep, 1 to 2 inches apart. Rows are 15-18 inches apart. Germinates in 7-10 days. Grows quickly. Make successive sowings every 3 weeks until **late summer.**

Matures in 8 weeks but leaves are harvested before that. Grows 18-24 inches tall. Transplants well.

Care: Likes full sun. Can take light frost. Likes soil pH of 6.5-7.5. Likes well drained soil.

Harvest leaves when 6-8 inches tall. Leaves become sparser and less flavorful after the plant blooms. Cut off flowers unless saving seed. Or allow seed to fall, and it will reseed itself.

Seeds (coriander) for eating are collected once fully formed but before they burst from the plant. Thresh by banging plants against container or by pulling seeds off by hand.

Uses: The seeds are used to flavor liquors and confections. The leaves can be eaten raw or cooked. The leaves are said to help remove heavy metals from the body.

Collard greens (brassicaceae family; Brassica oleracea). **Plant early spring and again late summer or early fall (8-10 weeks before first frost).** Best flavor after frost. Cold hardy to 20 degrees.

Biennial member of cabbage family. (See "Plant Families" in back of manual.) Grows similar to cabbage. (For **cabbage** see April Garden-Plant.)

It has high germination rate so sow thinly. Sow 1/4 to 1/2 inch deep. Germinates in 5-10 days. Thin seedlings to 6 inches apart. Let grow until begin to touch, then harvest whole plants to give 18 inches between plants. Can eat thinned plants. Matures in 80 days but can start harvesting at 40 days. Can pick individual leaves or cut whole plant.

Care: Likes full sun but can take some shade. Likes soil pH 5.5-7.5. Likes moderate water and lots of nitrogen. If it stays dry for a long time, the leaves may become bitter. Has disease similar to those of cabbage and broccoli.

Uses: Collard greens can be eaten raw or cooked but it is usually cooked for 45 minutes to an hour. If eating raw, choose small leaves. They are a good source of vitamin A and anti-oxidants.

(See August Garden-Plant and August-Greenhouse but details are in this section.)

 Divide and transplant mature **comfrey** plants (boraginaceae family; **Common Comfrey**: Symphytum officinale; **Russian Comfrey**: Symphytum x uplandicum; **Rough or Prickly Comfrey**: Symphytum asperum). There are other varieties too.

Plant crowns or part of root. One plant can be cut up into many pieces for propagation. Some varieties reproduce by seed.

The best time to transplant is in **March, April or May.** If you want to do it in the fall, the best time is **September.** However, it can be planted any time the soil can be worked (it is not frozen).

To **propagate**, drive a spade through the **crown** 3 inches below the surface. Cover up the cut crown with soil. The plant will grow back usually in 6 weeks. Divide the removed crown into 4-8 sections.

Replant in greenhouse for transplant outside later. Or plant outside in permanent location. Plant 3 feet apart in all directions. Cover each crown with several inches of soil. Put in a lot of fertilizer especially manure.

The plants live about **25 years**. The first 10 years are the most productive. For best leaf production, cut off flower stalks when they form. However, bees and other beneficial insects like comfrey so you may want to leave some flowers for them. The flowers are pretty. **Blooms late spring through summer.**

A member of the **borage family,** it is a **perennial** with hairy leaves and bell-shaped flowers. It is used as a vegetable, medicinal herb, animal feed and soil conditioner. Good at breaking up clay soil.

Care: Easy to grow. Grows 3-5 feet tall, 2-4 feet wide. It prefers full sun but does OK in partial shade. It likes moist soil. It is drought tolerant since it has a deep tap root. Likes a lot of nitrogen. Prefers soil pH 6-7 but is not picky. Very few diseases or pests.

Fertilize between November and March with the best time being March. Though it can be fertilized any time. Deep litter, raw poultry manure is best. Rock dusts can be added at any time. (See rocks dusts in "Garden Tips".)

Every year the plant gets bigger and wider. As plants age, they die in the middle. In **April or May** dig up roots away from the center, and replant in the center. The crown should be covered with soil.

Uses: Make first cutting when plant is about 2 feet tall. Comfrey is like a living compost. It pulls up minerals from deep in the soil. The leaves are high in nitrogen. It is great as animal feed and as a plant fertilizer. It is a compost activator because it is so rich it encourages the pile to heat up and decompose more quickly.

Comfrey can be used as **liquid fertilizer / compost tea**. Add leaves to a barrel or tub. Fill with water. For example, put 14 pounds of leaves in 20 gallons of water. Let sit for 3 to 5 weeks. It will smell bad. Use the liquid on your plants in the house, greenhouse and in the garden. Consult your herbalist.

Popular Varieties:

Common or True Comfrey (Symphytum officinale) is native to Europe. It reproduces by root cuttings and seed. It is used medicinally. It is used as a green manure. It produces less biomass than Russian Comfrey. Flowers are cream, creamy yellow, white or purplish. It can be somewhat invasive.

Russian Comfrey (Symphytum x uplandicum, Symphytum peregrinum, Symphytum asperum x officinale) includes varieties such as Bocking #4 and #14. It is a cross between Common Comfrey and Rough Comfrey. It is a sterile hybrid that does not self-seed. It is not invasive at all. Russian comfrey has purple, red or blue flowers. It has the greatest amounts of protein and allantoin. **It produces 100-120 tons per acre of biomass per year.** This is 3 times the amount that Common Comfrey produces.

Rough or Prickly Comfrey (Symphytum asperum) is native to Asia. The flowers are pink, blue and sometimes purplish. It is invasive. It sometimes produces seeds that are viable. It can be propagated by cutting up the root. It was traditionally grown in Russia as a silage crop. It is good for animals. It is a coarse, clumped plant.

Comfrey is sold at Nantahala Farm: www.nantahala-farm.com.

(See June/July/August/September Garden-Harvest with **harvest details in June**. For **replanting**, see April Garden-Maintenance.)

Creasy greens (brassicaceae family, Barbaraea verna, also known as upland cress, winter cress, early yellow rocket) is a **biennial mustard.** (For **growing mustard,** see February Greenhouse.) Two other major cresses are watercress (Nasturtium officinale) and garden cress (Lepidium sativum).

Sow seeds in **early spring (better fall planted)** when soil is cool and damp. Germinates in 7 days. Thin to 6 inches apart. Matures in 40-45 days. They grow in a rosette like dandelions.

Care: Tolerates partial shade. Easily grown even in clay soil. Does not need rich soil. Likes soil pH of 5.8-6.5. Unprotected plants survive all winter even in sub-freezing temperatures. They easily self sow.

Harvest spring, summer and fall. In greenhouse can harvest all year. Best picked before the yellow flower forms for less bitter taste.

Uses: Has peppery taste like radish but cooking lessens it. Can be eaten raw or cooked. High in vitamin A/C.

(See August Garden-Plant but all details are in this section. See September/October Greenhouse.)

Sow **dandelion** (asteraceae family; Taraxacum genus with hundreds of species; Common Dandelion: Taraxacum officinale). It is a very hardy, **short-lived perennial.** Commercial seed has wider and thicker leaves.

Sow **early spring through late summer.** Sow 6-9 inches apart in rows 12 inches apart. Germinates in 10 days at 55 degrees. Matures in 85-95 days.

Care: It grows in all soil types but rich soil improves growth and taste. Likes full sun but tolerates partial shade. Flowers are on stalks 2-18 inches tall. Self pollinates.

Uses: The plant has medicinal properties. Leaves are very nutritious, a great **spring tonic.**

Leaves are eaten raw or cooked. They are high in calcium, potassium and iron. Best eaten when young and tender especially in **early spring** before buds appear and again **late fall after frost.** Cooking reduces bitterness of leaves. Change water a few times while cooking.

Roots are up to 10 inches long. They can be eaten all year but are usually dug **late fall to early spring**. The outer skin is bitter so can be peeled. Best eaten cooked by changing water periodically.

Flowers can be eaten or made into wine. Collect **before mid-spring** when most flowers bloom. Use only the yellow parts because the little green leaves at the base are bitter. Eat raw or cooked. They are bittersweet.

Can **forage** for dandelion. (See July/August Greenhouse but all details are in this section.)

Dill (umbelliferae family; Anethum graveolens) is an easy to grow **annual** that is raised for the leaves and seeds. Cover seed lightly. Germinates in 7-10 days. Thin plants to 9 inches apart, in rows 12 inches apart. Grows from 18-42 inches tall depending on variety.

Care: Prefers full sun. Likes pH 5.5-6.5. Likes well drained soil. It readily self sows. If growing only for leaves, then remove flower stalk.

Harvest leaves any time. Young leaves taste best.
Harvest seeds after flower head has died and seeds are tan. Tie a group of stems together and hang upside down to dry. Put a container under them or a bag around them to catch seed.

  Plant **garlic** (amaryllidaceae family; Allium genus) **4-6 weeks before last frost in spring so that is March or early April**. It is **better to plant in the fall** because you get bigger bulbs. However, there is the risk of death by freezing over the winter.

(See April/September Garden-Plant with **planting details in September. For harvest and storage details**, see July Garden-Harvest. For **fertilizing** see April Garden-Maintenance.)

 Plant **horseradish** (brassicaceae family; Armoracia rusticana) roots in **early spring (March) or late September through October** (same time as perennial garlic and onions). It is a fast growing **perennial** that grows to 2-3 feet tall and 18 inches wide. Easy to grow.

Dig a hole 1 foot across and 6 inches deep. Place root at a 45 degree angle, around 6 inches deep for the small end, just below the surface for the top of the root. The bud faces up. Mound up a couple of inches of soil. Plants should be about 15-18 inches from each other.

Care: Prefers full sun but can tolerate some shade. Drought tolerant. Likes moderate nitrogen. Likes soil pH of 5.5-6.8. It is in the same **family as broccoli and cabbage**, so has similar diseases and pests.

One-year-old plants have the most flavor so can dig up and replant each year. (See April Garden-Harvest. For details about **digging up root to replant**, see October Garden-Harvest.)

In March sow **kale** (brassicaceae family; Brassica oleracea species, acephala group with many cultivars. Borecole.). It is a **biennial** member of the **cabbage family** that is green or purple. (See "Plant Families" in back of this manual.) It can be grown as a **forage crop** for animals. (See Pasture in "Garden Tips" in back of manual.)

The central leaves do not form a head. **Better sown in late summer** for fall harvest. **Sow 4-6 weeks before last frost in spring. Or sow 8-12 weeks before first fall frost. Cold hardy to 20 degrees.** Many sizes even up to 6 feet tall though these are not eaten by people.

Sow 1/4 to 1/2 inches deep, 1 inch apart in rows 18-30 inches apart. Thin seedlings to 8-12 inches apart. Fast growing. Matures in 55-65 days.

Care: Likes full sun. Likes well drained, moderately rich soil. Likes soil pH 5.5-7.0. Very hardy, few pests.

Kale varieties (cultivars) based on leaf type:

    1. Curly leaves (Brassica oleracea acephala)- Scotch or Scots type. It is the most popular. Types such as Curly, and Tall Green Curled.

    2. Plain leaves (Brassica oleracea acephala)- tall, coarser, pest resistant, very hardy and prolific. Eat in spring. Types such as Thousand Headed Kale and Cottagers.

    3. Rape (Brassica napus pabularia)- tender young shoots between March and May. Types such as Hungry Gap and Asparagus Kale. Sometimes called **canola** which can also be a field mustard (Brassica campestris or Brassica Rapa). Its seeds (rapeseed) are made into oil for people and livestock to eat. Can use oil for biodiesel.

    4. Leaf and spear- a cross between curly leaved and plain leaved. Types such as Pentland Brig.

    5. Cavolo nero (Brassica oleracea acephala) also known as black cabbage, Tuscan cabbage, Tuscan kale, Lacinato, and Dinosaur kale.

Harvest: Tastes better after frost. Younger leaves taste better. Can eat raw or cooked. Remove outer leaves and allow inside to grow.

Uses: Can eat kale a few leaves off of plant or harvest whole plant. Mature kale needs to be boiled a long time (10 to 20 minutes) to be tender. Young greens can be stir-fried, steamed, or sauted for a few minutes. Greens are high in vitamin C and calcium.

(See August Garden-Plant but all details are in this section.)

Sow **lettuce** (compositae family; Lactuca sativa) in succession every 3 weeks **March through May starting 2-4 weeks before last frost in spring.** Do not sow in June and July. Sow again in **early August or 6-8 weeks before first frost in fall.** Succession sow. **An annual hardy to 30 degrees.**

Seeds germinate better in light. Sow 1/8 to 1/4 inch deep. Rows are 12 inches apart, more or less depending on variety. Leaf lettuce is thinned to 8 inches apart. Crisphead is thinned to 12-14 inches apart. Does not transplant well.

Care: Likes 5 or more hours of sunlight but can tolerate partial shade. In hot weather partial shade is better. Likes soil pH 6.0-7.5. Does not need rich soil. Has shallow roots. Can take light frost.

It turns bitter in hot weather and when flowers form. It bolts (flower stalks form) in the summer.

There are 4 types of lettuce:

1. Leaf Lettuce- It is the most heat tolerant and easiest to grow. Leaves are loosely bunched instead of a tight head. Comes in many colors. Grows quickly. Harvest can be extended by cutting a few leaves at a time. Varieties include Black Seeded Simpson, Early Curled Simpson, Oak Leaf, and Red Sails. **Matures in 45-55 days.**

2. Butterhead Lettuce- Leaves form small, open heads that are not tightly bunched. Comes in many colors. Varieties include Buttercrunch, Little Gem, Tom Thumb, and Summer Bibb. **Matures in 55-75 days.**

3. Romaine (Cos)- Forms colorful, upright clusters of leaves. Varieties include Dark Green Cos, Green Towers, and Ideal Cos. **Matures in 45-55 days.**

4. Crisphead Lettuce- The traditional head lettuce that is only harvested once. Pick when head is firm and solid, and outside leaves turn a slight yellow green color. Varieties include Iceberg, Ithaca and Great Lakes. Less heat tolerant than other varieties. **Matures in 70-80 days.**

(For **late lettuce** see August Garden-Plant but all details in this section. See August/September Greenhouse.)

Sow **mustard greens** (succession plant) but better planted **late summer** for fall harvest. **Sow early spring (late March or early April) 3 weeks before last frost date in spring.**

(For **sowing/growing details**, see February Greenhouse. See August Garden-Plant.)

Sow **peas** (leguminosae family; Pisum sativum) in **late March and April, or 4-6 weeks before last spring frost.** Sow 3 succession plantings every 10 days. Sow when soil is 50 degrees or warmer. Best growing air temperature is 55-68 degrees. It is a **cool weather crop hardy to 25-28 degrees.** Peas do poorly in hot weather. Do not sow in June or early July. Sow again **late July and early August about 10 weeks before the first frost in the fall.**

Soak peas for 24 hours. Sow 1 inch apart about 1 inch deep with 24 inches between rows.

Grow better if you add inoculant when plant to improve nitrogen fixing. (For **inoculants**, see "Garden Tips".)

Care: Likes full sun. Likes well drained, medium rich soil. Likes soil pH of 6-7. Mature in 55-75 days. Most need trellis. Slight crowding is OK. Takes light frost. An **annual.**

Harvest: Pick peas frequently to encourage growth of more peas. Remove gently from vine.

Varieties: Peas can be green, purple or yellow.

English or Garden Peas The pods are not eaten.

Dwarf (bush) types grow 16 inches tall. Varieties include Frosty, Green Arrow, Knight, Maestro.

Tall types grow 3 feet tall. Varieties include Alderman, Freezonian, Lincoln, Thomas Laxton.

Edible-Podded Peas: Snow peas (sugar peas) are harvested when young, crisp and flat. But can also be eaten when the pod fills out. Most varieties mature in 63-72 days. Varieties include Dwarf Gray Sugar, Mammoth Melting Sugar, and Oregon Sugar Pod.

Sugar Snap Peas: Snap peas (sugar snaps) are grown like English peas and picked when pods have filled out. Pods are eaten. These wilt-resistant vines grow to a height of 2-6 feet. Varieties include Cascadia, Sugar Bon, Sugar Daddy, and Sugar Snap.

Early Season Varieties

Daybreak (54 days to harvest; 20 to 24 inches tall, good for freezing)

Spring (57 days; 22 inches tall; dark green freezer peas)

Main Season Varieties

Sparkle (60 days to harvest; 18 inches tall; good for freezing)

Little Marvel (63 days; 18 inches tall; dwarf)

Green Arrow (68 days; 28 inches tall; pods in pairs; resistant to fusarium and powdery mildew)

Wando (70 days; 24-30 inches; withstands some heat; tall; best variety for **late spring planting**)

(See July Garden-Plant but all details are in this section.)

Plant **rhubarb** (polygonaceae family; Rheum rhabarbarum). **Divide rhubarb root clumps** (crowns) when plant is dormant. Each clump should have one or more eyes or buds. Divide every 8-10 years by cutting into 3-4 pieces with shovel.

Plant crowns (divisions) early spring 3-5 feet apart. Prepare the soil well adding a lot of organic matter. Add a lot of fertilizer. Dig a hole and then make a mound in the hole. Place the roots over the mound. Cover with 2 inches soil. Cover with mulch. Rhubarb seed can be planted but it does not breed true.

Care: It does best in climates with a **long, cool spring.** Prefers soil pH of 5.5 to 6.5. Likes well drained soil. It has few pests. Rhubarb is very winter hardy. It is a perennial plant that lives for 20+ years. Fertilize in **spring and fall.**

Seed stalks form in the **spring** and should be removed since they take energy from the plant. Do not harvest the first year so the plant can gain strength. It is a **perennial.**

There are many different **varieties** but the main difference is the stem is either red or green.

(See April/May Garden-Harvest with **harvest details in April.** For **removing flowers,** see April Garden-Maintenance.)

Plant **sea kale** (brassicaceae family; Crambe maritima) **step 2** by taking root cuttings out of the refrigerator. By this time buds will be on the shoots. Rub off all but the strongest central bud. Plant the cutting 1 inch deep, 15 inches apart. It is a **perennial**.

Sea Kale seeds are sold at Nantahala Farm: www.nantahala-farm.com.

(For **propagating step 1** see November Garden-Maintenance. For **planting seeds and growing,** see May Garden-Plant.)

Sow **spinach (early spring)** (chenopodiaceae family; Spinacia oleracea) in March, **4-6 weeks before last frost in spring.** Sow every two weeks until last frost in spring. **Better planted late summer (August or September), 6-9 weeks before the first fall frost. Very cold hardy down to 15 to 20 degrees.**

Soak seeds overnight. Sow 1/2 to 1 inch deep with 12-15 seeds per foot, in rows 1 foot apart. Germinates well in cool soil. When 1 inch tall, thin 2-4 inches apart. Later thin 6-8 inches apart. Can eat thinned plants. Matures in 43-50 days.

Care: Likes full sun but tolerates partial shade. Is less likely to bolt if in partial shade. Does best with day temperatures around 60. Bolts (flower stalk forms) when weather gets hot. An **annual**.

Likes lime. Likes soil pH 6.0-7.0. Do not overdo it with nitrogen because it increases oxalic acid that is hard to digest. Likes well drained soil. Does not like being transplanted.

There are 3 types of spinach based on leaf type:
  1. Savoy spinach has curly, crinkly, dark green leaves. It is the type usually sold in the grocery store. Varieties include Bloomsdale Long Standing (48 days to mature; thick leaves) and Winter Bloomsdale (45 days to mature, tolerant to cucumber mosaic virus, good for winter).
  2. Flatleaf or smooth-leaf spinach has broad, smooth, spade-shaped leaves that are easy to clean. The stalks are narrow and tasty. Varieties include Giant Nobel and Olympia.
  3. Semi-savoy is a mix of the savoy and flat-leaf that has slightly crinkled leaves. Varieties include Five Star, Indian Summer, Melody, Tyee, and Vienna.

There are 2 types of spinach based on season:
  1. Spring or early spinach varieties: These are bolt resistant.
    Bloomsdale, Indian Summer, Space, Steadfast and Tyee.
  2. Winter or late spinach has robust and strong-tasting leaves that are coarser and usually curly.
    Fall planting varieties: Avon, Indian Summer, Melody, Razzle Dazzle and Tyee.
    Winter planting varieties: Giant Winter, Cold Resistant Savoy, Tyee and Winter Bloomsdale.

Uses: High in vitamins A, B and C. A pinch of baking soda in cooking water keeps spinach greener.

(For **late spinach** see August Garden-Plant but all early/late spinach details are in this section. For **early spinach harvest,** see May/June Garden-Harvest. For **all harvest details,** see October Garden-Harvest. See August/September/October Greenhouse.)

Sow **strawberry** in March or April. (For **seed sowing and growing details,** see April Garden-Plant. See May Garden-Harvest. See June Garden-Maintenance.)

Turnips (brassicaceae family) can be **sown in spring** but **better planted in midsummer** for fall harvest. Sow late March or early April. **Sow 4-6 weeks before last frost in spring.** (For details see August Garden-Plant. See October Garden-Harvest.)

🧺 Garden- Harvest 🧺 (March)

Vegetables: Burdock, jerusalem artichoke and stinging nettle.

 Harvest **burdock** leaves (asteraceae or compositae family; Arctium genus, Species: A. lappa, A. minus, A. minus nemorosum, A. pubens, A. tomentosum). It is a **biennial thistle.** (See "Plant Families" in back of manual.) Dark green leaves grow up to 28 inches long. Flowers July to October.

Leaves and stems (stalks) are edible. Peel stalks before eating. Leaves and stems are usually cooked like spinach. Stems can be eaten all year. **Leaves are best eaten in spring.** Consult your herbalist.

(For **sowing/growing,** see March Garden-Plant. For **harvesting roots,** see June Garden-Harvest.)

📖 For **foraging guidelines** read "Nature's Garden: A Guide to Identifying, Harvesting and Preparing Edible Wild Plants" by Samuel Thayer.)

Dig up **jerusalem artichokes** (compositae family; sunchokes) if ground is not frozen. (See April/October Garden-Harvest with **harvest details in October**. See May Garden-Plant, July Garden-Maintenance.)

Late March or early April harvest **ramps** (amaryllideacease family; Allium tricoccum) or wild leeks. Can eat entire plant. Grows in shady, rich, moist forests of Appalachia. One of the first plants to grow in spring.

By **late May**, ramp leaves begin to die back and a flower stalk emerges that blooms in **June**. Do not harvest every plant. Leave some for future years.

 Harvest **stinging nettle** (urticaceae family, Urtica dioica) leaves until **mid-June**. Early spring leaves do not sting at all or not as much. Can pick all summer but not as good tasting and more fibrous. During summer wear gloves to pick so do not get stung by leaves. If **stung,** apply crushed stems of jewelweed, or a poultice made from leaves of curly dock or common plantain.

Uses: Best harvested before it flowers. If harvesting after it flowered, eat young leaves. **Cooking a short time, wilting, or drying** gets rid of the sting in the leaves. Leaves can be eaten cooked or dried.

It is one of the highest **protein** sources of any plant. Comfrey is the highest.

The leaves can be make into a **tea**. This is especially good later in the summer when the leaves are tougher.

The **fibers** in the stalks are stronger than flax and can be spun and woven to look similar to hemp. The roots make a gold dye and are used for prostate problems.

It is very nutritious and medicinal: an astringent and blood builder. A good **spring tonic**. It has very high amounts of calcium, magnesium, iron, potassium, phosphorous, manganese, silica, iodine, silicon, sodium, and sulfur. It is laxative. Consult your herbalist and/or health care specialist.

Good to put in a compost pile to speed up decomposition. Can **forage** for stinging nettle in moist forests.

Nettle hay is good for ruminants such as cattle and goats. Can make silage from nettles and comfrey mixed.

(See April Garden-Plant.) 📖 For **foraging guidelines** read "Nature's Garden: A Guide to Identifying, Harvesting and Preparing Edible Wild Plants" by Samuel Thayer.)

🏠 Greenhouse, House, Hoop House or Cold Frame 🏠 (March)

Vegetables and Fruit: Arnica, asparagus, broccoli, broccoli raab, cabbage, celery, chrysanthemum (mum), ground cherry, kohlrabi, peppers, potatoes, sweet potatoes, tomatoes, and tomatillos.

📖 Read the book "Greenhouse Gardener's Companion" by Shane Smith and "Winter Harvest" by Eliot Coleman.)

  **Sow in greenhouse perennial herbs** such as chives, good king henry, oregano, parsley (biennial), sage, salad burnet, sorrel, thyme, valerian, and winter savory. Plant outside in **May**.

Most herbs have been less genetically manipulated by people so are **more like wild plants** than most fruits and vegetables. Therefore, they tend to be more resistant to disease and pests. Perennial plants usually take longer to germinate than annuals. Many can be **propagated** by dividing adult plants.

(For **oregano and most others** see May Garden-Plant.) (**Parsley**: For sow details, see April Garden-Plant. See August Greenhouse.)

Sow in Greenhouse in March:

Sow the herb **arnica** (asteraceae or compositeae family; Arnica montana or Arnica chamissonis) in **early spring** in greenhouse. Keep around 55 degrees. Transplant outside **after last frost**.

Varieties:

Arnica montana is also called mountain arnica, Leopard's bane, Wolf's bane and Mountain Tobacco. Likes very acid soil. Is slow growing. Likes growing in mountainous areas. Grows 18-24 inches tall, 36 inches wide.

Arnica chamissonis is called meadow arnica. It is much easier and more productive to grow in low elevations. It spreads by rhizomes creating a productive patch in a short time. Laboratory tests have shown this species of Arnica to be medicinally equivalent to Arnica Montana. Grows 6-8 inches tall.

Sowing and Care: Can stratify seeds. (For **how to stratify**, see "Garden Tips" at the back of this manual.) Sow on surface or 1/8 inch deep. Germinates slowly. Takes 16-40 or more days to germinate. Poor germination rates.

Both are a **perennial herb** good in hardiness zones 5-8. Can stand temperatures as low as minus 20 degrees. Likes partial shade or full sun. It likes well drained acid soil. Does OK in poor soil. Do not give much nitrogen.

Propagated by seeds or dividing rhizomes (roots) in early spring.

Harvest flowers when in full bloom. **Flowers bloom mid to late summer.** Dig up the **root** after the leaves have died in the **fall**. Dry the roots and then use like dried flowers. Roots also yield a small amount of essential oil. The entire plant can be cut and dried.

Used for medicinal purposes since the 1500s. The yellow daisy-like flower and rhizome (root) are dried. Can make into poultices, tinctures, powders or gels. Used externally as an ointment or compress for sprains, bruises, muscle pain, dental pain, and rheumatism. Used as an ingredient for salves and oils. Used for any trauma to the body. It is anti-inflammatory and a pain reliever. Do not use if pregnant or have high blood pressure. Do not use internally.

Make an herbal tincture by crushing 3 tablespoons of flowers. Then soak in 2 cups pure alcohol (not rubbing alcohol). Let sit for 2 weeks. During this infusion period, shake the jar 2 times every day. Best done around the full moon.

Dilute this solution to about 1 teaspoon of tincture per cup of warm water. Soak a small piece of clean cotton cloth in this diluted tincture solution. Place cloth directly on the external bruise or other problem. Leave on about 15 minutes. Repeat several times a day, and continue daily until the bruise or wound has healed.

Make an herbal ointment by grinding 2 tablespoons of dried flowers. Mix with 8 tablespoons of melted petroleum jelly (Vaseline) or Crisco shortening. To prevent the ointment from turning rancid, you can add tincture of benzoin or several drops of gum benzoin. Consult your herbalist, doctor and/or health care specialist.

(For general information on how to make an **herbal tincture, decoction or infusion,** see April Garden-Harvest.)

Sow **asparagus** (lilaceae family; Asparagus officinalis) seeds in greenhouse **12 weeks before last frost.** After 12 weeks (3 months), they should be about 12 inches tall and ready to plant outside.

(For **sowing seeds outside and planting details,** see April Garden-Plant. See April Garden-Harvest.)

Sow **broccoli** (brassicaceae family; Brassica oleracea) in greenhouse **6-8 weeks before last frost in spring.** Can also direct sow again in **early summer** for fall harvest. **Sow 12-14 weeks before first frost in fall.** (For "Plant Families" see back of manual.)

Sow 1/4-1/2 inches deep. Germinates in 6-10 days. Matures in 55-98 days. An **annual.**

Care: Likes full sun or a little shade. Likes soil pH of 6.2-7.2. Likes well drained soil. Grows best at 60-65 degrees. Can take frost to 25 degrees. May not produce heads if soil is low in calcium. Has diseases and pests similar to cabbage.

Transplant outside 2-4 weeks before last frost date so plant in **mid to late April**. Plant 18-24 inches apart in rows 24 inches apart.

Harvest: Cut broccoli head before the flowers open when the head is 4-6 inches across. You get a second harvest of smaller heads below where you cut off the first head.

(For **early broccoli or broccoli raab**, see April Garden-Plant, June Garden-Harvest. For **late broccoli or broccoli raab**, see June Garden-Plant, October Garden-Harvest. **Harvest details are in October.** Most planting details are in this section.)

Broccoli raab (brassicaceae family; Brassica rapa) is grown **similar to broccoli**. It is a leafy green in the **turnip family**. The stalks, florets (small flower that is part of a group of flowers) and leaves are edible.

(For **early broccoli or broccoli raab**, see April Garden-Plant, June Garden-Harvest. For **late broccoli or broccoli raab**, see June Garden-Plant, October Garden-Harvest. Most details are in broccoli above.)

Sow **cabbage** (brassicaceae family; Brassica olercea) in greenhouse. **Sow seeds indoors 6-8 weeks before last spring frost.** It can take light frost. Plant outside **late April**.

(For **sowing details and early cabbage** see April Garden-Plant. See June Garden-Harvest. For **late cabbage** see June Garden-Plant. For **harvest details,** see October Garden-Harvest.)

Celery (umbelliferae family; Apium graveolens) is a **biennial.** Start celery seed indoors **8 to 10 weeks before last spring frost.** Set transplants in the garden **2 to 3 weeks before average last frost** when seedlings have 5 to 6 leaves.

Sow 1/10 to 1/2 inch deep, 6-10 inches apart in rows 24 inches apart. Seeds need light to germinate. Takes about 21 days to germinate. Somewhat poor germination rate. When 6 inches tall, thin to 10 inches apart with 18 inches between rows. Eat thinned plants. Grows 12-18 inches tall. Matures in 98-130 days or even up to 5 months (150 days).

Care: Does best in soil 60-70 degrees. Likes full sun but tolerates partial shade. Likes soil pH of 5.8-6.8. Likes rich, moist soil. Keep well watered. Blanch by mounding soil around base to reduce bitterness.

Uses: Stalks, leaves, roots and seeds are edible. Celery, anise, caraway, carrots, chervil, coriander/cilantro, cumin, dill, fennel, parsnips, parsley and queen anne's lace are all members of the **carrot family** (apiaceae or umbelliferae). (See "Plant Families" in back of this manual.)

(For **transplanting,** see May Garden-Plant but details are here. See August Garden-Harvest.)

Sow **chrysanthemum** (asteraceae family; Chrysanthemum genus with 30+ species; mum, chrysanth) seeds in greenhouse in early spring (**late March**). It is a **perennial in the daisy family**. (See "Plant Families" in back of manual.) It can be planted anytime as long as it has time to get established before winter. Some varieties are **hardy** and some are somewhat **tender.** The old style of mums such as Chrysanthemum x rubellum 'Clara Curtis' and Chrysanthemum 'Mei-kyo' are hardy.

Cover seed lightly with soil. Likes air temperature around 72 degrees to germinate. Germinates in 1-2 weeks. Transplant outside 8-10 weeks after sowing or after last frost so that is **late May or early June**. Plant 18-30 inches apart. Grows 3 feet tall.

Care: Easy to grow. Likes full sun but will bloom in partial shade. Likes soil pH around 6.5. Likes rich, well draining soil. They are shallow rooted so keep well watered.

When 6 and 12 inches tall, **pinch back growth** to keep it bushy. Phosphorus encourages blooming. **Flowers** bloom mid-summer through fall in almost every color. They are 2-6 inches wide. Flowers can tolerate light frost.

Aphids sometimes bother them. Mulch in late fall. They can be **propagated** by cuttings or by dividing mature plants every 3-5 years. May need rooting hormone for cutting.

Food and Medicinal Uses: Yellow or white chrysanthemum flowers of the species C. morifolium are boiled to make a sweet drink in Asia. The tea is used to help recover from influenza. Consult your herbalist or doctor.

Insecticidal Uses: Chrysanthemum cinerariaefolium and Chrysanthemum coccineum (also called Pyrethrum) are natural insecticides. The flowers are pulverized. The active component pyrethrin is contained in the seed cases in the flower. This is applied as a suspension in water or oil, or as a powder. Pyrethrins attack the nervous systems of all insects. They are also an insect repellent and discourage mosquitoes from biting. They are harmful to fish, but are far less toxic to mammals and birds than many synthetic insecticides. They are biodegradable. They are among the safest insecticides for use around food. (Pyrethroids are synthetic insecticides based on natural pyrethrum, e.g., permethrin.) Consult your herbalist or County Extension office.

(For **more about pyrethrins**, see Products to Control Insects in "Plant Health" in back of manual.)

Sow ground cherry (solanaceae family; Physalis pruinosa or Physalis peruviana, cape gooseberry or husk tomato) in greenhouse 40-60 days **(6-8 weeks) before last frost**.

Sow seed 1/4 inch deep. Germinates in 10-14 days. Grows to 2-3 feet tall and spreads 2 feet wide. Thin to 24-36 inches apart.

Plant outside after last frost. Matures in 120 days (4 months), so matures **early September.**

Care: Grow similar to tomato. (See below.) Needs at least 8 hours of sun. Tolerates poor soil. Light feeder. Very productive. One plant can produce up to 300 fruit. Same pests and disease as tomatoes.

Aunt Molly is a good variety and grows 2 feet tall with 18 inch spread.

(For **transplanting** see May Garden-Plant but details are in this section. See July Garden-Harvest.)

Kohlrabi (brassicaceae family; Brassica oleracea) can be sown in **March** in greenhouse, then transplanted outside in **April**. Harvest **May or early June.** Can also be sown in **mid to late August** for harvest in late October. **Better planted in August.** Matures in 55-60 days. Good as animal fodder. (See pasture in "Garden Tips".)

(For **late kohlrabi and sow/grow information,** see August Garden-Plant. See October Garden-Harvest.)

Peppers (solanaceae family; Capsicum annum). Sow seeds in greenhouse **8-10 weeks before last frost**. Germinates in 2-3 weeks. Seedlings grow slowly. Space plants 18-24 inches apart in rows 24-36 inches apart.

Matures in 70-90 days depending on variety. Plant outside **late May or early June** after last frost.

Care: Likes full sun. Likes soil pH of 6.2-7.0. Likes well drained soil that is somewhat moist. Likes fertile soil.

Member of solanaceae or **nightshade family** which includes eggplant, potato, tobacco and tomato. They self pollinate. (See "Plant Families" in back of this manual.)

Varieties: Sweet green bell pepper (will turn red if allowed to mature), banana peppers with many hot varieties, round cherry peppers, and paprika. **Colors:** green, red, yellow, orange, purple.

(For **transplanting** see May Garden-Plant but details are in this section. See June Garden-Harvest.)

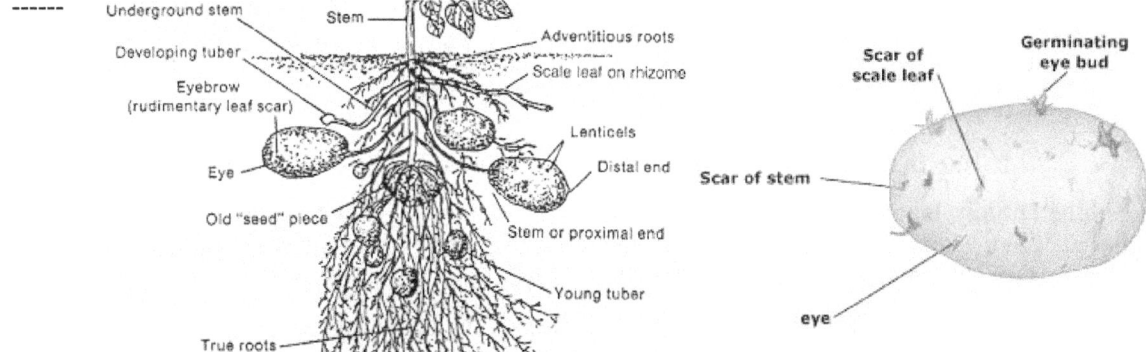

Pre-sprout **potatoes (white/red/purple)** (solanaceae family; Solanum tuberosum) in **late March or early April**. Do this 2-4 weeks before planting in **late April**. Some say do 6 weeks before planting. Potatoes must spend several months in cool storage before they will sprout. In **nightshade family.** (See "Plant Families" at back of manual.)

Potato tubers known as **seed potatoes** are planted. To reduce virus disease, it is best to use certified stock but you can use potatoes that you planted the previous year. Potatoes in the grocery store have been sprayed with anti-sprouting chemicals. You do not have to **pre-sprout before planting**, but if you do your yield and vitality are increased. They are less likely to rot in the ground and they mature sooner.

Optional, most people do not do this but you can experiment: Potatoes usually have more **eyes at the crown**, sometimes 3-5 in a cluster. On some roundish ones, they may be off center. (A tuber has one end called the **proximal end, stem or crown**, which is where it was attached to the plant. The opposite end is called the **distal end**. See above images.) If you let the crown eyes grow, they produce mainly small tubers. Using a potato peeler or knife, remove all eyes in the cluster by scooping about 1/8 inch deep. This stops all regrowth. Food reserves are directed to **shoulder and side eyes** to produce bigger potatoes. (end of optional)

If you get your seed potatoes out of **storage** and they have very long, thin sprouts, then remove them. Let new sturdy ones grow in their place.

Spread uncut potatoes (except for optional eye removal), eyes up, in **house** in indirect light, in 70 degree (always above 60 degrees) room to form sprouts (not in direct sun). This is called **chitting or greening**. If you do not remove eyes, then place crown end up. (See above image.)

Short, sturdy sprouts about 1/2 to 1 inch long are best. (Depending on how long you pre-sprout.) Throw away any diseased potatoes.

If sprouts are black or black tipped (but not the color from a red/purple potato), that means the sprouts got too cold and are dying. Remove the black sprouts and start again.

Potatoes are planted late April through June, and harvested June through October. Some people plant potatoes late March or early April but there is chance they will rot in the soil. **Earliest planting is 2-3 weeks before last frost in spring.**

In **late April plant** early/short season potatoes for harvest in June or July.

In **late April or May plant** mid season and late/storage potatoes for harvest in August through October.

In **June plant** early/short season potatoes for harvest in August or September.

(For **details about planting and growing**, see April Garden-Plant.)

(**Harvest:** For early/short season potatoes, see June/July Garden-Harvest. For mid-season potatoes, see August Garden-Harvest. For late/storage potatoes, see September/October Garden-Harvest. **October Garden-Harvest has most details about harvesting.**)

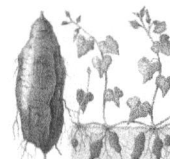 **Sweet potato slips (sprouts)** (convolvulacae family; Ipomoea batatas) are started **late March or early April** in house. Related to **morning glory.**

Put sweet potatoes in damp sand or in glass of water with half in water. (See second image above.) Or cut potato in half lengthwise, add water, and put potting soil over it. Keep in warm (75-90 degrees), sunny place. Shoots appear in 2-4 weeks.

When **shoots** are 4-6 inches, pull gently off potato. Put bottom of shoot in glass of water for a few days until roots form or plant in potting soil. (For general information about **potting mixes**, see "Garden Tips" in back of this manual.) Can get many slips from 1 potato.

Plant outside 2 weeks after last frost which is **early June**. If weather is too cold, plant in greenhouse.

Sweet potatoes can be orange, purple, red, white or yellow. Sweet potatoes **grow as vines** along the ground.

Sweet potatoes and yams are in different families. Yams are in the Dioscoreaceae family. (See June Garden-Plant, October Garden-Harvest.)

 Sow **tomatoes and tomatillos** in house **8-10 weeks before last frost in spring.** Some people start **5-6 weeks before last frost.** Move to greenhouse in a few weeks. Plant outside 2-3 weeks after last frost so that is **early June. Both are annuals.** Tomatoes are daylength neutral (they flower the same regardless of how many hours of light they receive). A member of the **nightshade family.** (See "Plant Families" in back of this manual.)

● **Tomatoes** (solanaceae or nightshade family; Solanum lycopersicum or Lycopersicon lycopersicum or Lycopersicon esculentum).

Sowing Seeds: (For details about **potting mixes and soil blocks**, see "Garden Tips" in back of manual.) Place 2-3 seeds in each pot. Cover with 1/4 inch of soil. Moisten soil and keep it moist. Seeds germinate in dark. Germinates in about 7 days. Germinates best when soil is 75-85 degrees.

They need 14-18 hours of light per day. If using fluorescent lights, place a couple of inches from light. Transplant when they get their first true leaves, and move into 4 inch pots about 2 weeks after that.

Brush plants daily with hand to strengthen stems. Or run a fan on them sometimes. When seedlings are several inches tall, they are old enough to be planted outside if there will be no more frost. If it is too soon to plant outside, then may need to transplant seedlings twice into bigger pots before plant in garden.

Planting Outside: Harden off seedlings before planting outside. (For **general hardening off details,** see "Garden Tips" in back of manual.)

Plant deeper than was in first pot (see below images). You can plant as much as half the plant underground if the plant is small. You can even bury all the way up to a few top leaves. The stem will develop roots. They self pollinate.

Care: Space plants 18-24 inches apart, in rows 30 to 36 inches apart. Needs good soil drainage. Likes rich soil. Needs full sun. Likes soil pH 5.5-6.8. Matures in 55-85 days depending on variety.

Fertilizers: Add soil amendments such as rock phosphate or bonemeal for phosphorus before planting. Early in the season the fertilizer should be high in nitrogen. During blossoming, fertilizers should be higher in phosphorus and potassium. If high in nitrogen during flowering, then you will get lots of leaves but not much fruit. Fruit set does not occur below 55 degrees or above 90 degrees. (For **general fertilizer details**, see "Garden Tips" in back of this manual.)

You can put **stake or trellis** in when transplant if you want. Use cotton cloth or twisty ties to tie up plant as it grows. If you stake/trellis them, the transplants can be put a little closer together such as 15 inches apart. A trellis is needed if weather conditions are moist since slugs and snails will eat tomatoes on the ground.

Mulch about 3-4 inches deep when soil warms later in season.

Water deeply and regularly while plants are developing. Irregular watering leads to **blossom end rot** and cracking. When fruit begins to ripen, use less water so fruit concentrates its sugars. But do not let plants wilt. Water at base of plant rather than on leaves to reduce fungus disease.

When the plants are about 3 feet tall, remove all leaves from the bottom 1 foot of stem. These are usually the first leaves to develop fungus problems.

Suckers (leafy shoots) develop in the crotch joint of two branches. If left on the plant, they keep growing and usually produce fruit. It is good to prune some suckers out so the plant does not get top-heavy. **Late in summer** remove suckers since they will not produce fruit before frost. The more suckers you remove, the less fruit you get but they are larger.

If plants are allowed to sprawl on the ground, then you may not want to remove any suckers. It is your choice as to how you want the plant to develop. To remove, pinch off with your finger. (See third image below.)

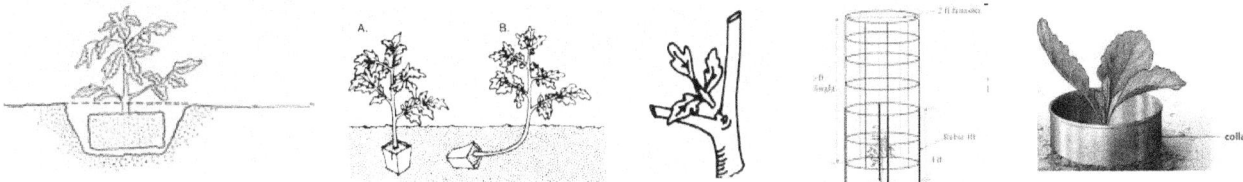

Pests:

    1. Cutworms: Put a paper, cardboard or metal collar around each plant to deter cutworms. Remove when weather gets warm. (See last image above though it is not a tomato plant.) (See Brassicaceae Family in "Plant Families".)

    2. Root-knot nematodes are caused by microscopic eelworms that live in the soil. Try rotating tomatoes with marigolds or grow near each other at same time.

Disease: Rock dusts reduce diseases. (For rock dusts, see "Garden Tips" in back of manual.)

    1. Blights and fungus infection can occur in high humidity. Early treatment with fungicides can help. Make sure there is good air circulation in the plants. Remove infected parts. Rotate crops each year.

    2. Blossom end rot is a round, brown, indented spot on the bottom of the tomato caused by uneven watering and/or a lack of calcium in the soil. Dolomite lime and bone meal add calcium to the soil.

    3. Cat-facing is when the tomato has irregular shapes and lines. It is caused by temperature shifts and incomplete pollination in cold weather. There is nothing you can do about it. The tomato is still good. Do not plant too early.

    4. Split skin or cracking happens when plants undergo accelerated growth. This is brought on by a sudden increase in water after being too dry. To prevent it, provide consistent moisture. Another reason is that the fruit is overripe.

Many cherry tomatoes crack with changes in weather and after heavy rains. There is not much you can do about it except to pick all ripe ones before it rains.

(See "Garden Tips" in the back of this manual. For more about **diseases**, see Solanaceae Family in "Plant Families" at the end of the manual. See "Plant Health" in back of manual.)

Types of tomatoes based on growth pattern:

    1. Determinate tomatoes only grow to a certain height and the fruit ripens at about the same time. They grow about 3 feet tall. They do not need to be caged or tied to a trellis.

    2. Indeterminate tomatoes keep growing taller and wider until frost kills them. Tomatoes ripen all summer.

Types of tomatoes based on use:

    1. Cherry tomatoes are very small, sweet, easy to grow, and produce early.

    2. Tomatoes for making **paste and sauces** are meatier, small to medium, and oblong such as Romas.

    3. Tomatoes for **slicing** are jucier, usually large and round such as Beefsteak.

    4. There is one variety called **Long Keepers** that is ready to harvest in 78 days and is semi-determinate. It is small 4-7 ounces, usually yellowish red-orange at maturity. It ripens 1 1/2 to 3 months after being picked. Harvest when has a pale, pink blush. To ripen store at room temperature. To slow ripening, store in a cool, dark place where they do not touch each other. They store up to 6 months. The quality is not as good as regular tomatoes.

Types of tomatoes based on ripening date:
1. **Early tomatoes** such as Early Girl, Early Pick, First Lady, Glacier, Native Sun, Oregon Spring and Tiny Tim. Matures in 78-98 days.
2. **Midseason** such as Better Boy, Big Beef, Big Boy, Big Girl, Celebrity, Delicious, Floramerica, Heatwave, Jelly Bean and Washington Cherry. Matures in 91-108 days.
3. **Late tomatoes** such as Beefmaster, Brandywine, Golden Boy, Homestead, Oxheart, Viva Italia, Wonderboy, and Supersteak. Matures in about 118 days.

(See May Greenhouse and June Garden-Plant but planting details are in this section. See August/October Garden-Harvest with **harvest details in August**.)

 Tomatillo (solanaceae or nightshade family; Physalis ixocarpa):

It is 1-8 feet tall depending on variety. They need two or more plants for pollination. Fruiting is day length sensitive (they flower more or less depending upon how many hours of light they receive). **Grow similar to tomato** (see above).

Plant seedlings 18-24 inches apart, in rows 3-4 feet apart. Matures in 90-100 days.

Used in Mexican cooking such as salsa. The fruit is green, about the size of a large cherry tomato. The inside is white and meatier than a tomato. Tomatillos grow inside of a thin paper-like husk.

(See May Greenhouse and June Garden-Plant but sowing/care details are in **tomato section above**. See August Garden-Harvest.)

 Farm Animals (March)

  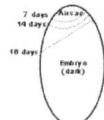   **Chicken, turkey, goose and duck** egg production is up. **Peak egg production is April and May**. Good egg production is **February through July.** Lowest egg production is **September through December.** (The longest day is in June. Molting is in September. The shortest day is in December.)

This is a good month to **start incubating poultry eggs** in an incubator such as a Hovabator. (See above second image.) Read the manual that came with your incubator. It is best if the incubator is in a room with a stable temperature around 70-80 degrees. Do not let sun hit incubator.

Before a fertile egg is incubated, the embryo inside is already developing. Collect eggs frequently. **Clean** lightly soiled eggs with a dry, soft cloth. It is better not to wash them since they have a bloom that keeps bacteria out. But if you must wash them, use warm but not cold water. Do not incubate eggs that are very dirty because bacteria may have gotten into the egg. Handle gently.

Storing eggs for incubation: It is best to incubate them within 1-2 weeks. If you only have a few hens and need to wait a few days or weeks to collect enough eggs for your incubator, then keep them around 50-60 degrees and 75% relative humidity. (Best hatch rate is with eggs stored less than 1 week.) Do not put in refrigerator. Put them in an egg carton with the big end of the egg up. Then 2-3 times a day lift up one end of the carton, then the next time the other end. Do not turn upside down. Let eggs warm up to room temperature before putting them in incubator.

If eggs are shipped, do not wrap in plastic because eggs need to breath.

Chicken eggs are incubated at 99 to 99.5 degrees. Chicken eggs incubate for 21 days.

Duck eggs are incubated at 99 degrees. Mallard and domestic duck eggs incubate for 28 days. Muscovy duck eggs incubate for 35 days.

Goose eggs are incubated at 99 to 99.5 degrees. They incubate for 28-32 days with smaller breeds being around 28, and larger breeds 30-32 days.

Turkey eggs incubate between 98-102 degrees with 99-100 degrees being best. They incubate for 28 days.

About 2/3 the way through incubation, the babies in the eggs start generating some heat. So you may need to turn the thermostat down if the temperature rises too much.

Chicken eggs need 50-55% humidity for days 1-18, and 65-75% for days 19-21.

Duck eggs like 55-65% humidity for days 1-24. From day 25 to first piping (ducklings start to crack egg, usually around day 28), humidity should be around 65%. When piping starts, increase humidity to 80-85%.

Goose eggs prefer 50-55% humidity for days 1-24. From day 25 to hatch, humidity should be around 75%.

Turkey eggs like 55-60% humidity for days 1-24. From day 25 to hatch, humidity should be around 75-80%.

You need a hygrometer (**humidity gauge**). Keeping the right humidity is very important. It is better to have higher than recommended humidity than lower especially when eggs are hatching. If it is too dry, the birds will have difficulty getting out of the egg.

It is better to have an electric **turning rack** but you can turn eggs by hand 3-4 times a day. Eggs are put in the turner with small ends down. Do not turn last 3 days. Remove turner and put eggs on side on incubator floor.

Open incubator only when absolutely needed such as adding water to maintain humidity. Water is usually added about 2 times a week. Opening the incubator changes temperature and humidity that can take hours to readjust.

You need to **candle eggs** which means looking at the eggs in the dark or in dim light with a flashlight touching the egg to see if the eggs are fertile and growing properly. You throw away infertile (clear) or dead (cloudy) eggs. (See above third and fourth images.)

Candle eggs after 7 days in the incubator. Candle again at day 14 if there were eggs that you were uncertain about. Otherwise, you do not need to candle again.

Hatching can take place over several days. For chickens, run the incubator for 24 days (3 days more than normal hatch time). For ducks and turkeys run the incubator for 32 days (4 days more than normal hatch time).

It is better **not to help a chick/duckling/poult crack out of its shell**. If you do, they almost always die (unless you follow the instructions in the next paragraph). Inside the shell are **2 membranes**. The inner membrane controls blood flow to the chick, and the outer membrane holds moisture. The chick breaks the shell slowly and controls the stopping of the blood flow. It can take 12 hours or more for the chick to come out of the shell once it first puts a little crack in the egg. It usually only takes an hour or so. Make sure humidity is high during hatching. You may need to spray eggs with a little water.

If you do want to help a hatch that has already started (chick has made a small hole in egg), then take egg out of incubator and go to a warm place. With dull tweezers, remove a little bit of the egg shell around the hole. Do not remove any of the inner membrane. Do this in a circle around the shell like the chick would do it. (See last image above.) Put the shell in the tweezer and pinch off small pieces. While you are doing this, keep the membrane moist with warm water in a dropper. The key to this is keeping the membrane moist. Do not drown chick. If you see blood, stop. After cracking around some or all of the egg, wrap a warm, very wet piece of washcloth around the cracked parts of the egg. Do not cover the chicks beak. Put back in incubator. The beak should be on its side or facing up.

Remove chicks, poults or ducklings from the incubator once a day. They can live for 48 hours without food or water. You only remove once a day so the incubator maintains stable temperature and humidity. Some incubators have a small plug you can remove to help reduce humidity while babies dry.

After the eggs hatch put the chicks (or ducklings or turkey poults) in a **brooder with a heat lamp.** This is also what you do with **1-day-old chicks** you get from a hatchery. You hang the heat lamp and then raise or lower it depending upon the behavior of the baby birds. Or change the wattage of the light bulb. Start with 100-125 watts. As they get older reduce to 75 and then 60 watts.

If they huddle under the light, then they need more heat. If they are walking around with no huddling under the lamp or crowded in a far corner, then they are happy. They have to have a place to get away from the heat too. They should not be in a breeze.

They need to be in a **cage** that rodents and other animals can not get into. The floor can be covered with small pieces of chaff left over from animals eating hay. You can also use straw, wood shavings/chips, or shredded paper. As time goes on you can put new material on top of the bedding. This is called **deep bedding**. The decomposing manure creates heat for the baby birds. It also provides natural protection from disease because it contains beneficial organisms that the chicks eat. (Chicks forage in the bedding.) It does not smell bad. If it smells, you need to add more bedding material.

Buy some **chick feeders and waterers**. Keep the water fresh. Feed baby chick feed. Ducks need extra niacin (vitamin B3). Feed greens with small leaves such as chickweed. After 3 days put some **sand (preferably coarse sand) or baby grit** in their feed. You can also put a dish of it out for them or sprinkle some on their bedding.

If the chicks get chilled, they may get **rear-end paste up** (gum up). This frequently happens when chicks are shipped. You have to remove the paste up or the chick will die. Just pull it off. Do not wash the chick's rear or it will get chilled again.

After a few weeks or more, **put them outside in a protected area** where predators and adult chickens can not get to them. They need more protection from the elements than adult chickens. Keep them enclosed in their shelter for about 2 days so they know where they are supposed to roost at night. After that their door can be open during the day. In time they can join the main flock of birds.

(For **chicken, duck and turkey breeds**, see February Farm Animals. For **broody hens** see April Farm Animals. For **molting**, see September Farm Animals.) (📖 Read the books "A Guide to Better Hatching" by Janet Stromberg and "Success with Baby Chicks" by Robert Plamondon.)

 Clean chicken, duck and turkey nest boxes. Put diatoms (**diatomaceous earth**) in nest boxes. It is a naturally occurring, silicon rich sedimentary rock made up of fossilized remains of diatoms. Diatoms kill insects in the box. It is a safe pesticide. (For **where to buy**, see fertilizers in "Garden Tips" in back of manual. For **more about diatoms**, see Dusts in "Plant Health" in back of manual.)

Also used as an **animal wormer internally**, but for worming must use food grade diatomaceous earth, not pool grade. Once every 3 months mix with feed at 5% diatoms. (For **natural wormer recipes**, see April Farm Animals.)

If needed, treat chickens, ducks and turkeys for **mites** by painting roosts and nest boxes with linseed oil thinned with turpentine, or paint with used motor oil thinned with kerosene. Or whitewash walls and perches with lime. The lime is antimicrobial. Repeat in 2 weeks.

Lime Whitewash Formula: Mix 2 gallons of water with 12 cups hydrated lime (calcium hydroxide, slaked lime, builders lime, pickling lime). Add 4 cups or 1 pound salt. Mix a little lime, then a little water, then lime, etc. If you mix it all together at once, it is very hard to stir. Let sit overnight before using.

In pioneer days homesteaders would soak **tobacco** leaves in water and then boil the mixture down. They would strain it and spray or paint it inside the coop to kill lice, mites and other bugs. When using any poison, always wear gloves and long sleeves. They also dried the tobacco leaves, then ground them into a powder and dusted the poultry and put some in the nest boxes. (For **tobacco** see April Greenhouse.)

If needed, treat for **lice and mites** by feeding a pinch of sulfur powder per bird monthly or more if bad infestation. Mix in feed. If badly infested bird, dust with wood ash, diatoms, insecticide, or rosemary/wormwood. Hold bird upside down and dust. Or place bird in a plastic bag with the birds' head out. Put dust in the bag and rotate/shake the bag to completely cover the bird with powder. Be sure to apply to vent (rear) and under wings. Repeat in 2 weeks.

Sevin (carbaryl) powder can be used but it is more toxic than the above methods. It kills bees and other beneficial insects. Or try Eprinex/Ivermectin pour on drops. Put 2-3 drops on a bantam, and 3-4 drops on a regular sized chicken.

Poultry need a place to take a **dust bath** in loose, dry, powdery soil. You can add wood ash, lime or sulfur dust to their dusting hole. The dust kills mites and lice in the feathers. Plus it is fun for them.

(For **garlic water,** see January Farm Animals.)

 **Clean barn, coop** and other animal sleeping areas. Put dolomite lime, gypsum, rock phosphate and/or rock dusts on floor. Then add straw, hay or other fibrous material for bedding. (For details see January Farm Animals. For more about **fertilizers**, see "Garden Tips" in the back of this manual.)

 **Honey Beekeeping in March**

The **beehive population is low** in March. They can die of starvation at this time of year. Be sure to feed them plenty of **sugar syrup or honey in the fall** so this does not happen.

With days growing longer, the queen steadily **increases her rate of egg laying**. More brood (embryo or eggs, larva and pupa) means more food consumed. The drones (male honey bees) begin to appear.

Early March on a mild day with no wind, look inside the hive. Do not remove the frames. If you do not see any sealed honey in the top frames, begin some emergency feeding with honey or sugar. Use 1 part sugar to 1 part water. Once you start, do not stop until they are bringing in their own food. Bees collect nectar and pollen on warm days.

At this time you can find out how many of your bees **died over the winter.**

Add Apistan strips for the treatment of varroa mite infestation (leave in hive for 45 days).

(For an **overview of beekeeping,** see January Farm Animals. For **harvesting honey**, see September Farm Animals.)

 # APRIL April 1 daylight is 12 hours 36 minutes. April 30 daylight is 13 hours 36 minutes.

Last of 3 "hunger" months. Lots of rain. Sometimes snow. Mud season. Last frost usually late April or early May.
Early April: Daffodils and forsythia bloom. April showers bring May flowers!
   Dandelion, mint, oregano, sorrel, stinging nettle, and lovage are still small but big enough to eat.
   Red maple buds turn red and then flowers open. (These form red, winged fruits known as helicopters).
   Pear and serviceberry (Saskatoon berry, juneberry, shadblow, shadbush) trees bloom.
   In greenhouse or protected porch rosemary blooms. Blackberry leaves start to grow.
Mid-Late April: The leaves of apple and poplar trees are the size of a dime. Oak tree leaves are size of squirrels ear.
   Dandelion, violet, rhubarb, wild comfrey, and everbearing strawberry bloom.
   Asparagus spears start to come up. Horseradish leaves start to come up.
   Apple trees, azelas, dogwood, peach, strawberries and walnuts bloom. Cherry bushes bloom.

Garden- Maintenance

As **comfrey plants** age, they die in the middle. In **April or May** dig up roots away from the center, and replant in the center. The crown should be covered with soil.
   (See March Garden-Plant. See June/July/August/September Garden-Harvest with **harvest details in June**.)

Last month to add nitrogen (manure) to **garlic and onions**. Do not add nitrogen when bulbing (**May to September**). Clip off seedstalks and flowerheads when they form unless you are saving seed. You can eat flowerheads. (For **garlic and perennial onions** see September Garden-Plant. For **spring planting of garlic**, see March Garden-Plant. For **onions and leeks**, see February Greenhouse.)

Put **fertilizers** such as azomite, dolomite lime, granite dust, greensand, gypsum, kelp, rock phosphate, rock dusts and/or other fertilizers on garden soil. Best if dug in. Do soil test first. Go to your **County Extension Office** for information about free soil tests. They also offer a lot of free information and advice about gardening and farming. (For **fertilizing**, see March Garden-Maintenance. For details about **natural fertilizers and where to buy them**, see "Garden Tips" in the back of this manual.)

Remove flowers on **rhubarb** plants so energy goes to leaves instead. Do this any time they form. Do not eat flowers or leaves. Eat only stalks. (For **dividing and planting crowns**, see March Garden-Plant. See April/May Garden-Harvest with **harvest details in April**.)

apply a band of duct tape *apply a coat of sticky material* *young caterpillars are caught*

Put **sticky tree bands** on fruit and nut tree trunks late April, or when leaves start to emerge to trap crawling insects.
   Wrap duct tape around the tree trunk about 3 feet from the ground. Push it into dips in the bark as much as possible so insects do not crawl under it. If you want larger band, use 2 strips of duct tape. Or make the band 10 inches wide. On thin-bark trees, place a layer of paper bag or butcher paper under the duct tape so do not damage bark.
   Can also use plastic wrap, garbage bags, foil, or any material that can block crawling insects.
   Then apply a thin coat of **sticky stuff** such as grease, petroleum jelly, or Tanglefoot to the duct tape or other material. (For how to make sticky stuff, see "Plant Health" at the end of this book.) Do not get sticky stuff on the tree.
   Crawling Pests: This prevents ants, cankerworms, cutworms, leaf beetles, and young caterpillars of codling moth/gypsy moths from crawling up the tree. This control method traps only insects crawling low on trees or that emerge from egg masses at a low location. Reapply more sticky stuff periodically.
   Some say this method does not make a significant dent in insect population to be effective. So experiment with

it. However, it is a good method to **monitor insect populations** to see how other control methods are working.

 **Whitewash trunks of fruit and nut trees** with white latex paint or other formulas. (For **whitewashing details and why**, see June Garden-Maintenance.)

Garden- Plant or Transplant (April)

Early April:
Bamboo, echinacea, garlic, grasses/cover crops, mullein, onions, parsnips, stinging nettle, and strawberry.

Mid to Late April:
Asparagus, broccoli, broccoli raab, cabbage, chickweed, hops, parsley, plantain, potatoes, and radish (summer).

Fruit/Nut Trees and Bushes:
Apple, apricot, Asian pear, blueberry, bramble (blackberry/raspberry), cherry, currant/gooseberry, elderberry, fig, grape, kiwi, peach/nectarine, pear, plum, rosa rugosa, and walnut.

Early April outside: Plant **hardy crops** when peach and plum trees are in full bloom.

 Plant Bamboo (poaceae/grass family; around 92 genera and 5,000 species)

Two types of bamboo: running and clumping. Both are **perennials**. If you want them in contained groups, choose the clumping type. If you want them for a hedgerow or to fill a large area, use the running types.

Running bamboo (Phyllostachys genus) spreads by sending out roots or rhizomes. Some are modest growers, while others can be extremely invasive. Most spread 3-5 feet per year with some 15 feet per year. Control running by cutting the ends of the rhizomes with a spade. Phyllostachys, or Giant Bamboo, are the tallest of the species and prefer sun or partial shade.

Clumping bamboo (Fargesia genus) is not invasive. New shoots occur on the outside of the plant. It is the hardiest of all bamboos. It is evergreen through winter and tolerates temperatures to minus 25 degrees. It is usually used to form a hedge with weeping foliage. It does best in partial shade.

Once per year, usually in **spring**, new growth occurs as a **bamboo shoot**. The shoot emerges from the ground and reaches its full height in 6-8 weeks.

Planting: Do not use fertilizer or manure when planting. Plant rhizomes 1-2 inches deep. Mulch around the plants, and keep the soil moist. At first it grows slowly but later it grows very fast. Varieties range in height from 7-33 feet. Be sure to keep it well watered for the first year.

Some bamboos die back to the ground in winter, but then send out culms (spears or shoots) in **May or June**. In **late fall** mulch with 4-6 inches of leaves, sawdust, straw or wood chips.

Others stay green all winter. It is a good food for animals.

Care: Easy to grow. Some varieties of bamboo prefer full sun, others prefer partial shade. They prefer rich, well draining soil with neutral pH of 7. Keep well watered.

Bamboo is **propagated** by rhizomes (roots) that multiply and spread rapidly. They are easy to transplant. Dig up rhizomes, divide them, and replant.

Some varieties:

Umbrella bamboo (fargesia murielae) grows to 10-14 feet tall with 1/2 diameter canes. Hardy to minus 20 degrees. It does not have running rhizomes so needs no containment to prevent spread. Flowers every 100 years!

Fargesia dracocephala Rufa grows to a maximum height of 10 feet with an average of 7 feet. Has cane diameter of 1/2 inch. Hardy to minus 15 degrees. Vigorous grower. Leafs out early summer. Prefers protection from hot afternoon sun. It does not have running rhizomes so needs no containment to prevent spread.

Phyllostachys nidularia grows 33 feet tall with cane diameter 1 3/4 inch. Running. Hardy to 0 degrees.

(See April Garden-Harvest.)

 Sow **echinacea** (asteraceae family; purple coneflower) seeds when soil is 55-70 degrees. Seedlings can take light frost. Sow 1/8 to 1/4 inch deep and 2 inches apart. Seeds need light to germinate. Germinates in 10-20 days. When an inch tall, thin seedlings to 18-24 inches apart.

Care: Prefers full sun or partial shade. Likes soil pH 6.0-8.0. Likes fairly dry soil. Drought tolerant. Native plant in the United States. **Blooms June through August.**

Echinacea angustifolia (narrow leaf purple coneflower) grows best in high elevations and cold winters. Hardy **perennial** good to zone 3. Least vigorous (slower growing) of all echinacea. Grows to 8-18 inches tall. Stratify seeds 21 days or more. (For how to stratify seeds, see "Garden Tips" in back of manual.)

Echinacea purpurea (common purple coneflower) grows best where winters are mild such as zone 6 or warmer. It is easier to grow. A **perennial**. Grows 24-36 inches tall. No stratification of seeds needed but can stratify for 7 days. Vigorous with large, purple-petaled flowers 4-6 inches across. Has fibrous roots that are easy to harvest (other species have taproots).

Can **propagate** either types by dividing plants in **late spring or summer.** Will self seed.

To save seed for planting: Choose fully ripened flower heads. Cut leaving a long stem. Hang upside down with the flower heads in paper bags. They release their seeds in the bag. Then remove chaff (plant debris) and spread the seeds on a newspaper for 10-12 days to dry.

Harvest leaves and flower tops just as flowers start to open. Dry them out of direct sunlight. Harvest **root in fall** of 3-4 year old plant.

Uses: Has anti-bacterial, anti-viral and anti-fungal properties. Good for immune system. Helps fight infections such as colds and flu. Make a tea from leaves, flowers or seeds. Can boil the root for 20-60 minutes and drink the water. Only use occasionally. Not good for people with immune disorders. Consult your herbalist, doctor and/or health care specialist. (See October Garden-Harvest. For how to make an **herb tincture, decoction or infusion**, see April Garden-Harvest.)

 Plant **garlic** (amaryllidaceae family; Allium genus) **4-6 weeks before last frost in spring**. Better planted in the **fall**. Harvest is larger if fall planted.

(See March/September Garden-Plant with **planting details in September.** For **harvest and storage details**, see July Garden-Harvest. For **fertilizing** see April Garden-Maintenance.)

 Continue to plant **grasses and cover crop legumes (cool weather).** (For **cool weather cover crops**, such as clover, fava beans, fescue, flax, oats and ryegrass, see March Garden-Plant. For general information about **cover crops**, see "Garden Tips" in back of manual.)

 Sow **mullein** (scrophulariaceae or figwort family; Verbascum genus with 250 species; velvet plant) **2-4 weeks before first frost in spring. Or sow 8 or more weeks before first frost in fall.**

(See August Garden-Plant for details.)

Plant or transplant **onions**. **Sow seed outside in April or 2-3 weeks before first frost date in spring. Or sow outside 8-10 weeks before first frost in fall.** (For **sowing/growing details**, see February Greenhouse.)

After daffodils bloom, sow **parsnips** (umbelliferae family; Pastinaca sativa) to eat fall through early spring. Start sowing same time as peas which is March. **Sow 2-4 weeks before last frost in spring. Sow from March 1 to late June.** It is best to time sowing so they **mature after first fall frost.** Matures in 100-150 days (4-5 months).

Seeds do not store well, usually only 1-2 years. Soak seed overnight before planting. Can pre-sprout seeds in a moist paper towel. Plant when rootlets are 1/4 inch long.

April p. 3

Germination is slow (2-4 weeks) and poor so **sow thickly**. Sow 1/2 to 3/4 inches deep, 2-4 inches apart in row 1 foot apart. Thin to 3-4 inches apart. When 2 inches tall, thin to 8 inches apart. Grows to 3 feet tall.

Can stay in the ground up to a year. Roots are 6-12 inches long. It is a **biennial.**

Care: Likes full sun and well drained soil. Likes soil pH 6.2-7.2. Do not use fresh manure or lots of nitrogen because that causes hairy, forked roots. Flavor is best after frost.

Make sure parsnips always have enough **water**, otherwise they get tough and may split.

Put **wood ashes** around the plants to keep away rust flies, carrot weevils, wireworms, and other carrot/parsnip pests. (For carrots see June Garden-Plant.)

To get big parsnip roots: Drive a crowbar into the soil 2 feet. Rotate the bar in a circular motion until the hole is about 6 inches across the top. Fill hole with a mixture of sand, peat moss and sifted soil, leaving a small depression at the top of the hole. Place 2 or 3 pre-sprouted seeds in the depression, then cover with 1/2 inch of sifted sphagnum moss and water. Space holes 8 inches apart. Later thin to 1 strong plant. Can mulch with straw or leaves.

Parsnips, anise, caraway, carrots, celery, chervil, coriander/cilantro, cumin, dill, fennel, parsley and queen anne's lace are all members of the **umbelliferae family.** (See "Plant Families" in the back of this manual.)

In **late fall** cover parsnips to be overwintered with 1 foot of straw, hay or leaves.

Varieties: There are **3 types of parsnip**- long/narrow, long/broad, short/stocky with rounded shoulders.

All America matures in 120 days. Broad-shouldered with white flesh and small core. Roots 12" long.

Cobham Improved Marrow has high sugar. Has 8 inch long roots. Better for heavy soils.

Harris Model matures in 120 days. Has 10-12 inch roots. Smooth white with few small branching roots.

Hollow Crown matures in 105 days. Roots are 12", tapered and somewhat free of hairlike side roots.

(See April/September/December Garden-Harvest with **harvest details in September**.)

 Sow or transplant **stinging nettle** (urticaceae family, Urtica dioica). It is easier to transplant a mature plant than to start from seed.

Seeds need to be stratified for 4-6 weeks. (For **how to stratify,** see "Garden Tips" in back of manual.) So it may be better just to plant the seeds in **fall** and let nature prepare the seeds for spring.

Barely cover seeds with soil. Germinates in 10-14 days. Thin to 8-12 inches apart. Grows 30-60 inches (3-4 feet) tall. Plant 24 inches apart. Plant away from areas where people walk a lot since they sting.

Care: Likes full sun or partial shade. Likes moist soil. They are **perennial.** Fast growing.

Propagation: They propagate by seeds and by underground runners that can grow up to 5 feet per year in good conditions. Somewhat invasive.

Uses: It is one of the highest protein sources of any plant. Comfrey is the highest. Nettle hay is good for ruminants such as cattle and goats. Can make silage from nettles and comfrey mixed. Good to add to compost pile to speed decomposition. Good to eat. For more uses, see harvest in March. Consult your herbalist.

(See March Garden-Harvest.)

 **Strawberry** (rosaceae family; Fragaria × ananassa, Fragaria x vesca with hundreds of varieties) **plants or seeds** are planted as soon as soil can be worked (**March/April**). They can also be planted in September. It is a hybrid.

Seeds germinate best at 60-65 degrees. They will not sprout if over 70 degrees. Germination takes 4 weeks. You may want to start them in a greenhouse. Do not plant in a frost pocket where flowers may be damaged by frost.

They are easy to propagate by their numerous **runners** (stolons) that grow from mother plants.

Care: Likes soil pH 5.5-6.5. Needs full sun and well drained soil. Likes rich soil. The root system is shallow so water on a regular basis.

Planting: Soak roots for a few hours before planting. Cut roots on transplants to 6 inches long before re-planting. Each plant needs a square foot of space. In the **matted row system** of planting, plants are planted 18-24 inches apart with 3-4 feet between rows. The plants are allowed to grow runners. After 1 year over 50 pounds of berries can be harvested from 25 plants grown in this system.

Remove buds from new plants **April through June** so develop good roots. Berries are harvested about 1 month after buds form. They are a very productive plant.

Can cover with row covers (Reemay) in **spring** to protect flower buds and berries from cold temperatures.

Life Cycle: A new plant grows from seed or runners from a parent plant. **Production peaks 2-3 years later,** and then the plant dies at about 4-6 years old. After 3 productive years it is best to replace old plants with new plants.

Pests and Disease: Remove any infested or diseased plants. Remove damaged berries. (See "Plant Health" in back of this manual.)

    1. Chipmunks, birds and other animals love strawberries so they may need to be protected with netting or traps. In peak season pick berries every day.

    2. Slugs and snails eat the berries. Spread wood ash, sand, gravel (chat), or diatomaceous earth around plants. Make sure there are no dead leaves or plants that give them shelter from the sun. Worse in wet weather. (For slug/snail traps see "Plant Health" in back of this manual.)

    3. Viruses attack strawberries with some producing symptoms such as mottling (spotting or blotching) of leaves while others produce no symptoms. Fruit yield is lowered, and plants can not be cured once infected. Most strawberry viruses are carried by aphids so control them first. (See "Plant Health" section in back of manual.)

There has been good success with helping sick plants recover from various health problems by sprinkling **rock dust** powder over the plants once a month. (See fertilizers in "Garden Tips" section in back of this manual.)

There are **4 types of strawberry plants:**

    1. June bearing strawberries (Fragaria x ananassa) start bud formation in fall. They produce a **large harvest in June** that lasts 2-3 weeks and that is it until next year. The berries are large. Then they produce runners.

    Varieties include Allstar, Annapolis, Brunswick, Chandler, Delmarvel, Earliglow, Honeoye, Kent, Lateglow, Mohawk, Sable, Seneca, Sequoia, Sweet Charlie and Winona.

    Earliglow is one of the most popular June bearing strawberry plants. The deep red berries are medium to large. They have an excellent, sweet flavor and are great for canning. They are resistant to many diseases.

    Honeoye are early season producers with large, firm, bright orange-red to red fruit. A heavy producer. Will grow in most soils but does best when grown in raised beds or lighter soils.

    Sweet Charlie has large sweet berries. Does well in heat/humidity. Bears early in the season.

    2. Everbearing strawberries (most are Fragaria vesca) start bud formation in the summer. They have smaller berries. They have a good crop in **spring,** then smaller **harvests through the summer until frost**. Strawberry plants from previous summer start blooming April.

    Varieties include Arapahoe, Eversweet, Fort Laramie, Gem, Geneva, Ogallala, Ozark Beauty, Red Rich, Rockhill, Quinault and Twentieth Century.

    Fort Laramie produces large, scarlet red berries. Produces blooms, berries and runners all at the same time. Very cold hardy. Berries have a great aroma and firm, honey-sweet flesh. Good for hydroponic growing.

    Ozark Beauty is the most popular everbearing strawberry plant. Large and sweet berries. Hardy and vigorous. Produces a large early crop and a second crop later in the season.

    3. Day neutral strawberries (Fragaria X ananassa) start bud formation at any time. They produce berries **all season**. They grow very few runners. Day neutral varieties need less sun than the other types.

    4. Alpine or wild strawberries (Fragaria vesca) are small and produce **June through summer**. They do not form runners. Will grow from seeds.

In the **mountains and Piedmont area of North Carolina** choose varieties that are red stele (fungus) resistant such as Earliglow and Sunrise. Other good varieties for the mountains are Apollo, Cardinal, and Tennessee Beauty.

(See May/June/July Garden-Harvest with **harvest details in May**. See March Garden-Plant but details are in this section. For **maintaining beds and fertilizing**, see June Garden-Maintenance.)

Mid to Late April sow or transplant outside:

Plant **asparagus crowns** (liliaceae family; Asparagus officinalis) **4 weeks before last frost in spring which is early to mid April** or when soil is 50 degrees or higher. A **perennial** that lives up to 15 years.

Asparagus crowns are the base and roots of one-year-old plants grown from seed.

Dig a hole about 10 inches deep. Make a 6 inch mound of dirt in the hole. Drape the roots over the mound. Fill in sides, and **cover crown** with 2-3 inches of soil. Plant about 15 inches apart in rows 3-4 feet apart in sandy soil.

Can also dig furrows 4-6 inches deep. Put crown in bottom and cover with 2 inches of soil. As the ferns grow, gradually fill in the furrow through the summer. (See above image.) (See Liliaceae Family in "Plant Families".)

Care: Likes soil pH 6.0-7.0 but is not picky about it. Needs a rich, well drained soil. Prefers full sun. Drought tolerant but better to keep well watered. Grows up to 5 feet tall and 3 feet wide (a group of ferns from same crown).

Varieties: There are green and purple varieties.

Harvest: Do not harvest spears the first year. One crown produces 1/2 pound of spears at peak production.

Sow **asparagus seed** (liliaceae family; Asparagus officinalis) when soil is 70 degrees or more. Or **transplant seedlings from greenhouse.** Can sow seeds **late April** or when apple blossoms fall (**May or early June**). (See Liliaceae Family in "Plant Families" in back of manual.)

Sow berry or seed 1-2 inches deep, several inches apart. (A berry contains about 6 seeds.) Germinates in 10-14 days. Thin seedlings by cutting off unwanted ones to 12 inches apart.

The top of the plant is **fern** like. The spears that are eaten are immature ferns. There are male and female plants. Male plants produce more spears than females. Some say kill all female plants. But if you want to **save seed**, save some female plants.

If saving seed, do not cut off flowers. Otherwise cut flowers off in the summer. Flowers turn into little **red berries**. The berries ripen and fall to the ground where new seedlings grow. Or can save berries and sow in spring.

(For **sowing seeds in greenhouse**, see March Greenhouse but details are in this section. For **fertilizing** see March Garden-Maintenance. See April Garden-Harvest. For **saving seed**, see August Garden-Maintenance.)

 Transplant **broccoli and broccoli raab** from greenhouse to outside. (For **sowing and growing** details, see March Greenhouse. See June Harvest.)

Transplant **cabbage** (European) (brassicaceae family, Brassica olercea) **from greenhouse to outside** after apple blossoms open. An **annual** that transplants well. Or sow directly outside **mid to late April** for early cabbage. (**For early cabbage sow seeds indoors or in greenhouse 6-8 weeks before last spring frost. Or sow outside directly 2-5 weeks before last spring frost.**)

For late cabbage sow 10-12 weeks (2 1/2 to 3 months) before first fall frost. That is June or early July.

Sow 1/4 to 1/2 inch deep. High germination rate so sow thinly. When 1 inch tall, thin plants to 3 inches apart. They are transplanted best when 4-6 inches tall. (See "Plant Families" in back of manual.)

Transplant or thin seedlings 12-18 inches apart in rows 18-36 inches apart. Gradually thin plants as they grow. Can eat pulled plants. Can take frost down to 20 degrees once hardened off (For hardening, see "Garden Tips" in back of manual.)

Early varieties are planted 12 inches apart in all directions. Late varieties up to 24 inches apart. Matures in 60-110 days but usually 90-95 days (about 3 months).

Care: Likes full sun but will tolerate some shade. Likes soil pH of 6.0-6.5. Keep soil moist. Likes rich, well drained soil. Likes boron, calcium and magnesium, particularly during early stages of growth.

Early varieties produce 1-3 pound heads. Late varieties produce 4-8 pound heads. Late maturing varieties are better for making **sauerkraut.**

Varieties: It can be green, purple or red. Most have smooth leaves but savoy cabbage has crinkled leaves.

    Green Cabbage- (Dutch White) Outer leaves are dark green, inner leaves are pale. **Good cooked.**
        Cheers (75 days to harvest; solid round heads; tolerant to black rot and thrips)
        Early Jersey Wakefield (63 days; pointed heads; resists splitting)
        King Cole (74 days; large; firm; uniform heads)
    Savoy Cabbage- Crinkly, thin with blue-green leaves. Good eaten raw. **Grown for cole slaw & salads.**
        Savoy King (85 days to harvest; dark green color; uniform)
        Savoy Queen (88 days; 5 pounds; deep green color; good heat tolerance)
    Red or Purple Cabbage- Smaller than green cabbage. Cook with vinegar to keep color. **Stores the best.**
        Red Meteor (75 days to harvest; firm; good for all seasons)
        Ruby Ball (71 days; 4 pounds; slow to burst; resists both cold and heat)

Pests: (See "Plant Health" section in back of this manual.)

    1. Aphids are black, brown, green, red, yellow or white insects that cause curled yellow leaves with a honeydew substance. To control use insecticidal soap or a strong stream of water or Sevin. Aluminum foil under plants reflects light to under leaves making them undesirable for aphids.

    2. Cabbage worms or cabbage loopers are the most common pest. They are green or greyish caterpillars that develop into white or yellow butterflies. The caterpillars feed on the underside of leaves, leaving holes. Deter them by using row covers in the earlier part of the growing season so moths do not lay eggs on plants. Can also pick off worms by hand. Spray with Bacillus thuringiensis (Bt).

    German winter thyme deters cabbage worms. (For **German winter thyme**, see thyme in May Garden-Plant. Anise hyssop repels cabbage moth. For **anise hyssop**, see mint in May Garden-Plant.)

3. Cutworms chew stems off just above ground level. They are 1 1/2 inch long caterpillars that are mottled green, brown or gray. When disturbed, they roll up in a ball. They like to stay at the moisture line in the soil. If the soil surface is dry, they will be found a couple of inches below the surface. Put cardboard or metal collar around new transplants that extends 1 to 2 inches above and below soil level. (See image for tomato cutworms in March Greenhouse.)

4. Flea Beetles chew tiny holes in leaves. They are 1/16 inch long, hard shelled, shiny, and dark-colored beetles. Slender, whitish, cylindrical larvae feed in or on roots but root damage is usually minimal. Dust with Rotenone. Keep plant debris removed. Rotate location of planting from year to year.

5. Slugs and snails like cabbage. (See "Plant Families" in back of this manual for more about control. See How to Make Traps and Deterrents for Slugs and Snails in "Plant Health" in back of manual.)

To reduce spread of disease, rotate non-Brassica crops with Brassica crops.

(For **early cabbage** see March Greenhouse but most details in this section. See June Garden-Harvest.)

Cabbage for storage should be planted in June. (For **late cabbage** see June Garden-Plant but most details in this section. See October Garden-Harvest.)

 Sow commercial **chickweed** (caryophyllaceae family; Stellaria media). It is a larger plant than the wild variety and better tasting. Sow 1-2 seeds per inch, 1/2 inch deep. Seedlings can survive medium frost. Thin seedlings to 4-6 inches apart. Matures in 37-50 days.

Does not like acid soil. Likes full sun or partial shade. A **hardy annual** that readily self seeds. Beneficial insects like the flowers. All above ground parts are edible. Mild tasting. Good in salads. Chicks and other baby birds love it.

Hops (cannabaceae family; Humulus lupulus) are a **perennial vine** propagated from rhizomes (their roots). Cut off a portion of root from a mature plant and plant a few inches deep and 3 feet apart. A few weeks later small, green shoots come up. A hop vine can grow 2 feet per week until the end of **June.** After that flowers form. Consult your herbalist or County Extension office.

Care: Likes soil pH 6.0-8.0. Likes full sun. Needs well drained soil. Likes rich soil. Does best on a trellis or fence 8 feet or more tall. Needs a lot of space.

After the first year, the earliest shoots that come up should be pruned off. The second set of shoots are sturdier growth. Choose 3-4 main vines and prune off all other vines so that all the energy goes to the main vines.

(See August Garden-Harvest.)

Sow or transplant **parsley** (umbelliferae family; Petroselinum crispum) **late April or early May. Sow 4-6 weeks before last frost in spring.** Soak seed in water 48 hours changing water twice. Discard water. Can stratify seeds. **Or plant late fall after a few frosts such as in October,** and let nature stratify them for you. (For how to stratify, see "Garden Tips".)

Germinates better in dark. Germination can take 3-4 weeks. Sow 1/4 inch deep, 1/2 inch apart in rows 10-12 inches apart. Thin plants to 6-8 inches apart. Slow growing. Matures in 10-13 weeks. **Leaves are hardy to 25-28 degrees.**

Care: Likes soil pH 6.0-7.0. Tolerates poor soil but does better in good soil. Prefers full sun but tolerates partial shade. Needs at least 6 hours of sun a day. Mulch in late fall.

A **biennial**. No flowers the first year. Will come up again next spring and bloom in summer, and then die in the fall. Cut off flower stalk unless you want to save seeds. Collect seeds the second year. Can eat some from it second year but leaves are not as developed.

Varieties: flat and curly leaf.

Parsley, anise, caraway, carrots, celery, chervil, coriander/cilantro, cumin, dill, fennel, parsnips and queen anne's lace are all members of the **carrot family** (umbelliferae or apiaceae). (See "Plant Families" in back of this manual.)

Swallowtail butterflies use parsley as a host plant for their larvae. Their caterpillars are black and green striped with yellow dots. They feed for 2 weeks and then turn into butterflies. Let them live.

Harvest: Can harvest up to 26 weeks, sometimes more. For harvest eat fresh or dry the leaves. Can eat leaves the second year but they are more bitter than first-year leaves.

(See October Garden-Plant. See March/August Greenhouse but details are in this section.)

 Sow broad and narrow leaf **plantain** (plantaginaceae family; Plantago major). It is a **perennial** that grows in a rosette form with large oval, dark green leaves 4 to 10 inches long with ribbed veins. Flower stalks are tall and slender with dense flower spikes. Flowers **May through July.**

Can also **forage** for it. Harvest roots in **late summer or fall.**

Care: May take weeks to germinate. Likes sun or partial shade. Average soil is OK. Does OK in most soil pH.

Uses: Plantain leaves and seeds are antibacterial, astringent, anti-inflammatory, antiseptic, diuretic (increase urination), and expectorant (coughs). Young leaves are edible raw or cooked. After the middle of spring, leaves are very fibrous so should be cooked. Crush plantain leaf and apply to **bee or other stings** to relieve pain and itching.

Seeds can be sprouted and eaten. The seeds have up to 30% mucilage which acts as a bulk laxative. The seeds are used in the treatment of parasitic worms. Gather seeds to dry after flower spike forms.

Chew plantain **root** to relieve toothache. A decoction of the roots is used to treat diarrhea, gastritis, irritable bowel syndrome, hemorrhoids, cystitis, sinusitis, coughs, and hay fever. (A decoction is made by making a tea and then boiling down so most of the water is gone.) (For how to make a **tincture, decoction, or infusion,** see April Garden-Harvest.) Consult your herbalist, doctor and/or health care specialist.

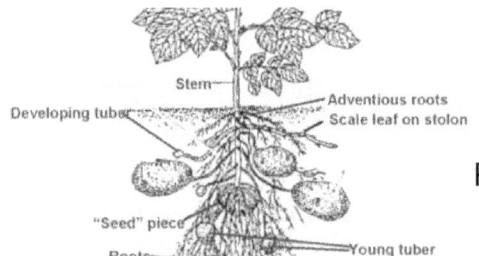 Plant **potatoes** (white/red/purple) (solanaceae family also known as nightshade; Solanum tuberosum with 4,000 varieties) in **late April** when soil is at least 45 degrees or **2-3 weeks before last frost in spring.** Plant potatoes when the shadbush (serviceberry, Juneberry), daffodil and dandelion flowers. They are high in protein and vitamin C. (See "Plant Families" in back of this manual.)

Potatoes are the **most productive crop** you can grow. They are easy to grow and drought resistant. They do well even if soil is not that good. They like cold, wet weather.

Care: Potatoes like soil pH 5.2-6.8. Likes full sun but tolerates partial shade. Likes well drained soil. Likes average soil. Soil should not be too hard.

There are less problems with potato scab if pH is low and there is not too much nitrogen in the soil. It is best to rotate the land where you plant potatoes with other crops such as with corn or squash.

Water early in the day to reduce odds of getting fungus infection. They need the **most water** right before they flower. After flowering, less water is needed. Do not overwater as the tubers may be watery and not store as long. You may not want to water them towards the end of the season unless very dry.

Potatoes are planted late April through June, and harvested June through October. Some people plant potatoes late March or early April but there is risk they will rot in cold soil. **Earliest planting is 2-3 weeks before last frost in spring.**

In **late April plant** early/short season potatoes for harvest in June or July.

In **late April or May plant** mid season and late/storage potatoes for harvest August through October.

In **June plant** early/short season potatoes for harvest in August or September.

(For **sprouting to plant in April,** see March Greenhouse.)

(**Potato Harvest:** For early/short season potatoes, see June/July Garden-Harvest. For mid-season potatoes, see August Garden-Harvest. For late/storage potatoes, see September/October Garden-Harvest. **Harvest details are in October Garden-Harvest.**)

How to Plant Seed Potatoes: Potatoes the size of a chicken egg are planted whole. Larger potatoes are cut so each piece has 2-3 eyes or sprouts. The larger the seed piece, the larger the individual potatoes and greater yield overall.

Do not plant diseased tubers. Lightly dust cut potatoes with **agricultural sulphur** (sulfur) to stop fungal diseases. Put 1 tablespoon of sulphur in a paper bag and gently toss pieces around. Let dry for a few hours or a day out of the sun.

Seed potatoes rot if the soil is too cold and wet. The best soil temperature to plant is between 50-70 degrees. **Plant 2-3 weeks before last frost** so that means planting late April. Tuber pieces can freeze and die so do not plant too early or plant a little deeper.

Plant 3-4 inches deep with cut side down and sprout side up. Plant tubers 12-18 inches apart in row. Plant early potatoes in rows 18 inches apart, mid-season 24 inches apart, and storage potatoes 30 inches apart. Leaves emerge in about 2 weeks.

As potatoes grow, **hill up** around it with dirt or mulch. When the plant is 8 inches tall, hill up 3 inches of soil around plant. Do again every 2-3 weeks. You end up with 8-12 inches or more of soil over the seed potato. Tubers form between the seed potato and the leaves of the plant. (See above image.) **Harvest is 10-25 pounds per 1 pound planted.**

Colors of potato skins: russet (brown skin), red, white, yellows (Yukon) and purple (blue).

Flesh (inside) can be white, yellow, blue, red, pink or purple.

Three types of potatoes based on season:

1. Early or short season mature in 60-80 days such as Bison, Bliss, Caribe, Charlotte, Dazoc, Early Ohio, Early Rose, Irish Cobbler, Mountain Rose, Norland, Purple Majesty, Purple Viking, Red Dale, Red Norland, Red Pontiac, Russet Norkotah, and Yukon Gold.

- **Irish Cobbler**- round white potato. Heirloom that bruises easily. Plant as soon as ground dries.
- **Norland** has red skin and white flesh. Excellent boiled, fried, or mashed. Stores well.
- **Purple Viking**- drought resistant, high yield. Purple skin, white inside. Gets sweeter in storage.
- **Red Pontiac**-easy to grow, red skin/white flesh. Good for new potatoes. High yield, stores well.
- **Yukon Gold** is large with yellow skin/flesh. Excellent storage. Good baked, boiled, or mashed.

2. Mid-season mature in 80-100 days such as Adirondack Blue, Agria, All Blue, All Red, Atlantic, Colorado Rose, Dakota Rose, Desiree, Durango, Huckleberry, Kennebec, Keuka Gold, Kerrs Pink, Maris Piper, Pink Pearl, Red Gold, Red La Soda, Rio Colorado, Rose Gold, Sangre, Valisa, Viking Red, White Rose, and Yellow Finn.

- **Desiree**, a red potato, very high yielding. Resistant to diseases. Good for all around cooking.
- **Kennebec** does well under harsh conditions. Has light tan skin and white flesh. Good storage. Excellent producer, large potatoes. Great for baking or frying.
- **White Rose**- good producer. Good for boiling & potato salad. Fair for baking. Does not store well.

3. Storage or late season mature in 100-130 days such as Bintje, Carola, Garnet Chile, German Butterball, Inca Gold, Lemhi Russet, Nicola (some say mid season), Ranger Russet, Rio Grande, Romanze, Russet Burbank, Russet Norkotah, Russet Nugget, Shepody (some say mid season), Silverton Russet, Snow White, Superior.

- **Carola (Carole)**- a yellow potato with some scab and disease resistance. Excellent for storing.
- **Russet Norkotah**- an excellent baking potato. Excellent producer, large potatoes. White flesh.
- **Silverton Russet** is medium size, dual purpose with smooth, long tubers and white flesh.
- **Superior**- buff skin and white flesh. Washes easily. Good for long storage. Resistant to scab.

Potato Pests (Insects and Animals):

Choose varieties that are resistant to blight, blackleg, scab, potato cyst eelworm, and other problems. Rotate crops on a 3-4 year rotation with plants not in the **nightshade family**. (See "Plant Families" in back of manual.) Good soil quality helps plants fight disease. (For more about pests and disease, see "Plant Health" in back of this manual.)

1. Colorado potato beetle is the worst pest of potatoes. Adults are 1/2 inch long, yellow with black stripes. Larvae are plump, hump-backed, copper colored grubs up to 3/5 inch long. They feed on leaves. Eggs are bright yellow-orange laid on underside of leaves. Hand pick adults and egg masses if possible. Can use Sevin but use sparingly to avoid spider mite build up.

2. Flea beetles are tiny, and black or brown. They chew small holes in leaves. Cover ground with fabric row covers when you plant the tubers. Keep populations low through crop rotation and by having high soil organic matter. (See Brassicaceae Family in "Plant Families" in back of manual.)

3. Voles, moles and gophers eat potatoes. Trap them with snap traps. (For **how to make trap boxes**, see September Farm Animals. For animal pest control, see Animal Pests in "Plant Health".)

Potato Diseases: Weather has a lot of influence on what diseases your plants get.

1. Common scab (Streptomyces scabies) is a bacterial disease that produces patches of corky tissue on the surface of tubers. Damage is usually superficial, and infected tubers are edible. To control, improve soil moisture holding by adding organic matter. Manure should be aged. Scab is much worse in alkaline conditions so do not lime soil. Not usually a problem if soil pH is less than 5.4. Make sure soil does not dry out. (See Chenopodiaceae Family in "Plant Families".)

2. Late Blight is caused by **downy mildew fungus**. It caused the Irish crop failures of 1845-1846. Water-soaked areas on leaves turn brown and black. The leaf dies. It occurs during cool, wet weather and may spread rapidly if the weather warms up. Plants can die and tubers can be seriously affected, especially when stored. Plant certified seed and use a potato dust fungicide. (See Amaryllidaceae Family in "Plant Families".)

3. Mosaic Virus is spread by **aphids**. It makes leaves curl and look 2-toned (light and dark green). It reduces harvest, but does not kill the plants. Kennebec and Katahdin varieties have some resistance to certain kinds of mosaic. (See Leguminosae Family in "Plant Families". For aphids see Brassicaceae Family in "Plant Families".)

 **Saving and Planting Potato "True" Seeds (not tuber):** Yes it is possible and it is worth doing due to the great importance of potatoes for calories. **If you are having disease problems, this is a way to solve it.**

Potato plants flower and produce seed. An average plant produces dozens of berries, each of which has hundreds of tiny seeds about the size of tomato seed. Each seed is **genetically different** from all the other seeds which is what you want.

When the **potato berries** are soft like a ripe tomato, harvest them. Put them in a blender, cover with water, and blend just enough to break them up. Put this mixture in a bowl to ferment 1 day. The seeds sink. Rinse several times to get clean seeds. Drain in a coffee filter or other method. Then spread on a flat surface to dry. They store for several years.

Sow these in a nursery area **3-4 weeks before the potato planting season**. They produce small tubers (tuberlets) that are then replanted in the field the same as regular potato tubers.

Most of these plants are vigorous and disease free. Plant some seeds and see what happens. Weed out any **seedlings** that have symptoms of disease. Your goal is to do your own breeding and select for the best plants. This can continue over the years by letting insects pollinate your new plants and create **cross breeding** with plants best suited for your area.

To create a **wider range of breeding possibilities** get as many different varieties of potatoes as you can. Ask your neighbors for a few tubers if they have types you do not have. And go online to buy more. In an ideal situation you would have 20 varieties or more. (There are 4,000 varieties of potatoes.) Plant a row of each variety.

The seedlings do produce less yield the first year than regular tuber plantings. But if you are having problems with diseased tubers this may solve the problem. Then the next year you can plant the tubers created from the seedlings and yield will be back to normal.

(For more information about this method read the Adobe pdf "**The Amateur Potato Breeder's Manual**" by Raoul A. Robinson.)

Sow **radish (summer)** (brassicaceae family; Raphanus sativus) **2-3 weeks before last frost in spring** in succession every 2 weeks until May. Do not sow June and July. Sow again **August and September**.

Sow 1/4 to 1/2 inch deep, one inch apart in rows 1 foot apart. **Spring radish** matures in 20 to 30 days; **winter radish** matures in 50 to 60 days.

Care: Likes full sun but will grow in partial shade. Likes soil pH of 5.5-6.8. Tolerates light frost. Few pests or diseases.

Summer radishes are small and can be red, pink, purple, white or red/white. They are globe-shaped or elongated, very hot or mild. An annual that goes to seed in the summer. **Winter radish** is larger than summer radish so sow further apart. Winter radish can grow 8-9 inches long.

Uses: The leaves and bulbs of all radishes are edible. Harvest bulbs when they are small before they become tough and bitter.

(For **more succession planting of summer radish and varieties/storing winter radish**, see August Garden-Plant but sowing/growing details are in this section. See September Garden-Plant. See August/September Greenhouse.)

In April plant or transplant new fruit/nut trees and bushes.

Apple, apricot, Asian pear, blueberry, brambles (blackberry/raspberry), cherry, currant/gooseberry, elderberry, fig, grape, hazelnut (filbert), kiwi, peach/nectarine, pear, plum, rosa rugosa, and walnut.

(For **hickory and oak**, see September Garden-Harvest.)

(Read the books "Fruits and Berries for the Home Garden" by Lewis Hill, "The Backyard Berry Book" by Stella Otto, "The Grape Grower" by Lon Rombough, and "The Apple Grower" by Michael Phillips. Read "Peterson Field Guides: Field Guide to Eastern Trees" by George A. Petrides and "Botany in a Day: The Patterns Method of Plant Identification" by Thomas J. Elpel.)

🍒 Fruit Bearing Age, Plant Longevity, Chill Hours, & Annual Fruit Yields 🍒

Fruit	First Fruit	# Years Live	Chill Hours	Annual Yield per Plant or Row
Apple (standard)	5 to 7 years	35 to 45 years	400-1000 hours	10-15 bushels/tree (1 bushel=42-48 lbs)
Apple (semi-dwarf)	5	20 to 30	400-1000	6-10 bushels/tree (1 bushel= 42-48 lbs)
Apple (dwarf)	3	15 to 20	400-1000	3-6 bushels/tree (1 bushel= 42-48 lbs)
Blackberry (erect)	1	5 to 12	200-500	40 quarts per 100 foot row
Blackberry (semi-erect)	1	8 to 10	200-500	4 to 10 quarts per 100 foot row
Blueberry	2 to 3	20 to 30	800	4 to 8 quarts per plant
Cherry- Tart	3 to 5	15 to 20	700-800	60 to 80 quarts per tree
Grape	3	20 to 30	100+	50 to 100 quarts per 100 foot row
Peach	2 to 4	10 to 15	600-800	3 to 6 bushels/tree (1 bushel= 48 lbs)
Pear	5 to 8	35 to 45	600-800	3 to 5 bushels/tree (1 bushel= 56 lbs)
Plum	4 to 6	15 to 20	800-900	3 to 5 bushels/tree
Raspberry (fall red)	5 to 6 months	5 to 12	700-800	100 to 150 pints per 100 foot row
Raspberry (summer)	1 month	5 to 12	700-800	150 pints per 100 foot row
Raspberry (black)	1 month	5 to 12	700-800	1 quart per plant
Raspberry (purple)	1 month	5 to 12	700-800	1 1/2 quarts per plant
Strawberry (June-bear)	1 month	4 to 5	200-300	50 to 100 quarts per 100 foot row
Strawberry (everbear)	3 to 4 months	2 to 4	200-300	50 quarts per 100 plants
Strawberry (neutral)	2 to 3 months	2 to 4	200-300	45 to 90 quarts per 100 plants

Not listed in chart are **chill hours** for: Almond 500-600 hours, Apricot 500-600, Chestnut 400-500, Currant 800-1000, Fig 100-200, Gooseberry 800-1000, Kiwi 600-800, Mulberry 400, Persimmon 200-400, Pomegranate 100-200, Quince 300-500, Walnut 600-700.

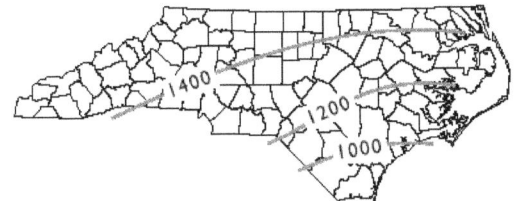

Average Chilling Hours in North Carolina

🌳 Fruit and Nut Tree/Bush Dormancy and Chilling Hours

In the fall as the days get shorter, **perennial plants go into dormancy.** Photosynthesis slows down and the plants create a type of anti-freeze to get through the winter. Plants stay in this dormant condition until there are a certain number of **chilling hours below 45 degrees from November to mid February.**

Temperate zone (between tropics and polar regions) fruit and nut trees need a minimum number of chilling hours if they are going to bloom and produce fruit or nuts. Every variety has a different number of chilling hours to go through before they **break dormancy in spring**.

The **lower the chilling requirement,** the earlier the tree begins growing in spring. Once the requirement is met, any warm period during late winter causes trees to prematurely bloom. The blossoms will be killed at next frost. For **high chilling requirement** varieties, if the chilling is not met, trees bloom very poorly with little or no fruit or nut setting in spring.

In **North Carolina**, varieties with a chilling requirement of 750 hours or greater are recommended. (See above table.)

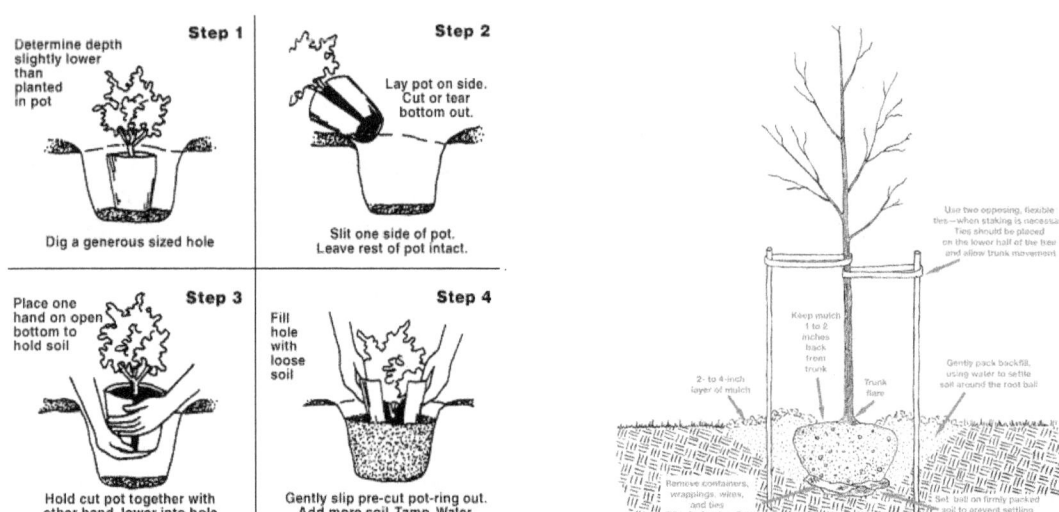

(**Above images:** Remove containers, wrappings, wires and ties from root ball. Set it on firmly packed soil to prevent settling. Gently pack dirt, using water to settle soil. Put down 2-4 inches of mulch. Keep mulch 1-2 inches from trunk.)

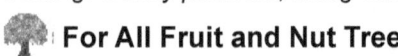 For All Fruit and Nut Trees

When you receive a tree from a nursery, **soak the roots** in water for 24 hours before planting. Do not soak longer than that. To plant a young tree, **dig a 3 foot wide hole** or dig a hole that is 2 times the width of the roots. Dig the hole at least 2 feet deeper than the roots. It is important to dig a good hole so the tree gets off to a good start. **Prune** very long roots rather than curl back.

Do not put **fertilizer** in the planting hole and do not fertilize right after planting. Fertilize in the spring. If planting in spring, fertilize next spring. Do a **soil test.** Contact your County Extension Office for a free soil test. If you overfertilize with nitrogen, it makes trees more susceptible to fire blight and other problems. (For information about fertilizers, see "Garden Tips" section in back of this manual.)

Keep **bud/graft union** 1-3 inches above soil line. (A bud/graft is where 2 different varieties are grown together. This is where the root system meets the trunk.)

All fruit trees sold commercially consist of 2 parts that are **grafted together**. The scion is the top part with branches, and the rootstock is the bottom. The scion determines the variety of fruit. The rootstock determines whether the tree is dwarf or standard size.

Put **peastone gravel** (3/8 inch diameter) several inches deep from the trunk to about 1-2 feet away from the trunk, depending on size of tree. Gravel should not touch the trunk. This is to keep the area around the trunk weed free. This makes it harder for the apple borer and rodents to damage trees. It reduces disease due to excess moisture. It also reduces nutrient and water competition from weeds. (Some quarries do not sell pea gravel. But they do sell **chat**, a small but not round gravel, that is the closest to it.)

Plant **comfrey** along dripline (outer most leaves on a tree define the drip line) of fruit trees just past the peastone gravel. Do not cut comfrey leaves, or if cut use as mulch by trees. Do cut off flowers. Comfrey is high in nitrogen. It brings up minerals deep in the soil. If used as mulch, keep it at least 12 inches away from trunk to keep rodents from chewing trunk in winter and to reduce disease. (For comfrey see March Garden-Plant.)

All fruit and nut trees need full sun, well drained soil and good air circulation.

Select a site where the tree will not be in a **frost pocket** which is where cold air settles in low-lying areas. Low spring temperatures can kill blossoms or developing fruit. Trees need good air drainage, especially during early spring frosts. Choose a higher site with a slope so cold air flows down away from the trees (hot air rises). Do not plant close to a fence row, wooded area, or at the bottom of a hill since air drainage will be poor.

Plastic milk containers or buckets can be used to slowly **water** trees. Put small holes in bottom of container and fill with water. The water gradually seeps out and waters the tree. Water trees deeply so they develop deep roots.

Trees and bushes with good soil fertility are less prone to disease and pests. Too much nitrogen makes plants more prone to disease and frost injury. Never spray insecticides when trees are in bloom. It kills the **bees** that pollinate the blossoms. (For **whitewashing trunks, bagging fruit, and setting maggot traps** see June Garden-Maintenance. For **protecting trees from crawling insects and tree borers**, see May Garden-Maintenance. For **dormant oil and fungicide spraying,** see February Garden-Maintenance.)

(For **more planting information and pruning non-brambles** see February Garden-Maintenance. For **pruning brambles** see October Garden-Maintenance. For **thinning fruit**, see June Garden-Maintenance. For information about **disease and pests**, see "Plant Health" section in back of this manual.)

 ## Dwarf, Semi-Dwarf and Standard Fruit and Nut Trees

Dwarf: Plant about 8 feet from other trees. They range in height from 4-10 feet. They are easy to prune and harvest. The fruit is normal size, but the yield is less. Dwarf trees are not as long-lived as larger fruit trees. They begin bearing fruit in 3-4 years. They live 15-20 years.

Semi-dwarf: Plant about 15 feet from other trees. They range in height from 10-16 feet. They need annual pruning to keep the height down. Very productive. Sometimes trees will rest a year and produce little or no fruit. Most fruit trees planted today are semi-dwarf. They begin bearing around 4-6 years old. They live 20-30 years.

Standard: A huge tree. Standard trees need more space and are a big job to prune and harvest. They grow to 25-30 feet or taller if not pruned. Most standard trees begin bearing in 6-10 years. They live 50-100 years.

Fruit and Nut Tree/Bush Pollination and Blooming

Many fruit trees and bushes need **pollination** (transfer of pollen from a male to a female flower) from another variety of tree or bush to produce fruit. (Many do not self pollinate.)

Peaches, European plums, tart cherries, most apricots, brambles, strawberries and walnuts are **self-pollinating**. They only need 1 plant to produce fruit or nuts.

Blueberry, Japanese plum, most sweet cherries, pear, most apple varieties, and hazelnut (filbert) need **cross-pollination** from another plant by insects or the wind to produce fruit. There are exceptions to this.

Most **flower blossoms** are killed if temperatures go below 25 degrees. It is worse if a cold snap follows a warm period. Extreme cold can kill buds before they open (blossom). Dead blossoms or buds turn brown.

When trees and bushes bloom:
  Late February to Early March: apricot, Asian pear, blueberry
  Mid March to Late March: apricot, Asian pear, blueberry, currant/gooseberry, hazelnut (filbert), peach/nectarine, pear, plum
  Early April to Mid April: apple, currant/gooseberry, peach/nectarine, pear, plum, walnut
  Late April to Early May: apple, blackberry, grape, hickory, walnut
  Mid May to Late May: apple, blackberry, cherry bush, grape, hardy kiwi, hickory, oak, walnut
  June: elderberry, oak, rosa rugosa

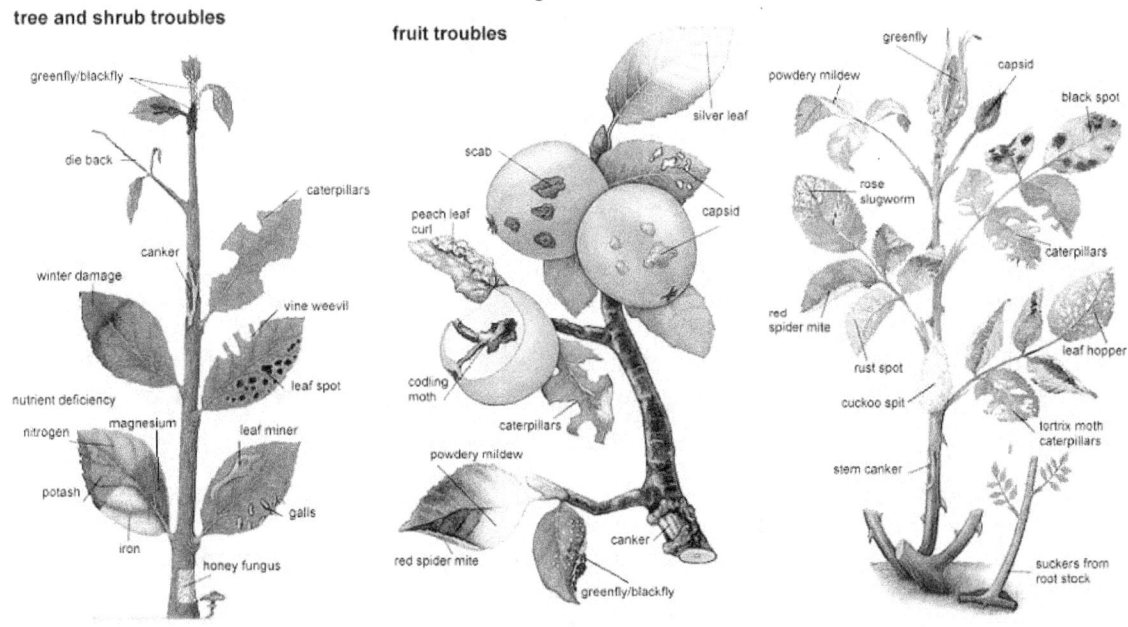

Plant Apple Trees in April: (rosaceae family; maloideae or spiraeoideae subfamily, maleae tribe, Malus domestica)
(For **pruning** see February Garden-Maintenance. For **thinning** see June Garden-Maintenance. See August/September Garden-Harvest with **harvest details in August.** For how to plant, see "For All Fruit and Nut Trees" above.)

(For **whitewashing trunks, bagging fruit, and setting maggot traps** see June Garden-Maintenance. For **protecting trees from crawling insects and tree borers**, see May Garden-Maintenance. For **dormant oil and**

fungicide spraying, see February Garden-Maintenance.)

Apple seed does not breed true. Most apple trees are grafted onto a good rootstock.

Care: Likes soil pH 6.0-6.5. Needs full sun and well drained soil. They tolerate a wide range of soils as long as water, nutrients and pH are good. Fertilize in the spring. Do a soil test.

Apples bloom April through May depending on variety. Apple blossoms can tolerate temperatures down to 27 degrees. Standard trees take 7-10 years to produce fruit and can live over 100 years. Dwarf trees take 3-4 years to produce fruit and live around 15-20 years. Trees should be staked at planting to prevent leaning. (For an image of a staked tree, see above in April Garden-Plant.)

It is better to have **ground around apple trees** growing with plants such as raspberries, broadleaf weeds, buckwheat, plantain, clover, and flowering herbs rather than sod such as fescue. (Sod competes too much for nutrients.) Or can use peastone gravel with comfrey under tree. (See above "For All Fruit and Nut Trees".)

Pollination: Most apple trees need another variety of apple tree to pollinate. All apple trees produce more fruit if there are other varieties nearby. Some apple trees only pollinate certain varieties but not others. For cross pollination they need to flower at the same time. Crab apple trees can also be used as pollinators.

There are 7,500 varieties (cultivars) of apples.

1. Early apple varieties ripen as early as **August** such as Centennial, Empire, Granny Smith, Idared, Jonagold, Liberty, Lodi, Paulared, Pristine, Spartan, Sunrise, Redfree and William's Pride.

2. Mid season varieties usually ripen in **September** such as Arkansas Black, Cortland, Fuji, Gala, Ginger Gold, Golden Delicious, McIntosh, Mutsu (Crispin), Red Delicious and Stayman.

3. Late varieties ripen as late as **October** such as Boskoop, Enterprise, Freedom, Goldrush, Gold Star, Idared, Melrose and Rome. Late varieties are **best for storage.**

Varieties recommended for North Carolina: Crispin (Mutsu), Empire, Fuji, Gala, Ginger Gold, Golden Delicious, Jonagold, Rome, Red Delicious, and Stayman.

Arkansas Black ripens late in season. Heavy bearer. Dark red to black skin. **Stores well.** Tart. Cross pollinates with Yellow Delicious, Stayman Winesap and Jonagold. It has a long bloom time so is **good pollinator.**

Fuji is green with red stripes and is good eaten fresh. It cross pollinates with Rome.

Gala is yellow-orange to red and good eaten fresh. Ripens early season. **Stores well.** It cross pollinates with Golden Delicious and Granny Smith.

Golden Delicious is yellow green to light yellow and is good eaten fresh or cooked. It cross pollinates with Empire, Gala and Red Delicious. It is a good mid season pollinator.

Granny Smith is green, medium to large. Firm and juicy with tart taste. Ripens early in the season. Cross pollinates with Gala.

Jonagold is extra large with light scarlet red skin. Flesh is crisp. Ripens late September. Cross pollinates with Fuji, Gala, Granny Smith and Red Delicious. Though some say it's pollen is sterile so it will not pollinate other trees.

Red Delicious is red and good eaten fresh. It cross pollinates with Gala and Golden Delicious.

Rome is blush to red and is good eaten fresh or cooked. It cross pollinates with Fuji.

Stayman Winesap is blush to red and is good eaten fresh or cooked. It cross pollinates with Arkansas Black, Gala, Golden Delicious, Jonagold, Red Delicious and Yellow Delicious. However, it's pollen is sterile so it will not pollinate other trees.

Yellow Delicious is very popular. Firm and sweet. Excellent for cider. Good fresh and for pies. Ripens mid season about 1-2 weeks after Red Delicious. It cross pollinates with Arkansas Black, Stayman Winesap and Jonagold.

Apple Tree Pests: (See apricot pests and disease below. For **bagging fruit and whitewashing trunks,** see June Garden-Maintenance. See "Plant Health" section in back of manual.)

Fungicide such as sulfur, lime-sulfur, or Bordeaux mix can be sprayed **late winter or early spring while tree is dormant.** Horticultural oil can be sprayed to suffocate scale insects and to kill mite and aphid eggs. Do not use sulfur on trees within 3-4 weeks of applying horticultural oil. Do not spray when tree is flowering. After flowering, spray every 10-14 days in the summer. (For **dormant oil and fungicide spraying,** see February Garden-Maintenance. For **crawling insects**, see April Garden-Maintenance.)

1. Aphids and Mites (See plum pests below. For more about aphids, see "Brassicaceae Family Disease and Pests" in the back of this manual.) Spray apple trees with a biological soap to treat rosy aphids, apple aphids and apple grain aphids. Spray in summer but before petals fall. (For **dormant oil spraying,** see February Garden-Maintenance. For **nasturtium insect repellant**, see "Plant Health" in back of manual.)

2. Apple Maggot (railroad worm)

It causes pitting and dimpling in fruit. Trap adult flies on red sticky balls by hanging them in trees after petal fall. Throughout the year remove any fallen fruit. (For **how to make trap**, see June Garden-Maintenance.)

3. Apple Tree Borers
(For **how to control borers,** see May Garden-Maintenance.)

4. Codling Moth
Moths lay eggs on leaves and twigs when petal fall begins. After petal fall, moth larvae enter young fruit through the blossom end and tunnel inside the fruit. Wormy apples and pears are caused by the codling moth.

Codling moth traps lure male adult moths with female pheromones and trap them. If your moth problem is small, this method works. For larger codling moth problems, spray the trees with Bt (Bacillus thuringiensis). Fish oil added to the spray mix helps stick the bacterium to leaves and slows its breakdown. Do 2 spray applications in **June**, and 2 more between **August and mid-September**. Or spray beginning 15 days after petal fall, spray 3 times, once every 5 days.

Another Codling Moth Trap: In **spring** put out some traps to check for the presence of male Codling Moths. Dissolve 1 part sugar and 1 part molasses into a jar of water. Cover with 3/16 inch mesh that traps the moths but not bees. Check daily. If you see males, then selectively spray blossom clusters with a Pyrethrum based spray.

5. Curculio Beetle
Found in the Eastern United States, this beetle is hard to get rid of with either sprays or traps.

An effective method is to put sheets on the ground in the morning during blossom time. Then shake the apple tree. The curculios fall into the sheets. Then kill them.

Try a repellant spray that contains 2 tablespoons garlic extract, 2 tablespoons liquid seaweed, 1 tablespoon neem oil, and 1 tablespoon fish oil per gallon of water. The beetles migrate away from the fruit trees.

6. Tentiform Leaf Miner (See Solanaceae Family in "Plant Families".)
Use Pyrethrum or Sabadilla sprays between bud-break and apple harvest. Rake up and remove all dead leaves in the fall to eliminate leaf miner pupae habitat.

7. Leaf Roller (See apricot and blueberry below.)
Spray with Bacillus thuringiensis once in June, and again between mid-July and mid-September.

Apple Tree Diseases: (See "Plant Health" section in back of manual.)

1. Cedar-Apple Rust
It is a fungus that looks like 1/4 inch rust-colored blisters. It is controlled with sulphur spray.

2. Fire Blight (See Asian pear and pear below. See fire blight in June Garden-Maintenance.)

3. Powdery Mildew
It is a fungus that attacks many types of plants. It is a white to gray, talcum-powdery growth that thrives in warm, dry climates but needs high humidity to germinate its spores. The best way to control powdery mildew is by pruning to let air circulate freely through the tree. Sulphur spray also controls powdery mildew. (For more about powdery mildew, see Cucurbitaceae Family Disease and Pests in the "Plant Families" section at the end of this manual.)

4. Scab (See peach below.)
It is a fungus that is the most widespread and damaging of apple diseases. Control scab by spraying trees with sulphur near the time buds begin to turn pink.

Plant Apricot Trees: (rosaceae family; Prunus armeniaca) (For **pruning** see February Garden-Maintenance. See August Garden-Harvest. For **thinning** see June Garden-Maintenance. See "For All Fruit and Nut Trees" above for how to plant.)

Apricots will grow where peaches grow. Apricots are the first fruit trees to **bloom** in spring usually **late February to late March** so are very susceptible to frost damage. They bloom about 3 weeks before peaches. (Peaches bloom late March to mid-April.) They are a subtropical tree so may have trouble getting fruit in zone 5. Less hardy than apple trees.

Most require **cross pollination** from another variety of apricot. All will produce more fruit if another variety is near by. They produce fruit around age 3-5 years. They live up to 50 years. Standard varieties grow to 30 feet tall. Prefers full sun but will tolerate partial shade. Likes well drained soil. Likes soil pH of 6-7.

Varieties: Moongold ripens mid July. Sungold ripens late July.

Apricot Tree Pests: (See "Plant Health" section in back of book. See apple tree diseases and pests above.)

1. Aphids (See apple tree diseases and pests above. For **dormant oil spraying**, see February Garden-Maintenance.)

2. Fruit tree leafroller lays eggs in a gray or brown casing that hatch into green worms with black heads. The eggs become a moth with brown and tan top wings, and white to gray bottom wings. Adult fruit tree leafrollers are 3/4 to 1 inch long with a wingspan of 5/8 to 7/8 inch. Rollers feed on leaves and roll leaves together that kills the leaves.

The pesticide Bacillus thuringiensis kills and prevents leafroller infections. (See blueberry below.)

3. Peach tree borer worms are fatter than fruit tree leafroller. They range in color from pale cream to dark amber. Adult moths are 1 inch long and black with bands of yellow. Borers make burrows into trunks, damaging wood and causing sores around the upper branches. Permethrin and esfenvalerate are effective for peach tree borers. (For more on **how to control**, see May Garden-Maintenance.)

Apricot Tree Diseases:

1. Bacterial Spot

Apricots are severely affected by bacterial spots, a common bacterial disease. It affects fruit, leaves and twigs. It can spread to other trees. There is no cure. Keep good orchard sanitation. Have good air flow. Do not overwater.

Apply copper-based fungicides weekly at first sign of disease to prevent its spread. It does not kill leaf spot, but prevents the spores from germinating. Applications of Fire Blight Spray also control the disease.

2. Brown rot and coryneum blight. Can use a dormant oil spray. Do not apply lime-sulfur on apricots. It burns the fruit buds and shoots.

3. Crown Gall

Crown gall is a bacterial infection affecting a plant's roots and stems. A gall forms and turns very hard and dark. The plant's leaves turn colors, and the plant wilts. Infected plants should be removed.

4. Plum Pox

Plum pox is a virus that frequently affects trees of fruits with stones such as apricots. Apricots may develop rings on them or leaves may have yellow veins. Over time, the tree no longer produces healthy fruit. There is no cure. Keep aphids under control so it does not spread. Remove infected branches.

5. Powdery Mildew (see apple tree diseases and pests above).

It is a fungus that eventually causes affected leaves to wilt and drop off, but the tree usually survives.

6. Ripe Fruit Rot

Apricots are very susceptible to ripe fruit rot (fungus) which forms dark brown hard spots on ripening fruit. The fruit most likely will not drop and will stay on tree. Remove the fruit. It is less likely if the fruit has not been damaged.

Plant Asian Pear Trees: (rosaceae family; Pyrus pyrifolia and Pyrus serotina) (For **pruning** see February Garden-Maintenance. See August Garden-Harvest. See "For All Fruit and Nut Trees" above for how to plant.)

Asian Pears are crisp and sweet somewhat like a cross between a pear and an apple. Japanese varieties are round. Chinese varieties are shaped more like a standard pear.

They are self pollinating a little but do much better with 2 or more varieties. **Blooms in March.** The flowers are frost resistant. Plant 15 feet apart. Likes full sun. Likes well drained soil. Likes soil pH of 5.9-6.5. Hardy from 0 to minus 30 degrees depending on variety.

Thin fruit twice. When tree is in bloom, remove about half the flowers. When fruit is the size of a cherry, thin to 6 inches apart.

Asian pears are susceptible to **fire blight** so must be grafted onto a fire-blight-resistant rootstock such as Pyrus calleryana or Pyrus communis. Korean Giant and Shinko are varieties that have shown some resistance to fire blight disease.

Varieties recommended for North Carolina: Chojuro, Nititaka (pollen source), Shinseiki (New Century), and Twentieth Century (Nijisseiki).

Chojuro is a buttery Japanese pear with a caramel sweetness.
Kosui has a vanilla undertone.
Shinseiki has a crisp texture and flavor like honey and walnuts. It ripens late August.
Starking Hardy Giant ripens mid September.
Twentieth Century is a crisp pear that tastes like apples and citrus.

Asian Pear Disease and Pests: (See above apple and apricot tree diseases and pests. See "Plant Health" in back of manual.)

1. **Bacterial canker** can occur in areas with cold springs. Control with good sanitation and good air flow. Remove infected branches. Disinfect pruning tools after each cut (1 part bleach to 9 parts water).

Whitewash trunk to reduce bark-damaging temperature fluctuations. (For **whitewashing trees**, see June Garden-Maintenance.) Apply copper fungicides to the trunks and lower branches.

2. **Codling moth** is severe on Asian pears, requiring 3 to 4 well-timed sprays of the insecticides used on apples and European pears. Thinning clusters to a single fruit reduces infestation. (For more about codling moth, see apple tree above.)

3. **Fire blight**, a bacterial disease, is a problem with all Asian pear cultivars except Shinko. Asian pears are as susceptible to fire blight as most European pears. Spray with antibiotics such as Streptomycin and Terramycin, or spray with copper during the bloom period and later in spring when average daily temperatures exceed 60 degrees and there is rain. Continue through April and May and after harvest. Remove diseased (blighted) branches. (See pear below. See fire blight in June Garden-Maintenance.)

4. **Pear psylla** are insects that cause sticky fruit and requires at least one delayed dormant spray. (For **dormant oil spraying**, see February Garden-Maintenance.)

5. **Two-spot spider mites** are serious on Asian pear trees especially if trees do not get enough water. Mite spray before harvest.

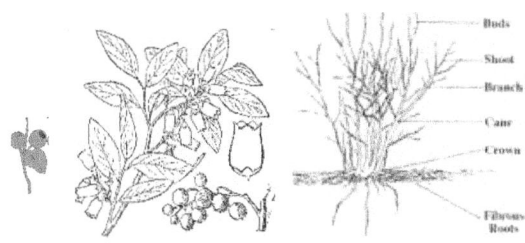

Plant Blueberry Bushes: (ericaceae family; Vaccinium cyanococcus, corymbosum and other varieties) (For **pruning** see February Garden-Maintenance. For **propagation** see March/December Garden-Maintenance. See July Garden-Harvest.) (**Above image on right** has text going from top to bottom: buds, shoots, branch, cane, crown, and fibrous roots.)

When planting **2-year-old** bare root plants, prune 50-60% of wood. For blueberries in dirt ball, prune 30% of wood. Remove at least half of the height, and thin to 1 to 3 strong canes per plant. Remove all weak/twiggy growth. Do not fertilize when planting.

Soak peat moss several days before planting to add to hole. **Dig a wide hole**, add peat moss and some dirt/leaf litter from other blueberries to add beneficial fungus. Aged sawdust is good mulch for blueberries. Can put 4-6 inches of sawdust mulch around plants now and in the future. (Some cedars and walnut can be toxic to plants. Hickory, maple and oak may tie up soil nitrogen more readily than sawdust from evergreens.)

Blueberries can grow 3-15 feet tall. Plant 4-5 feet apart. Rows should be 8-12 feet apart. Remove flower buds the first 2 years after planting. Thin flower buds the third year. Do some **bud thinning** every year. Flowers are white or pink.

Care: Blueberries like acid soil, 4.5-5.5. Prefer full sun and well drained soil. Shade reduces yield. They have a shallow root system. Needs a growing season of 160+ days with at least 5 months no frost.

Blooms early spring in March. Need at least 2 varieties for pollination.

You get larger and more fruit if more than one variety is planted. Can get 7-10 pounds of fruit per plant in **June through August.** Is mature at 5-8 years. Can live 50 years. No serious pests or diseases.

Mowed sod around plants is OK. **Fertilize** in early spring as the buds start to open, then again a month later. Magnesium is a common deficiency with blueberries.

Blueberries are **propagated** by hardwood and softwood cuttings. (For general information about herbaceous, softwood, semi-hardwood and hardwood cuttings, see "Garden Tips" section in back of this manual.)

Varieties:

Highbush blueberries (Vaccinium corymbosum) are best for western North Carolina. Highbush bears fruit late June-July in the mountains. Berries are small and somewhat sour. Pick every 5-7 days. One plant produces 10 pounds of berries a year. **Grows to 5-9 feet tall.** Plant 4-6 feet apart.

Other varieties for western North Carolina are Berkeley, Bluecrop, Blueray, Duke, Earliblue, Jersey, Sunrise, Toro and Weymouth.

Rabbiteye blueberries (Vaccinium ashei) do not usually survive the low winter temperatures in the mountains but are good in other parts of North Carolina. Rabbiteye bears fruit July and August. **Grows about 15 feet tall.** Plant 8 feet apart.

Tiftblue is a rabbiteye blueberry that ripens mid to late season with large berries.

Powder Blue is a rabbiteye blueberry that ripens mid season with medium to large berries
Premiere is a rabbiteye that ripens early to mid season with large berries. Very productive.
Wild blueberries (Vaccinium angustifolium or **lowbush**) have smaller fruit. Usually produce fruit in August. **Grows about 3 feet tall.** Plant 2 feet apart.

Blueberry Pests: (See "Plant Health" section in back of manual.)

1. Aphids Lacewing larvae (family Chrysopidae) feed on aphids. (See apple tree diseases and pests above. For **dormant oil spraying**, see February Garden-Maintenance.)

2. Birds love blueberries. Put netting over bushes or put up a scarecrow or hang pie plates nearby.

3. Blueberry Maggot (Rhagoletis mendex)

They are the chief pest of blueberries. The adults are small flies about 3/16 inch long with dark patterns across their wings. The flies lay eggs in the fruit. Each egg hatches into a small white larva (maggot), which feeds on the fruit. Infested fruit falls to the ground and the maggot enters the soil, pupates and overwinters. (For **how to make traps**, see below.)

4. Leafhoppers (family Cicadellidae) (See Solanaceae Family in "Plant Families".)

They are under 1/4 inch, found on stems or undersides of leaves. They feed by piercing the plant and sucking juices. Most are green or brown, but some are green and red. The young look like adults but are smaller and wingless. These insects do little damage by feeding on plants. They do some damage when they make a slit in the stem to lay their eggs. However, some leafhopper species transmit blueberry stunt mycoplasma, a serious disease.

5. Leafrollers (family Tortricidae) (See apricots above.)

They are the larvae or caterpillars of a few species of small moths. They roll leaves for shelter. Early in the growing season, these brown or green worms tie together blossoms and feed on them. When full grown at 1/2 to 3/4 inch long, the larvae seal up the leaf shelter, form a cocoon (a glossy brown case) and undergo metamorphosis. One to two weeks later, they emerge. The adults vary from brown to yellow. They are 1/2 inch long. Small numbers of leafrollers are not a concern. Remove them by hand and kill them. Some files (Tachinidae, Sarcophagidae) and wasps (Braconidae) are parasites of leafroller larvae. (See control below.)

Control of the above blueberry pests

Many pesticides are more toxic to beneficial insects than to the pests they are used against. Therefore, **natural control methods** are often better. Beneficial insects survive better if you provide a diverse habitat. Ladybird (ladybug) beetles (family Coccinellidae), praying mantids (mantis) and lacewings can be bought from some nurseries or plant dealers. Ordering eggs rather than adults is preferred, because many of these insects fly away as adults.

Traps: Leafhoppers and aphids are attracted to goldenrod yellow and can be caught on a yellow surface covered with a sticky substance. Sticky material (e.g., Tangletrap) is available from garden centers. Paint pieces of plastic or plywood (1 foot square by 3/8 inch) goldenrod yellow, cover with tangletrap or sticky stuff. Suspend them in or near blueberry bushes with one trap for every two bushes. (For **how to make your own sticky stuff,** see "Plant Health" in back of this manual.)

Use **bait traps** to trap adult blueberry maggots. Ammonium carbonate is the attractant. The flies are caught on surfaces coated with sticky stuff. Start with a cottage cheese or similar container. Place 1 teaspoon of ammonium carbonate on paper and fasten it to the bottom of the container. Coat the inside walls with tangletrap or sticky stuff. Hang trap upside-down from the foliage by a string at the bottom of the container. You can paint the trap goldenrod yellow.

Culture: Cultivation techniques affect both pests and beneficial insects. Exposed earth under bushes leave few places for insects to live but this affects both good and bad insects. (See "For All Fruit and Nut Trees" above. For **bagging fruit,** see June Garden-Maintenance.)

Pesticides: Botanical pyrethrins, synthetic pyrethroids, and Bacillus thuringiensis may work. They have low mammal toxicity but rapidly kill pests.

Malathion provides good control of leafroller larvae and blueberry maggot adults. If it is applied to the undersides of leaves, malathion effectively controls aphids. Malathion is relatively low in toxicity to humans and other animals. Sevin is effective against leafrollers but does not kill aphids and is highly toxic to bees. Therefore, it should never be used during periods of bloom.

Blueberry Disease (See "Plant Health" section in back of this manual and see apple tree above.)

1. Crown gall is a bacterial disease that forms bumps on the branches and roots of blueberry bushes. Infected bushes are usually stunted and weak. Treat with a suspension of hypervirulent Agrobacterium. Remove diseased plant parts and spray with a solution of Bordeaux mixture (copper sulfate and lime).

2. Fungal diseases are controlled by removing infected plant parts and spraying with a fungicide. Gray mold causes leaf spots and blossom blight (deformity). Powdery mildew coats stems, leaves and flowers with a white layer. Fusicoccum and Phomopsis cause cankers on stems, then they wilt and die. Other fungal diseases cause shriveled or watery fruit and root rot. (See Fungi Disease in "Plant Health" in back of manual.)

3. Viral diseases commonly affect leaves. Symptoms include reddish-brown spots on leaves or bright yellow and green mottling of leaves. Leaves may be cupped, twisted or elongated. Berries with shoestring virus turn purplish-red instead of blue. Blueberry bushes infected with a virus should be removed and burned. (See Viral Disease in "Plant Health" in back of manual.)

   Plant Brambles in April: (rosaceae family; Rubus genus)

Brambles are **perennial** plants that live for about 10 years. They include **blackberries and raspberries**. (See details for each below.) Plants with no virus problems live longer. Brambles must have good drainage and good air circulation. They have shallow root systems. Prefer full sun. Brambles are self pollinating.

When planting, cut back canes to ground level. Do not fertilize for several weeks after planting. Grow grass between plants and keep mowed. Mulching is not needed.

Make sure brambles have enough water when fruit is forming.

Bramble Disease and Pests: (See "Plant Health" section in back of manual.)

To reduce disease use **good sanitation** (removal and destruction of infected plant parts); keep plants in good vigor through weed control / good fertility / and moisture management (mulching and irrigation). Use resistant varieties that are certified disease free. Remove floricanes (a cane bearing flowers) soon after they bear fruit. Grow red and black raspberries 700 feet or more apart. (For **pruning brambles** see October Garden-Maintenance.)

Bramble Diseases
1. Anthracnose (Elsinoe veneta)
It is a fungus that can be a serious disease on blackberries and black raspberries. It is less of a threat to purple and red raspberries. Most thornless blackberry varieties have resistance. Anthracnose can be controlled organically with a single delayed-dormant spray of liquid lime-sulfur. Under severe conditions, additional later sprays using bordeaux, burgundy mix, or a fixed copper may help.

2. Blackberry Rosette (Double Blossom, Witches Broom)
It is the most serious disease affecting blackberries in North Carolina. It is a fungus that distorts flowers and shoots. Berries do not grow from infected branches. Remove and destroy infected branches. Use fungicide sprays. Arapaho, Humble and Navaho varieties are resistant.

3. Crown Gall
It is a bacterium which enters through wounds in the roots or crowns. Plant healthy stock on a clean site. If a site has had crown gall, wait 3-5 years before replanting brambles. (See blueberry disease above.)

4. Gray Mold
It is a serious disease of bramble fruit caused by fungus. It can be devastating if rainy weather occurs when fruits are at their ripest. Raspberries are more susceptible than blackberries. No organic fungicides are available for controlling this disease. Choose a location with good soil and air drainage. Good control includes trellising, removal of spent (old) floricanes, thinning, weed control, cover crop mowing, and picking when fruit is ripe (so none rot).

5. Leaf Spot
It is a fungal problem in humid climates. Most blackberries are resistant, but raspberries can be hard hit. Practice good sanitation, especially the removal of old floricanes. Make sure branches have good air flow and sun penetration. Overwintering fungus can be reduced with a single delayed-dormant spray of lime sulfur. There is some resistance among raspberry varieties Heritage, Southland, and Fall Gold.

6. Orange Rust (Gymnoconia peckianus)
It is a fungus that affects blackberries and black raspberries. It causes rust-colored lesions on leaves. It usually infects the whole plant, stunting some plants seriously. Control measures include planting clean stock and removing infected plants including roots. Improve air circulation. Blackberry cultivars Arapaho, Chester, Eldorado, Hull, Raven, Snyder, and Ebony King appear resistant.

7. Spur Blight (Didymella applanata)
It is a fungal disease of red raspberries and sometimes of purple raspberries that reduces yields and the life of the plant. Control procedures are the same as those for anthracnose above. The cultivars Amity, Festival, Haida, and Prestige all have some resistance.

8. Verticillium Wilt

It is a fungus. Wait at least 3-4 years before planting in a field where potatoes, tomatoes, peppers, eggplant, tobacco or other plants in the nightshade family have been grown. Soil solarization can greatly reduce verticillium inoculum. (For how to **solarize soil,** see potting mixes in "Garden Tips" in the back of this manual.) Red raspberries Cuthbert and Syracuse, and blackberries Himalaya and Evergreen are resistant. (See Leguminosae Family in "Plant Families".)

9. Viruses

Raspberries can be very hard hit in areas where aphids are spreading viruses. Plant resistant varieties.

Bramble Pests

1. Aphids

Besides spreading viruses, aphids can be so bad that leaves curl. High aphid populations are often an indication of excessive nitrogen. If you do not spray, naturally occurring enemies of the aphids will eventually control them. Or use insecticides such as soap sprays, rotenone, or pyrethrum. (See apple tree diseases and pests above. For **dormant oil spraying**, see February Garden-Maintenance. See Brassicaceae Family in "Plant Families".)

2. Beetles (See Gramineae family in "Plant Families" in back of manual.)

Japanese and June beetles can be major pests of bramble fruit if their adult emergence coincides with ripening. Hand picking, trapping, exclusion with row covers, and reducing grubs in the soil with tillage, and/or beneficial nematodes can be used.

Buckets of water with rotting fruit attract June beetles. Scoop out the beetles and kill them. Change the fruit brew if it has lost its yeast-like smell or is no longer attracting the beetles. Place traps away from brambles so you do not draw pests to them.

Surround WP is a kaolin-clay-based insect repellent effective against both June and Japanese beetles. However, the clay is nearly impossible to wash off from the seed-bearing parts of the berry.

3. Borers

This includes the raspberry crown borer, raspberry cane borer, and red-necked cane borer. All 3 attack blackberries as well as raspberries. Control is best achieved by removing and destroying infested plants. (For **how to control borers,** see May Garden-Maintenance.)

 Blackberry: (rosaceae family; Rubus fruticosus) (For **trellis types and pruning brambles** see October Garden-Maintenance. See July Garden-Harvest.)

Plant **erect blackberries** 3-4 feet apart, and **trailing blackberries** 5-6 feet apart. Rows should be 8-10 feet apart.

Care: Likes full sun. Blackberries like 5.5-7.0 soil pH. Needs good air circulation.

Blackberry plants are larger than raspberry plants. Trailing blackberries need a grape-type trellis. (For **trellis system,** see grapes in February Garden-Maintenance.) Blackberries prefer warmer climates than raspberries so are not as frost hardy. It has **biennial fruit production** meaning fruit grows on canes that are 2 years old.

Blackberry winter is a brief period of cold weather that happens at the same time blackberries are in **bloom, usually early to mid May.**

Varieties good in the mountains of North Carolina: Arapaho (erect thornless), Chester (semi-erect thornless), Cherokee (erect thorny), Cheyenne (erect thorny), Kiowa (erect thorny), Navaho (erect thornless), Shawnee, and Triple Crown (semi-erect or trailing thornless).

Arapaho is thornless and disease resistant. Excellent quality. Ripens 11 days before Navaho. Hardy.

Navaho is thornless and ripens late. Good overall quality. Hardy to minus 9 degrees.

Varieties good for all of North Carolina: Apache (erect thornless), Arapaho, Chester (semi-trailing thornless), Hull (semi-trailing thornless), Natchez (erect thornless), Navaho, Ouachita (erect thornless), and Triple Crown

Harvest: Plants reach full production at 3 years. Blackberry produces about 5-7 pounds or 2-4 quarts of fruit per plant. Berries are seedy. The berries ripen **June through September** depending upon the variety, climate, and how much sun they get. Pick often so they do not get overripe.

Uses: Young leaves around 1 inch long can be eaten raw or cooked. Leaves are dried to make a tea and gargle.

Blackberry Propagation: (For **propagation methods,** see "Garden Tips" section in back of this manual.)

Leafy Stem Cuttings: This is the best way to propagate large quantities of plants. During **late spring or summer,** get 4-6 inch cuttings from tip of cane when it is succulent but still firm. Place cuttings 2 inches deep in a perlite/peat or peat/sand mixture. Cuttings should be misted especially in the 2-4 week period before roots form. Can use rooting hormone. (For details about potting mixes and making rooting hormone, see "Garden Tips" in back of book.)

Root Cuttings: Fastest way to propagate blackberries. When **dormant**, cut roots 1/4 to 1/2 inch in diameter into 6 inch strips. Put in a pot, cover with 2-4 inches of soil. Grow in pot for up to 1 year.

Suckering: Easiest way to propagate blackberry. Suckers naturally form from the roots. In **late spring or summer**, cut the sucker with the roots from the mother plant, and replant.

Tip Layering: Semi-erect and trailing blackberries can be propagated this way. Good for propagating a few plants. Put first year vegetative shoots (canes) in contact with soil. Cover with 3 inches of soil.

A more efficient method of tip layering is to cut off the shoot tip to induce lateral (side) branching. During the **summer** dig a 3 inch deep hole and put middle and end of shoot in hole. Cover with 3 inches of soil. By fall new rooted shoots will have developed at each lateral branch. Transplant in **spring**.

  Raspberry: (rosaceae family; Rubus fruticosus idaeus) (For **propagating red raspberry**, see February Garden-Maintenance. For **propagating black and purple raspberries**, see July Garden-Maintenance. For **trellis types and pruning brambles** see October Garden-Maintenance. See July Garden-Harvest. For **propagation methods**, see "Garden Tips" in back of this manual.)

Plant **red raspberries** 2-3 feet apart; **black raspberries** 3-5 feet apart. Rows are 6-8 feet apart.

It has **biennial fruit production.** Blooms April through June depending on variety. Bears fruit in 2-3 years.

Care: Raspberries like 6.0-6.8 soil pH. Like full sun and well drained soil. Needs good air circulation.

Red raspberry is very frost hardy. Of the brambles, it is the first to produce in the summer. Red raspberry produces about 1-2 pounds of fruit per plant. Yellow raspberries have the same characteristics as red raspberries.

Purple or black raspberry (blackcaps) produce about 3 pounds of fruit per plant. Purple raspberries are a cross between red and black raspberries. Black raspberries are susceptible to anthracnose, crown gall, orange rust and viruses. (See bramble diseases above.)

Harvest is at its peak the third year after planting. It slowly declines in following years. Most growers replant a site after 10 years. Get about 2-4 quarts per plant.

Raspberry propagation: Raspberries are propagated from suckers, root cuttings and by dividing roots of plant. (See blackberry propagation above.)

In the **mountains of North Carolina** floricanes (canes with fruit) of raspberries can be damaged by very cold temperatures and by large fluctuations in temperature in the winter. The variety Heritage is the best one for the mountains but it is somewhat susceptible to disease.

To have an almost continuous harvest of raspberries from **late June until early October,** plant these varieties:
Late June and July Harvest:
Floricane (second year cane) fruiting- Citadel, Latham, Mandarin, and Reveille.
Primocane (first year cane) fruit- Autumn Bliss, Cherokee, Heritage, Red Wing, Ruby, Summit.
August and September Harvest: Primocane fruiting- Heritage and Nordic.
October Harvest: Primocane fruiting- Heritage and Nova.
Varieties recommended for the mountains and foothills of North Carolina:
Allen- fruits in June. Erect, large black berries.
Bristol- fruits in June. Erect, susceptible to anthracnose. Large black berry.
Cumberland- fruits in June. Erect, large black berry.
Dormanred- fruits in June. Trailing, large berry, vigorous. Glossy red. Canes are prone to cold damage.
Heritage- fruits June through August (everbearer). Erect. Large deep red berries. Disease resistant.
Royalty- fruits June and July. Erect, vigorous. Sweet purple berry.
Southland- fruits in June and mid August. Erect. Light red.
Varieties recommend for North Carolina Coastal Plain & Piedmont: Southland, Dormanred and Mandarin.

Plant Cherry Bush (rosaceae family; Prunus pumila) (sand cherry).

(See July Garden-Harvest. For how to plant, see "For All Fruit and Nut Trees" above.)

Sand cherry is a 5-10 foot tall, 5-6 foot wide bush. Dwarf is 5 feet tall. White flowers **bloom in spring usually May.** Plant 2-5 feet apart. When plant, cut branches so half as tall.

Care: Drought resistant; frost hardy. Likes full sun or partial shade. Likes well drained soil. Likes soil pH 4.5-7.5.

Will produce more fruit if have more than 1 plant. Produces 1/2 inch fleshy, purple-black fruit in **fall**. Does not need much pruning. Will produce a good hedge or tall groundcover.

Varieties include prunus pumila var. besseyi such as Dwarf Flowering Cherry and Hansen's Bush Cherry.

Harvest: Can get 16 quarts of sweet cherries from a 3-year-old Hansen bush.

Plant Currant and Gooseberry Bush in April: (Currant: grossulariaceae family; Ribes nigrum/black, Ribes rubrum/red, Ribes glandulosum/white) (Gooseberry: rossulariaceae family; Ribes uva-crispa)

(For **pruning** see February Garden-Maintenance. See June Garden-Harvest. For **propagation methods**, see "Garden Tips" in back of manual.)

Can plant in **early fall or early spring.** Plant 3-5 feet apart in rows 8-10 feet apart. When transplant cut back to 6-10 inches tall for 1-2 year old plant. Gooseberries can be trellised to increase air flow.

Care: Both like soil pH 5.5-6.5. Likes full sun but will produce fruit in partial shade. Can adapt to wet soil.

Blooms in March or April. Remove flowers the first year so rest of plant develops better. Grows to 3-6 feet tall. Very winter hardy. Will produce for 10-15 years. Produces 10-12 pounds of berries per bush each year.

They are **self pollinating** so do not need other varieties to bear fruit. They bear fruit at the base of one-year-old wood, and on spurs of two- and three-year old wood. Currants bear fruit mid to late summer. They ripen mid to late summer.

Propagating currants and gooseberries is easy.

Take **hardwood cutting** from mature wood, 10-12 inches long. Put the cutting 3/4 of the way into the ground. New roots grow along the stem. In cold climates, cover cuttings with dirt or straw to prevent winter heaving. In early spring transplant rooted cuttings.

Another way to propagate is by **tip layering.** In fall or spring bend a low-growing branch to the ground and cover with soil. Keep under soil by weighing it down with a rock. When roots are established, cut branch off and transplant. (For images of tip layering and for details about herbaceous, softwood, semi-hardwood and hardwood cuttings, see "Garden Tips" in the back of this manual.)

Currant and Gooseberry Disease and Pests: (See "Plant Health" in the back of this manual.)

1. Currant borer damages the inside of the cane. The larvae feeds on the pith, then emerges as a moth in spring. A sign of borer damage is the tip of canes wilt and drop leaves. Cut the cane back, then burn the infested wood. (For **how to control borers**, see May Garden-Maintenance.)

2. Currant fruit fly (gooseberry maggot) may infest fruits. The maggot is the larval stage of the currant fruit fly. Adult flies are about 1/5 inch long, yellow to orange in color, and have dark bands across the wings. In spring, flies lay eggs under the skin of fruit. Maggots burrow and feed in the berries. Berries turn reddish and may drop. Adult fruit flies emerge in mid- to late April and May.

On uninfested plants, use a floating row cover before adult flies emerge. Place at petal fall after pollination has occurred. If placed after fruit formation begins, they may not be completely effective. Pick and destroy infested fruit daily. Place tarps beneath bushes by **mid-June** to prevent larvae from entering soil to pupate.

3. Currantworm can defoliate bushes in a few days. When you see the leaves being chewed, search for the tiny green worms. A dusting of rotenone is effective, as is hand-picking, or giving the bush a good, hard shake.

4. Powdery mildew is a major disease. (See apple tree diseases above. See diseases in Cucurbitaceae Family in "Plant Families".)

Plant Elderberry Bush/Tree: (adoxaceae family; Sambucus canadensis, Sambucus nigra)

(For **pruning** see February Garden-Maintenance. See August/September Garden-Harvest with **harvest details in August**. For **propagating,** see March/July Garden-Maintenance. See "For All Fruit and Nut Trees" above for how to plant.)

New canes grow from plants and reach full height the first year. The second year they develop their branches. Flowers and fruit develop on the current years growth and continue strongly into the next year. After that production drops on those branches.

Plant between **October and March**. Plant 5-7 feet or more apart. After plant it, cut back main shoots to 3-6 feet tall with an outward facing bud.

It is a large bush or small tree. Grows in clumps of canes 4-15 feet tall with a width of 6-10 feet. Very winter hardy.

Care: Easy to grow, vigorous, fast growing. Prefers full sun. Likes soil pH of 5.5-6.5. Likes well drained but moist soil. They have shallow roots.

Blooms in June with 10-inch clusters of tiny cream-colored flowers.

Harvest: Do not eat red elder berries because they are poisonous. Eat only dark purple, dark blue, or black fruit. Cook before eating. They mature **late August through early September.** Harvest for each bush lasts 2 weeks. You get more fruit if there is more than one variety. Full fruit production in 3-4 years. Get 12-15 pounds of fruit per tree.

The **cultivars** Adams and York are heavy fruit producers. There are wild red and purple elders. Wild varieties produce less than commercial varieties.

Self-propagating by above ground suckers. Can also sow seeds. Stratify for 4-6 weeks first. (For **how to stratify,** see "Garden Tips" in back of manual.)

Varieties:
European elderberry (Sambucus nigra) is a tall shrub growing almost 20 feet tall. It has dark black berries that are safe to eat if cooked.
American elderberry (Sambucus canadensis) is a smaller bush, usually growing only 8 to 10 feet tall. It has dark black berries that are safe to eat if cooked.
Red elderberry (Sambucus racemosa) is **very toxic and should not be eaten at all**. It has red berries. (For more about **cooking and toxicity**, see August Garden-Harvest.)

Elderberry Disease and Pests: Few pest problems. (See "Plant Health" section in back of manual.)
1. Birds are the major pest. Control by prompt harvesting of ripe fruit and netting trees.
2. Leaf spotting fungi, root rots, thread blight and verticillium are the less common diseases.
3. Powdery mildew can affect canes and berries in late summer and early fall. It results in a grey appearance on the berries, but does not lower the quality of the juice. (See apple tree diseases above.)
4. Stem and twig cankers are fungus diseases controlled by pruning and burning infected canes.
5. Tomato Ringspot Virus is among the most serious diseases affecting elderberries. It is spread by nematodes and through pollen transfer. It results in weakened plants, reduced productivity and plant death. Soil should be tested for nematodes prior to planting, and fumigated if necessary. (See **Insects and Nematodes** in "Plant Health".)

 **Plant Fig Tree early April or March:** (Common fig: moraceae family; ficeae tribe, Ficus carica) (For **propagating,** see July Garden-Maintenance. See August Garden-Harvest. For **covering fig,** see November Garden-Maintenance. See "For All Fruit and Nut Trees" above for how to plant.)

Figs like warm climates. They can not survive outside in zones 5-6 but must be put in a greenhouse or enclosed porch in the winter. If it is a very cold night, cover with a sheet or frost blanket even if in enclosed area.

Care: Needs at least 8 hours of sun. Fertilize in **March or April.** Likes well drained soil. Likes soil pH of 6.0-6.5.

Blooms early spring. The flower is not visible. Common figs do not need pollination from other figs. Brown turkey fig (a common fig) bears fruit on branches produced the same summer. In protected areas such as a greenhouse, fig fruit starts maturing in **early August** and continues through October until frost.

Varieties recommended for North Carolina: Celeste, Brown Turkey, Brunswick/Magnolia (for preserves), Greenish, and Marseille. Brown Turkey fig is hardy to 10 degrees.

Fig Disease and Pests: Figs are relatively pest and disease free. (See "Plant Health" in back of manual.)
1. Birds are pests. Figs can be netted.
2. Borers, mealybugs and scale insects attack fig trees. The best remedy is good sanitation.
3. Fig rust is a yellowish-orange fungus disease that attacks young leaves causing them to fall off. Control by using good sanitation methods. Prune and burn infested wood and fruit. Do not allow piles of leaves and fruit to accumulate that give breeding areas for insects. Also controlled with neutral copper sprays. One or two applications made in **May or early June** usually keep trees in good condition until after fruit ripens.
4. Mosaic is a viral disease. The main symptom is mottled leaves. Some varieties show dwarfed leaves and fruit; others are scarcely affected. Mosaic is incurable but is rarely a reason to remove plants. (See Leguminosae Family in "Plant Families".)
5. Root-knot nematodes
If you see tiny galls or swellings on the roots, you have root-knot nematodes. The best control is to destroy infected plants and not use that site for figs again. Nematodes contribute to premature fruit drop. (See **Insects and Nematodes** in "Plant Health".)

Plant Grape Vine in April: (**European common grape vine:** vitaceae family; Vitis vinifera with 2,000 varieties. **Fox grape:** Vitis labrusca. **Muscadine grape:** Vitis rotundifolia or Muscadinia rotundifolia with 300 varieties. **Wild summer grape:** Vitis aestivalis with 4 varieties.)

(For **trellis types and pruning,** see February Garden-Maintenance. See September Garden-Harvest. For **propagation methods,** see "Garden Tips" in back of this manual.)

Bare root plants should be planted in **spring** as soon as soil can be worked. Soak roots for several hours before planting. Plant 4-8 inches deep. Do not plant in low-lying frost pockets. Distance between plants depends on vigor of variety, varying between 5-8 feet apart with 5-6 feet between rows. Muscadines can be planted up to 20 feet apart.

Trellis wires can be 2 1/2 to 3 feet off the ground for the first wire, and 5-6 feet off the ground for the second wire. (For more trellis information, see February Garden-Maintenance.)

Care: Likes soil pH 6.0-7.5. Needs full sun and good air circulation. Likes well drained, moderately rich soil.

Do not over fertilize with nitrogen because there will be too much leaf growth and not enough grapes. Do not need other varieties for pollination. **Blooms in May.**

Growing season is about 150–180 days (5-6 months) for standard varieties. Will produce for 50+ years. May need to put nets over grapes to keep out birds. May need to use fungicides.

Harvest: Bears in 2-3 years. Six grapevines are usually enough for a family. A mature vine yields 10-15 pounds.

Varieties:

1. European Common grapes (Vitis vinifera, wine grapes) such as **Chardonnay, Cabernet Sauvignon, Cabernet Franc, Merlot and Viognier** are harder to grow than other types of grapes. They need spraying and are more sensitive to fluctuating winter temperatures and more sensitive to spring frost. They are bunch (grow in clusters) grapes.

2. Fox grape (Vitis labrusca, American grape) such as **Catawba, Concord, Delaware, Isabella and Niagra** are bunch grapes. A type of fox grape is found **wild in Appalachia**. It is high climbing (up to 30 feet high) with very large, smooth leaves. It has large, 3-lobed leaves with teeth on the edges. Grapes are small and sweet. They get ripe in late August and September.

3. Muscadine (Vitis rotundifolia) such as **Carlos and Noble**. One variety is a common native grape with 1-2 grapes per **cluster** (not a bunch grape). Can climb up trees 100 feet. The bark is white speckled. The leaves are small and glossy on both sides. The grapes are large and thick skinned. One variety is **Scuppernong**.

4. Summer Grape (Vitis aestivalis) is a **wild grape**. Similar to fox grape. Woody vine can climb over 30 feet. It has large, 3-lobed leaves with teeth on the edges. Bark is gray to reddish brown. Likes full to partial sun but needs sun to make grapes. Blooms late spring. Fruit ripens late summer and fall. One variety is Cynthiana (Norton).

Varieties recommended for **Piedmont and western North Carolina**: ("Early" ripens mid summer.)

Variety	Season	Color	Size	Use	Notes
Alwood	Early to Mid	Purple	Medium grape	Fresh or juice	Resembles Concord
Catawba	Late	Red	Medium	Fresh, juice or wine	All purpose grape
Cabernet-Sauv.	Very Late	Black	Small	Wine	Vinifera, excellent
Chambourcin	Late	Purple	Medium	Wine	French hybrid, red wine
Chardonnay	Late	White	Small	Wine	Vinifera, excellent wine
Concord	Early to Mid	Black	Medium	Fresh or juice	Not for Piedmont
Delaware	Mid	Red	Small	Fresh, juice or wine	Good wine, table grapes
Fredonia	Early to Mid	Purple	Large	Fresh, juice	Heavy producer
Himrod	Very Early	White	Small to Med	Fresh	Seedless, prone to diseases
Lakemont	Early to Mid	White	Small to Med	Fresh	Seedless, good quality
Leon-Millot	Late	Black	Small	Wine	French hybrid, red wine
Moored	Early	Red	Med-Large	Fresh	Fruity, productive
Niagara	Mid	White	Medium	Fresh or juice	Popular, fruit, hardy
Seyval	Late	White	Small to Med	Wine	French hybrid, good
Suffolk Red	Early to Mid	Red	Medium	Fresh	Seedless, good quality
Venus	Mid	Black	Med to Large	Fresh	Seedless

More about varieties:

Catawba (Vitis labrusca × Vitis vinifera) grows 4-10 feet tall and 8 feet wide. Medium-large red fruit keeps well and is widely planted for their sweet, rich taste. Self-fruitful but does better with 2 or more varieties.

Canadice (Vitis) has medium-size red grapes that grow in large, well-shouldered, compact clusters.

Moderately vigorous vines. Has good winter hardiness. Ripen weeks before Concord. A little susceptible to mildew.

Concord (Vitis labrusca) consistently produces good crops of large blue-black grapes. A mid-season bearer, it is widely planted for its sweet, rich taste.

Himrod (Vitis) is one of the hardiest and best quality of white grapes (pale green to yellow). Make great raisins. Seedless. Ripens 28 days before Concord.

Niagara (Vitis labrusca) has hardiness and productivity of Concord but ripens earlier. Good dessert grape.

New Himrod (Vitis labrusca himrod) is a golden amber grape noted for its sweet fruit and is similar to grapes found at the grocery store. Ripens a month ahead of Concord. Disease resistant.

Ontario is an early white grape.

Steuben (Vitis) is a mid season purple grape that is excellent for making juice.

Vanessa (Vitis) is black seedless with large sweet fruit. Very disease resistant. Resists cracking. Winter hardy. Ripens 28 days earlier than Concord.

Grape Disease and Pests: (See the "Plant Health" section in the back of this manual.)

1. **European common grapes,** with varieties such as Vitas vinifera cultivars, are planted in areas where they are not native. Therefore, they are very susceptible to all American grape diseases and pests such as black rot, downy mildew, Phomopsis leaf spot, powdery mildew, and phylloxera (a root-feeding, aphid-like insect).

States with **dry, Mediterranean climates** (warm to hot, dry summers and mild, wet winters) are good for growing the European wine grape, and organically acceptable fungicides are adequate for controlling most disease problems. However, in non-Mediterranean climates such as North Carolina there are problems where fungicides may not work.

2. **Fox grapes (American grape)** (Vitis labrusca) and **Summer Grape** (Vitis aestivalis) vary in their susceptibility to diseases. Concord (Vitis labrusca), for example is very resistant to anthracnose but susceptible to black rot. Ives (Vitis labrusca x unknown) is relatively resistant to black rot but highly susceptible to downy mildew. Edelweiss (Vitis labrusca) and Cynthiana (Vitis aestivalis, also known as Norton) have significant resistance to most of the major grape diseases.

3. **Muscadine grapes** (Vitis rotundifolia), good in the southern United States, are very resistant to most grape diseases and pests.

Organic Control: Make sure plants have good air circulation. Trim out diseased parts right away. Keep area mowed and remove fallen fruit. Plants resist diseases and pests better when there is good soil fertility.

Where varietal resistance, sanitation and other cultural controls do not work, an organic grower has to rely on acceptable mineral fungicides (various sulfur and copper formulations), microbial-based fungicides, compost teas, and vegetable and mineral oils used as dormant applications or on foliage (depending on weather). **Compost teas** have been successfully used as a combined foliar feed and disease suppressive. You can make a compost tea with comfrey. (For **comfrey** see March Garden-Plant.)

Pests include caterpillar, grape berry moth, grape phylloxera, leafhoppers, mealy bugs, mites and nematodes. Plant disease-resistant varieties and use certified disease-free stock.

Diseases: black rot, botrytis, downy mildew, Pierce's disease, phomopsis, powdery mildew, root rot.

Viruses in grapes are managed by using clean planting stock. Viruses spread from one plant to a neighboring plant, but the spread is generally slow.

Fungicides: (For more about **fungicides & pest control,** see "Plant Health" in back of this manual.)

Sulfur and sulfur-containing **fungicides** can be disruptive to beneficial insects and other arthropods such as spiders and mites. Another problem with the use of sulfur is tissue injury or phytotoxicity. Some cultivars, especially those of Vitis labrusca such as the Concord, are highly susceptible to sulfur injury.

In regions where it rains a lot during the growing season, wettable sulfur or flowable sulfur formulations are preferred. Flowable formulations are less damaging to predatory mite populations.

Bordeaux mix (copper sulfate mixed with hydrated lime) is less likely to be phytotoxic than sulfur. However, damage can still occur on sensitive cultivars, especially during high temperatures. (See **How to Make Organic Fungicide Sprays and Solutions** in "Plant Health".)

There are **organically acceptable alternatives** to mineral-based fungicides. There is a new generation of microbial fungicides such as AQ-10 (for powdery mildew control) and various formulations of the bacteria Bacillus subtilis (Serenade, Epic, and Kodiak).

 **Plant Hardy Kiwi Vine:** (actinidiaceae family; Actinidia deliciosa) (For **trellis types and pruning** see February Garden-Maintenance. For **more pruning** see June Garden-Maintenance. For **propagation** see July/October Garden-Maintenance. See September Garden-Harvest.)

Care: Likes soil pH 5.0-6.5. Likes full sun but partial shade is OK. Likes moderately rich, well drained soil.

There are male and female plants. Need about 1 male plant to every 6 female plants. Place plants 8-15 feet apart. Need strong trellis. A **2-arm cordon (trellis)** 6 feet tall such as used with grapes is good. (For **trellis system**, see grapes in February Garden-Maintenance.)

Wrap trunks with cloth, or paint with white latex paint. (For **whitewashing trunks**, see June Garden-Maintenance.)

Growth starts very early in spring and is frost sensitive so do not plant in a low-lying frost pocket. Sometimes new growth is killed by late frost.

Blooms late May. No fertilizer after July 4.

Plants start bearing fruit in 4-5 years, and are in full production at 8-9 years. Can live 50+ years.

Harvest: Bears fruit **July through September**. One plant will produce 50-100 pounds of fruit a year.

Hardy Kiwi Disease and Pests: (See "Plant Health" in back of this manual.)

Pests include Japanese beetles, root knot nematodes, and spider mites. Diseases includes phytophthora crown and root rot.

 **Plant Hazelnuts** (betulaceae family; Corylus genus with 14-18 species, also known as filberts). A member of the **birch family** that is a large bush or small tree. **Flowers very early spring usually in March.** They grow about 10 feet tall and 10 feet wide. Plant 20-40 feet apart. They live 40 years.

Care: (See "For All Fruit and Nut Trees" above.) Likes full sun. Likes soil pH of 6-7. Likes well drained soil. Likes moderately rich soil. Grows in zones 4-9. Drought tolerant.

Pollination: Hazelnuts have male and female flowers on the same plant that form during the prior year and remain dormant through most of winter. Male flowers are called catkins. Female flowers are red and small. They produce the nuts. Pollination is by wind. They require other varieties for pollination. Nuts form in clusters called burrs with 1-12 nuts per burr.

Pests and Disease:
Eastern filbert blight can kill trees unless sprayed. Geneva, Slate and Grimo are resistant varieties.
Leafroller moths lay eggs then the caterpillars roll up in the leaves damaging them. Cut them off. (See blueberry pests above.)

Pruning: A lot of suckers form at the base. Cut these off and let one central leader (trunk) grow instead. But overall not much pruning is needed.

Varieties: American hazelnut type is more cold tolerant than the European type.
American Filbert (Corylus Americana) is native to the eastern United States. It grows 8-12 feet tall and 10-15 feet wide. Nuts mature September to October.
Beaked Filberts (Corylus Cornuta) is native to the United States and Canada.
Common Hazel (Corylus avellana) is the most popular worldwide.
Filbert (Corylus maxima) is grown in southeastern Europe and southwest Asia.

Propagation by 2 Methods:
Method 1: Remove the husk (shell) from the nut or use a file to scratch a small notch in the husk. Sow nuts several inches deep in potting soil in a pot outside in the fall. Put wire mesh or hardware cloth over the pot so animals do not eat the nut. They should germinate early spring. Let them grow 6-12 inches tall and then transplant them. Nuts are ready to harvest in 3-5 years.
Method 2: Dig up runners or starts with roots from established bushes. Cut with a spade in late fall when the bushes are dormant. Transplant to new location.

(See September Garden-Harvest.) (Read "Peterson Field Guides: Field Guide to Eastern Trees" by George A. Petrides. Read "Botany in a Day: The Patterns Method of Plant Identification" by Thomas J. Elpel.)

Plant Peach or Nectarine Tree: (Peach: rosaceae family; Prunus persica) (**Nectarine** is a cultivar group of peach.) (For **pruning** see February Garden-Maintenance. See August Garden-Harvest. See "For All Fruit and Nut Trees" above for how to plant.)

Peach trees are grown from seed (pit). It takes 3-4 years to go from a seedling to a fruit-producing tree. Trees from a nursery are usually a few years old. Full production starts after 5-6 years.

Plant at least 15 feet apart. Grows 15-30 feet tall and 20 feet wide. May need spraying.

Care: Likes full sun. Likes soil pH 6.0-6.5 with good drainage. Needs good air flow.

Peach trees flower early, so are more prone to frost damage than most fruit trees. Do not plant in low-lying frost pocket. Usually **blooms late March to mid-April.** Self fertile (self pollinating) but some may need hand pollinating.

Thin developing fruit early in the growing season, leaving about 8-10 inches between peaches and nectarines.

Varieties recommended for North Carolina: Carolina Belle (white-fleshed), Contender, Cresthaven, Encore, Legend, Norman, Redhaven, Summer Pearl (white-fleshed), and Winblo. Elberta **ripens early August**.

Peach and Nectarine Disease and Pests: (See "Plant Health" section in back of this manual. See **bagging fruit** in June Garden-Maintenance.)

All fruit should be picked before becoming over-ripe. No fruit should remain on trees after ripening. All fruit should be removed from the ground. This reduces disease and insect problems for later-ripening fruit and the following year.

Rigorous cultural and sanitation practices reduce the number of sprays needed. Trim out diseased parts right away. Make sure the soil is fertile so plants stay healthy.

Fungus Disease
   1. Brown Rot
   It affects blossoms, shoots and fruit. It causes blossom blight. Remove mummified fruit and infected plant tissue. Handle good fruit gently. Can use a dormant oil spray. (For **dormant oil spray,** see February Garden-Maintenance.)

   2. Leaf curl can be controlled with a single application of Bordeaux mixture, fixed coppers or lime-sulfur. Some cultivars are less susceptible to leaf curl than others. The spray must be applied during the dormant season before buds swell. Peach and nectarine foliage is very sensitive to copper and can be injured. Do not spray if there are leaves. (For **dormant oil spraying**, see February Garden-Maintenance.)

   3. Peach Scab (See apples above.)
   It infects fruit just after shuck split (when small growing fruit splits open the old flower). Symptoms appear mid to late May as small, shallow, greenish black spots. The fruit may crack. Apply fungicide 4-6 weeks after shuck split.

   4. Rhizopus Rot
   Happens after harvest. Do not let fruit get overripe before picking. Handle gently. Store in cool location.

Pests
   1. Mites
   They feed on the leaves which then become bronzed or blackened. Defoliation and abnormal fruit can occur. Populations are larger in warm, dry weather. Use a miticide. (For **dormant oil spraying**, see February Garden-Maintenance.)

   2. Nematodes (See Insects and Nematodes in "Plant Health".)
   They are microscopic wormlike organisms that can transmit viruses. Types include ring nematode and root rot nematode. Plant resistant varieties and remove diseased plant or parts. Plant marigolds. (For **marigolds** see April Greenhouse.) Solarize the soil. (For how to **solarize soil**, see potting mixes "Plant Health" in back of this manual.)

   3. Oriental fruit moth, peach scab, plum curculio, and white peach scale are controlled with 2 dormant oil sprays at 2-week intervals before green tissue emerges. When flower petals begin to drop, apply the combination spray and repeat at 2 to 3-week intervals until 3 weeks before harvest. Most insecticides should not be used within the 2 to 3 week period before harvest.

   4. Scale insects
   Includes white peach scale. They are often a serious pest of fruit trees that can kill young trees. Usually damages shoots and large branches. Use a dormant oil such as Volck at 5 tablespoons per gallon of water (8 fluid ounce per 10 gallons) just before green tissue appears (early spring before bud break). Oil also helps control mites and aphids. (For **dormant oil spraying**, see February Garden-Maintenance.)

Plant Pear Tree in April: (rosaceae family; maloideae or spiraeoideae subfamily, pyrus genus with about 30 species) The most common species is the **European pear** (Pyrus communis subspecies communis).

(For **pruning** see February Garden-Maintenance. See August Garden-Harvest. See "For All Fruit and Nut Trees" above for how to plant.)

Pear seeds do not breed true. Varieties are grafted onto a good rootstock. Plant 20-25 feet apart for standard varieties. Standard pear trees grow 30 to 40 feet tall and 25 feet wide.

Pears need other varieties to **pollinate** them. They should be within at least 40-50 feet of each other. Seckel and Bartlett do not pollinate each other. Blossoms can tolerate temperatures down to 26 degrees.

Care: Likes full sun and well drained soil. Likes soil pH of 6.0-6.5. Likes good air flow. Does OK in clay soils.

Blooms late March to early April, 2 weeks before apples. Standard pear trees take 10-20 years to bear fruit. Dwarf trees take much less than that. Soon after setting fruit, **thin fruit** to prevent limb breakage and to get larger fruit.

Varieties recommended for North Carolina: Kieffer, Harrow Delight, Harrow Sweet, Harvest Queen, Moonglow (a good pollinator, ripens mid August), Magness (not a pollen source), and Seckel (ripens mid September). Delicious ripens early September.

Pear Disease and Pests: Overall they have fewer diseases and pests than most fruit trees. (See "Plant Health" section in back of this manual. For **bagging fruit and whitewashing trunks,** see June Garden-Maintenance. For **dormant oil sprays including killing pear psylla,** see February Garden-Maintenance.)

All fruit should be picked before becoming over-ripe. No fruit should remain on trees after ripening. All fruit should be removed from the fruit-growing area. This reduces disease and insect problems for later-ripening fruit and the following year. Use of rigorous cultural and sanitation practices can reduce the number of sprays needed.

Pears are affected with many of the same diseases as **apples** with the exception of pear scab, which has not been reported in North Carolina, and cedar apple rust, which does not occur on pear. (See apple trees above.)

1. Codling Moth Wormy apples and pears are caused by the codling moth. (See apple tree pests above.)

2. Fire blight (See Asian pear above. See fire blight in June Garden-Maintenance.)

Some sources say fire blight is a **fungus**, other sources say it is a **bacteria**. It is more severe on pear than apple trees. It can kill large limbs and even entire trees. It is one of the most serious diseases of pear trees. Very few trees are completely resistant and those that are usually produce poorer fruit. Shoots infected turn brown or black and look scorched. Blossoms wither and die. Reddish water-soaked bark lesions appear, and on warm days they ooze an orange-brown liquid. Later they become brown and dry.

It can be controlled with **correct fertilization** (not too much nitrogen) and pruning (for good air flow). In the **winter** branches with cankers can be pruned off a few inches below the cankers. On limbs too large to remove, pare away diseased tissues, sealing wounds with tree paint.

In the **summer,** look for blackened leaves, stems or fruits. Cut them off 12 inches below the disease. Cut off vigorous suckers growing up along branches and the rootstock. These are very prone to fire blight. **Disinfect pruning tool** in a bleach solution (1 part bleach to 9 parts water) after each cut. Hold in the solution for at least 2 seconds.

Pear tree varieties resistant to fire blight: Asian varieties and Comice, Dawn, Douglas, Duchess d'Angouleme, El Dorado, Fan-stil, Harvest Queen, Harrow Delight, Lincoln, Luscious, Mac, Magness, Maxine, Moonglow, Orient, Seckel, Starking Delicious, Sugar, Sure Crop, Waite, and Winter Nelis. **Disease-resistant rootstocks:** Kieffer, Moonglow, Old Home, Oriental (Pyrus bitulafolis), Oriental (Pyrus communis), Oriental harbin pear (Pyrus ussuriensis), Seedling, Stark Honeysweet, and Starking Delicious.

3. Pear leaf spot (fabraea leaf spot) can be a problem on some cultivars. Captan is not registered for use on pears. Use benomyl (Benlate) 1 teaspoon per gallon of water (3 tablespoons per 10 gallons).

Sprays for pear scab, such as Bordeaux mixture, will usually control pear leaf spot. Surround, a kaolin-clay-based insect repellant, is labeled for suppression of leaf spot.

4. Pear psylla look like very small cicadas (1-2 inches). Overwintering adults are dark reddish brown. In the summer adults are tan to light brown. Adults overwinter in ground litter and in cracks in bark. They become active between 40-50 degrees. The whitish or yellow eggs are laid in a line when 50-60 degrees prior to bud burst. They hatch and feed on new growth. Honeydew from the feeding damages the leaves and tree. (For **dormant oil sprays that kill pear psylla,** see February Garden-Maintenance.)

Plant Plum Tree: (rosaceae family; prunus genus with 20 species) (For **pruning** see February Garden-Maintenance. See August Garden-Harvest. See "For All Fruit and Nut Trees" above for how to plant.)

Standard plum trees grow up to 20 feet tall and 25 feet wide. Plant at least 15 feet apart. Plums do not produce fruiting spurs as apples and pears do. **Blooms late March or early April.**

Care: Likes soil pH 6.0-6.5. Needs full sun and well drained soil. Likes moderately rich soil. Needs good air flow. The wood tends to be brittle.

Types of plums:
1. American plums (Prunus americana) are a wild plum native to North America.

2. European plums (Prunus domestica Lindl.) are blue. Eaten fresh or made into prunes. Varieties include Bluefre, Damson, Earliblue, and Stanley. European plums usually do not need 2 varieties for cross-pollination, but there are exceptions. All produce more fruit with more than one variety. European plums will not pollinate Japanese plums. European plum trees are adapted to conditions throughout most of the United States. The plums store longer than Japanese plums.

3. Hybrid plums from European and Japanese plums. Good for areas that have weather extremes.

4. Japanese plums (Prunus salicina Lindl.) are red, yellow or green and less hardy than blue plums. Varieties include Ozark Premier, Redheart and Santa Rosa. They must be cross-pollinated from other Japanese plum varieties. Japanese plums are better able to withstand summer heat. They bloom earlier than European plums. They do not like cold, damp springs. They need more pruning and fruit thinning than European plums.

Varieties recommended for North Carolina:
Japanese: Burbank, Byrongold, Ozark Premier (may bloom early), and Methley (self-fruitful). Japanese Shiro is a gold plum that ripens early August. Japanese Superior is a red plum that ripens early August.

European: Bluefre, Shrophire (Damson), and Stanley.

Plum Diseases: (See "Plant Health" in the back of this manual. Also see apple tree diseases above.)
1. Black Knot

This **fungal** disease occurs on branches of cherries and especially plums. This is most common when trees are grown near wooded areas that contain wild cherry. Prune out these knots and spray a fungicide such as Benltate starting early bloom with 2 to 3 additional sprays 7 to 10 days apart.

2. Brown Rot

This **fungal** disease is commonly found on cherries and plums. Apply the first spray when the flower petals begin to drop and repeat at 2 to 3 week intervals until 3 weeks before harvest. If weather conditions are wet, apply 1 to 2 fungicide (myclobutanil) sprays during the 3-week preharvest period.

3. Canker

Bacterial canker often attacks plum trees. Look for depressions in the bark at the base of limbs in the fall. They grow large in spring. They quickly cover the entire branch and the branch dies. Remove disease by cutting the branch off and cutting away diseased wood. Buy a canker paint for plum trees and paint on the wound. Spray the tree in August, September and October with copper based fungicide.

Plum Pests:
1. Aphids

They are a common pest on plum trees. These tiny green and black insects feed on leaves, buds and stems. Aphids do not kill a plum tree but they weaken it. Manage by hosing tree with a high-pressure sprayer. Do this every other day. Or mix 25 percent dish soap with 75 percent water in a hand-held sprayer and spray the aphids. The dish soap suffocates them. (See apple tree pests above. See Brassicaceae Family in "Plant Families". For **dormant oil spraying**, see February Garden-Maintenance.)

Peach Tree Borer

Can cause serious damage to trunks of peach, nectarine, cherry and plum trees. Can cause tree death. To control, apply endosulfan 50W (Thiodan) at 2 tablespoons per gallon of water (7 ounce per 10 gallons) to the lower limbs, trunk and base of the tree the first week in September. (For **control methods**, see May Garden-Maintenance.)

Plum Sawfly

It lays eggs on flowers during the spring. Once the flowers produce fruit, the eggs hatch and maggots tunnel into the fruit. Look for tiny holes in new fruit. The holes may ooze a dark fluid. Kill adult sawfly by digging up the soil around the plum tree in the fall. Expose the pupae to cold nights so they die. Birds eat exposed pupae.

Red Spider Mites

They are common on plum trees. Leaves start to turn dry and brown. Look for fine webbing on the leaves and tiny red spiders on underside of leaves. Spider mites are resistant to virtually all insecticides. Spray the tree every other day using a hose with a strong water sprayer. This can easily manage spider mites. (For **dormant oil spraying**, see February Garden-Maintenance.)

Plant Rosa Rugosa Rose Bush: (rosaceae family; Rosa rugosa) (For **pruning** see February Garden-Maintenance. For **eating petals and leaves** see April Garden-Harvest. For **harvesting fruit**, see September Garden-Harvest. For details about herbaceous, softwood, semi-hardwood and hardwood cuttings, see "Garden Tips" in the back of this manual.)

Thorny, hardy (cold tolerant), and vigorous woody shrub in the **rose family.** Plant is 3-6 feet high and wide. Good for erosion control.

Care: Likes full sun but can take partial shade. Likes acid soil with pH of 5.5-7.5. Likes well drained soil. Tolerates poor soil. Likes good air circulation. Requires little care.

Harvest: Petals, shoots and leaves are edible in **early spring.** Flowers are deep pink, rose pink or white in **late spring (June) and early summer** (flowers may bloom sporadically throughout summer and into early fall). They may be single or double; very fragrant.

Produces large red-orange rose hips that make good tea, jelly, etc. Rose hips are 1 inch orange to red when ripe.

Can be **propagated** by seed, softwood cuttings in **July-September,** or hardwood cuttings in **November-February.** For hardwood cuttings, store at 35-40 degrees in sand or peat. Then in spring plant cuttings outside with top 1 inch protruding from soil. (For **herbaceous, softwood, semi-hardwood and hardwood propagation**, see "Garden Tips".)

Rosa Rugosa Disease and Pests: It has few diseases and pests. (See "Plant Health" in back of this manual.)
Less troubled by the common rose diseases (**black spot and powdery mildew**), than other roses.
Japanese beetles can be a problem. (For how to control, see bramble pests above.)

Plant Walnut Tree in April: (juglandaceae family; juglandoideae subfamily, juglandeae tribe, Juglans genus with 21 species) (For **pruning** see February Garden-Maintenance. For **harvest and processing**, see September Garden-Harvest. See "For All Fruit and Nut Trees" above for how to plant.)

Can grow from seed but squirrels may dig it up so cover with wire mesh. Plant seed 3 inches deep. It can take 5 months to a year for it to germinate because they need cold to sprout. They grow 1-2 feet the first year. Harvest is in 10 years for standard varieties.

If you buy a young tree, then cut back top of tree to about 3 feet tall where there is a dormant bud. This bud will be the main trunk.

Different varieties grow from 30-150 feet tall and up to 70 feet wide. Hardy.

Care: Likes full sun. Likes well drained soil. Likes soil pH of 5.0-7.5. Walnuts raise the pH of the soil around them. They kill plants growing under or near them by poison in the roots. Likes moderately rich soil. Drought tolerant.

Harvest is usually in **September.** The nuts are edible and enclosed in a thick green husk about 2 inches across. They turn brown as they dry.

Self fertile but **pollination** is better with 2 or more walnut trees. **Flowers April to May** so is frost sensitive. Do not plant in low-lying frost pocket.

Varieties:

Eastern black walnut (Juglans nigra) is a common species native to eastern North America. The bark is brown, and has furrowed ridges that form a diamond pattern. The nuts are edible but the **Persian walnut** (Juglans regia) is the type found in grocery stores. Black walnut wood is very valuable. Grows to 150 feet tall. Lives 200 years or more.

English Walnut (Juglans regia) is a European walnut that grows about 60 feet tall. The bark is light gray with flattened ridges. Nuts are edible.

Stark Northern Prize (Juglans regia Domoto) produces nuts 5-6 years after planting. Grows to 20 feet.

Lake English (Juglans regia Lake) produces nuts 4-5 years after planting. Grows to 30 feet.

Walnut Disease and Pests: (See "Plant Health" in back of this manual.)
Walnuts are prone to several diseases/pests but most do not threaten life of the tree. But they may ruin nuts.
  1. Walnut Blight
    Also known as **bacterial** blight. It causes black spots on leaves, holes and blotches on fruit, and dieback of shoots. Nuts are usually damaged extensively, especially when male catkins (a flowering spike of trees) are affected. It affects mainly young growth and nuts. It is most damaging during cool wet periods, and around flowering time. The bacteria overwinters in buds.
    Cut out damaged parts (cutting too much is better than cutting too little) and burn them. Bordeaux mixture may help. May prevent it by having a soil with pH above 6, avoiding excessive nitrogen and wetness, and having good air flow by pruning properly.
  2. Walnut Leaf Blotch (Walnut anthracnose)
    This widespread **fungus** causes nuts to turn black and then fall. It creates brown blotches on leaves and fruit. Leaves fall off. It spreads in wet weather, usually during May and early June. It overwinters on leaf litter. The only treatment is burning fallen leaves and using a Bordeaux mixture.

Garden- Harvest  (April)

Vegetables and Fruit: Asparagus, bamboo, forsythia, Jerusalem artichoke, lambs quarters, matteuccia ferns, mushrooms, parsnips, petasites, poke, rhubarb, rosa rugosa, salsify/burdock, and yellow dock.

Harvesting Herbs (When using herbs, please contact your herbalist, doctor and/or health specialist.)
All herbs should be harvested when oils responsible for **flavor and aroma** are at their peak. Herbs grown for their **leaves** should be harvested before they flower. Herbs grown for their **seeds** should be harvested when seed pods change from green to brown to gray but before they shatter (open). Herbs grown for their **flowers** should be collected just before full flower. Herbs grown for their **roots** should be dug up in the fall after the leaves have died.
For **leaf harvesting** most herbs can have up to 75% of their foliage cut without damaging the plant. It is best to cut in the morning. **Annual** herbs can be harvested until frost. **Perennial** herbs should be harvested until late August so plants have time to store energy for winter. Also late pruning of perennials encourages new growth that may be damaged by frost.
Read "Encyclopedia of Herbal Medicine: The Definitive Home Reference Guide to 550 Key Herbs with all their Uses as Remedies for Common Ailments" by Andrew Chevallier and "Prescription for Herbal Healing: An Easy-to-Use A-Z Reference to Hundreds of Common Disorders and Their Herbal Remedies" by Phyllis A. Balch.)

 ### How to Make a Decoction with Herbs

Decoctions are best when the herb is hard and woody such as **roots, rhizomes, twigs, bark, nuts and some seeds.** More heat is needed to process them. An infusion is better when the herb is a leaf or flower.
When using a woody herb with a lot of **volatile oils**, powder it as finely as possible and then use in an infusion rather than a decoction so the oils do not boil away.
Use a glass, ceramic, earthenware or enameled pot. Do not use aluminum. For every 1 ounce (1 cup) of chopped herb, add 4 cups of water. Bring to a boil and **simmer for 45-60 minutes**. Do not remove the lid while simmering. Turn off heat and let sit to cool with the lid on for about 2 hours. Transfer to glass jar and **let sit for 8-12 hours**. Then strain out herb. Refrigerate it. It lasts a few days. Consult your herbalist, doctor and/or health care specialist.

 ### How to Make an Infusion with Herbs

Infusions are made using the **leaf, flower or any soft parts of a plant.**
Water Infusion: In a quart jar put 1 ounce (1 cup) of chopped herb. Add 4 cups (or how much to fill the jar) of boiling water. Put the lid on the jar. **Let sit for 8-12 hours**. Strain out the herb and refrigerate. It lasts a few days.
Oil Infusion: Put 9 ounces of dried or 1 1/4 pounds of fresh herb with 2 cups of a light oil such as sunflower, safflower or sweet almond in a glass bowl. Put the bowl over a pan with simmering water. Or use a double boiler. Cook

at a low heat for about two hours. Strain into another container. Will last in the refrigerator for 3 months.

Consult your herbalist, doctor and/or health care specialist.

(For how to make **horehound cough syrup**, see May Garden-Plant. For **how to use wormwood tea**, see May Garden-Plant.)

 ## How to Make a Tincture with Herbs

Tinctures are **concentrated extracts of herbs** made with vinegar or drinkable alcohol such as wine, brandy or vodka. They are not heated.

Fill pint jar with 2-3 ounces of dried herb. Or if fresh herb, completely fill the jar. Then fill jar with 8 ounces of 100 proof vodka. For fresh herbs use higher proof alcohol. Never use rubbing alcohol since it is not safe to eat. Put the lid on and shake. Shake every day for 2-4 weeks. Then strain liquid and store in bottle in dark, cool place. This method can be used for any herb.

Using: The mixture stays potent for several years. Start with 10 drops three times a day. The maximum dose for most tinctures is 30 drops or 1 teaspoon three times a day. How much depends on strength of tincture.

Consult your herbalist, doctor and/or health care specialist.

(For **making an herbal tincture and ointment with arnica,** see March Greenhouse. For **how to test calendula tincture,** see May Garden-Plant. For making **walnut tincture,** see September Garden-Harvest.)

 ## Harvest Perennials

Bronze fennel, chickweed, chives, comfrey, dandelion, garlic/onion leaves, good king henry, horseradish leaves (young) and roots, lambs quarters, lovage, mint, oregano, plantain, poke, purslane, ramps, rosemary, sage, salad burnet, sea kale, sorrel, stinging nettle (best before flowering), thyme, violets, and winter savory. Some of these are found by foraging.

Leafy greens are best in April and May...a spring tonic. Once temperatures are 85 degrees or higher, many turn bitter. (For **horseradish** see October Garden-Harvest. For **perennial herbs** such as mint see May Garden-Plant.)

Harvest **asparagus** (asparagaceae family; Asparagus officinalis) until spindly or until spear diameter is 3/8 inch or less (at least as thick as a pencil), for up to 2 months. They should be at least 7 inches tall. Cut or bend till it snaps. Leave 2-3 inches of plant.

Do not harvest all spears. Leave some to grow to strengthen plant for next year. Do not harvest the first year it is planted. Harvest lightly the second year.

Freezing temperatures below 28-32 degrees can damage the spears so cover with cloth if cold.

One technique for **lengthening the harvest time** is to let 3 large spears per plant grow to ferns during the harvest season. This way the crown is being fed so more spears can be harvested. After harvest season, new spears in addition to those 3 spears form into ferns.

Uses: Asparagus is good quick sauted on high heat with butter or olive oil. It is done when it turns bright green.

(For **planting seeds and crowns,** see April Garden-Plant. For **fertilizing,** see March Garden-Maintenance. For **salt and production,** see July Garden-Maintenance.)

Shoots from most **bamboo** (poaceae/grass family; bambusoideae subfamily; bambuseaeare tribe) are edible but some varieties contain hydrocyanic acid which can be removed by boiling. Shoots should be dug **as soon as the tip emerges from the soil.** When it starts getting sunlight, it turns bitter. Look for mounds of soil being pushed up by the shoots.

A shoot that is fairly stout, light yellow or light brown, purple on the root buds, and white at the bottom will be tender, fragrant, and tasty. A slender shoot with purple black skin, reddish root buds, and a dark colored cut base will be tough.

Uses: Do not eat raw. Boil freshly cut shoots for 10 minutes with 1 or 2 changes of water. Then peel off the skin and eat.

Stalks and leaves are good as **fodder for animals.** Good to use for bean poles or trellis. Can be used to make furniture, small buildings, bridges, baskets, clothing, firewood, fishing rods, hats and toys.

(See April Garden-Plant.)

**Forsythia** (oleaceae family; Forsythia suspensa) bush **blooms in early spring** with yellow flowers before the leaves emerge. The flowers bear fruit that go from green to yellow.

Medicinal: Chinese medicine uses green forsythia fruit as a remedy for respiratory problems. It is used to treat colds and viral infections with a fever. Usually combined with **honeysuckle** flowers. Honeysuckle flowers are highly antibacterial and antiviral. Consult your herbalist, doctor and/or health care specialist.

Harvest: Collect the fruit when it is green in spring. Use it right away or dry it to use later. To make forsythia tea, put the fruit in boiling water, let stand for a few minutes, strain and drink. (Not for pregnant women.) You can also use stems and bark in a tea.

Continue digging up **Jerusalem artichokes.** (See March/October Garden-Harvest with **harvest details in October.** See May Garden-Plant and July Garden-Maintenance.)

Forage for **lambs quarters** (chenopodiaceae family; Chenopodium album, white goosefoot, giant goosefoot, fat hen) through **spring, summer and fall.** It is a highly nutritious **annual.**

It is fast growing and weedy. It has little or no odor. Leaves are alternating, almost triangular, with a blunt tip and jagged edges. Leaves may develop a white tinge.

Harvest young shoots up to 10 inches tall. Tender new growth can be eaten any time. Eat raw or cooked. Leaves taste somewhat like spinach. Steam leaves 5-10 minutes or cook like spinach.

Flowers are edible. The tiny black seeds that form in the fall are tasty but a lot of work to collect.

"Nature's Garden: A Guide to Identifying, Harvesting, and Preparing Edible Wild Plants" by Samuel Thayer.)

Harvest young **fiddlehead ferns** (onocleaceae family; Matteuccia struthiopteris; ostrich fern). Collect before the round fiddlehead has started to unfurl. Wash to remove fur or scales. Cook for 12 minutes before eating. Must cook before eating. Consult your herbalist, doctor and/or health care specialist.

Forage for **morel mushrooms** (morchellaceae; Morchella esculenta) after good rain. Soak in salt water brine, then cook. Must be cooked. When picking mushrooms, be absolutely sure you have the correct species. Some are **poisonous.** Some people react poorly to all mushrooms. Ask an expert to help you identify it.

"Nature's Garden: A Guide to Identifying, Harvesting and Preparing Edible Wild Plants" by Samuel Thayer.)

Harvest **Shiitake Mushrooms**

Logs inoculated with Shiitake spawn/spores over a year ago, usually in February, have mushrooms to eat **April and May**. Consult an expert before eating any mushrooms. (For many details, see February Garden-Plant.)

Harvest **parsnips** planted last April. Parsnips taste better when been through hard frosts. (See April Garden-Plant and September Garden-Harvest.)

Harvest stalks and flowers of **petasites** (asteraceae family; Petasites japonicas, Japanese coltsfoot, butterburr). Has giant green leaves that can be 16 inches wide. Has fragrant white daisy-like flowers that grow on spikes. Likes deep to partial shade. Likes moist soil. Can be invasive.

The edible flowerbuds in **spring** are somewhat bitter. Also cut young stalks in spring when 2 feet or more long. Cut into 6 inch pieces and boil for 10 minutes. Dip in cold water for 1 minute and peel.

Hard to peel so after boiling and cooling, cut into bite size pieces and put in vinegar. Use small diameter stalks.

 **Poke- American** (phytolaccaceae family; Phytolacca americana) Also known as Virginia poke, American nightshade, cancer jalap, coakum, garget, inkberry, pigeon berry, pocan, pokeroot, **pokeweed**, pokeberry, redweed, scoke, red ink plant, and chui xu shang lu.

It is a **hardy perennial** that grows wild usually in fields. Native to eastern United States. It grows up to 10 feet tall. It dies back each winter. Do not feed to animals.

Young leaves and shoots are eaten in spring. Cooked and eaten like spinach except should be boiled twice (blanched) with the first water being discarded. It is poisonous if eaten raw. Native Americans ate them cooked.

Consult your herbalist, doctor and/or health care specialist.

Berries are usually not eaten. Some Appalachian old timers eat a few berries a day for arthritis. Sometimes berries are cooked into a jelly or pie with the seeds strained out. Cooking is believed to inactivate toxins and others believe toxicity is in the seeds within the berries so if strain out seeds it may be OK.

Roots are dried and used medicinally as a laxative.

Juice from the berries can be used to make ink and dye.

Late April harvest **rhubarb** (polygonaceae family; Rheum rhabarbarum) stalks for 4-6 weeks. **Do not eat the leaves** because they contain poisonous levels of oxalic acid. Cut stems or twist off outer stalks when 1 inch diameter.

Do not harvest the first year so plant can build good root system. In the second year harvest 2 stalks only. The third year and after, **harvest 3-4 stalks per plant** if there are also at least that many left on the plant. High in vitamin A and C.

Cooking: For 2.2 pounds rhubarb, use 1 1/3 cups sugar. Cut the stalks into pieces 1-2 inches long. Put stalks and sugar in pan and cover with a little water. Cook 10 minutes. Can cook with **strawberries** since they ripen at the same time. Can use less sugar. Can also use honey.

(For **dividing crowns**, see March Garden-Plant. See May Garden-Harvest but harvest details are in this section. For **removing flowers,** see April Garden-Maintenance.)

 **Rosa Rugosa** rose petals, shoots and leaves are edible in **early spring.** (See April Garden-Plant, September Garden-Harvest.)

 Harvest biennial young leaves, stems and flowers of **salsify and burdock**. Eat **salsify** raw or cooked. **Burdock** stalks are peeled, and eaten raw or boiled. Young leaves are eaten raw or cooked.

(**Salsify:** For sowing see May Garden-Plant. For digging up roots, see September Garden-Harvest.)

(**Burdock:** For sowing see March Garden-Plant. For digging up roots, see June Garden-Harvest.)

Harvest/forage for **yellow dock** (polygonaceae family; Rumex crispus, curly dock, narrow dock, sour dock). It is a **perennial** found in fields and roadsides. The roots are 8-12 inches long. The leaves are 6-10 inches long. It is closely related to sorrel and rhubarb. It is a good **spring tonic** like dandelion, stinging nettle, and burdock. It is a blood purifier.

Medicinal: In colonial days dock **leaves** were used to treat sores, sore eyes, and glandular swellings. To cure itchy skin, the leaves were bruised, mixed with butter or lard, and placed on the problem area. It was also used for diarrhea and constipation.

It is high in nutrients. It is used for digestive problems, liver diseases and skin disorders. It is an astringent (causes contraction), cholagogue (stimulate bile), hepatic (liver healer), and laxative. Can make **root** into a tea or poultice. Consult your herbalist, doctor and/or health care specialist.

Eating: Young leaves are eaten cooked. Do not eat raw. If the leaves are too bitter, then parboil, wash, add clear water, and cook until tender. Since the leaves contain oxalic acid (similar to spinach), they should not be eaten frequently in large amounts. The **seeds** are ground and used as flour.

Greenhouse, House, Hoop House or Cold Frame (April)

Early to Mid: Cauliflower, cucumber, fennel, globe artichoke, sweet potato, tobacco, watercress, and winter squash.
Late: Basil, eggplant, okra, marigold, melon/watermelon, and tarragon.

Early to Mid April in Greenhouse:

 Sow early **cauliflower** (brassicaceae family; Brassica oleracea) in greenhouse. Start about **4-8 weeks before last frost in spring. If sowing outside start 2-3 weeks before last spring frost.**

It is finicky; requires more care and attention than cabbage. Germinates in 7-10 days. Sow 1/2 inch deep. Grows best at 57-68 degrees. Matures in 50-100 days (2-3 months) so harvest June or July.

Will withstand light frost down to 29-32 degrees. Cauliflower and chard are more sensitive to cold than broccoli, collards, kale, kohlrabi, or mustard.

Fall planting is better. Sow outdoors 10-14 weeks (2 1/2 to 3 1/2 months) before first fall frost.

Care: Does not like acid soil. Likes soil pH 6.0-7.0. Needs full sun and well drained soil. It does not like hot weather. The soil should not be hard. It has shallow roots so water on a regular basis.

In **early May** when seedlings are 3-6 inches tall, transplant outside 18-24 inches apart in rows 24 inches apart. Harden off before planting. (For **how to harden plants,** see "Garden Tips" in back of manual.)

You can **blanch** cauliflower (tying up the leaves around the head to prevent light from reaching it so it stays white). Blanching is optional.

Harvest heads when they are full but before sections begin to loosen. Start checking plants every day when heads reach 3-4 inches across.

Colors: Cauliflower can be white, yellow, green, orange or purple.

(For **early cauliflower,** see May Garden-Plant. For **late cauliflower,** See June Garden-Plant but details are in this section. See September Garden-Harvest.)

 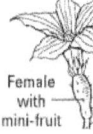  Sow **cucumber** (cucurbitaceae family; Cucumis sativus) in greenhouse. Germinates best at 90 degrees. Will not germinate below 60 degrees. Frost will kill the plants. (See "Plant Families" in back of manual.)

Can sow in degradable pots **3-4 weeks before last frost** since they do not transplant well. **Transplant outside in May after last frost.** They also can be **planted again in summer 10 weeks before first fall frost.**

Sow 1 inch deep, 2-3 inches apart. Germinates in 3-10 days.

Plant vining varieties 2 feet apart. Plant bush types 18 inches apart. When planted in hills and allowed to run, grow 3-4 plants spaced 1-2 feet apart. Matures in 48-80 days.

Care: Likes soil pH 6.0-7.0. Needs full sun and well drained soil. They like a lot of fertilizer. Do not let the plants wilt from lack of water.

Pollination is better if there are at least 3 plants.

There are 4 types:

    1. English cucumbers (burpless) are sweeter with a thinner skin. They are easier to digest and have a pleasant taste. They grow up to 2 feet long. They are nearly seedless. (Seeds and skin of other varieties may give people gas.) Varieties include Burpless Hybrid and Sweet Slice.

    2. Pickling (picklers) cucumbers are uniform in the length-to-diameter ratio (about 3-4 inch long and 1 inch wide) and do not have voids in the flesh. Picklers are usually short, thick, less regularly shaped with bumpy skin (tiny white or black-dotted spines). Color can be creamy yellow to pale or dark green. Common ones are Kirby and Liberty. Others include County Fair, National Pickling, Pickle Bush, Regal and Saladin.

    3. Slicing cucumbers are eaten fresh while they are green and not ripe. Ripe cucumbers are usually yellow, bitter and sour. Unripe cucumbers are pickled when they are small. Slicers are usually long and smooth with a tough skin. Varieties include Dasher II, Greensleeves, Marketmore 76 & 80, Raider, Slice Master, Straight Eight and Sweet Success.

4. Specialty cucumbers include round, yellow cucumbers such as Lemon Cucumber. Others such as Sweet Armenian are pale green or white with big ribs. Boothbys Blonde are light yellow with black spines. Suyo Long from Asia have a curl at one end of the cucumber.

Harvest: Cucumbers mature quickly so harvest often to keep plants productive. **Harvest while still small,** between 3-4 inches for most pickling varieties such as dill. Sweet pickles or gherkins for pickling can be harvested when about 2 inches long. And harvest at 6-9 inches for slicers. Green varieties should be bright, dark green. Seeds for all should be soft. If you will be storing them for a while, leave 1 inch stem on it to prevent rot.

(See May Garden-Plant but details are in this section. For **insect repellant that deters cucumber beetles and squash bugs**, see nasturtium in "Plant Health" in back of manual.)

 Sow **fennel** (umbelliferae family; Foeniculum vulgare) mid April in greenhouse. Cover seeds with 1/4 inch of soil. Germinates in 12-18 days. Thin seedlings 10-12 inches apart, in rows 18-24 inches apart. Grows 3-5 feet tall and 2 feet wide depending on variety. About 100 days (14 weeks, 3 1/2 months) to maturity for annual variety.

Care: Likes full sun and well-drained soil. Likes soil pH of 5.5-7.0.

It is aromatic with anise-like flavor. It is related to caraway. It has feathery leaves with yellow umbrella-like flowers.

Plant annual fennel such as **Florence fennel in April and again mid summer** to harvest in fall. It is **hardy to 25 degrees** and can be eaten after frost. **Bronze fennel** (foeniculum vulgare dulce rubrum) is a perennial.

Harvest fennel when bulb is 4 inches across. Leaves, bulbs, stems and seeds are edible. Seeds are mature in **late August.** If overripe, the seeds fall out of the flower so harvest before then.

 Move **globe artichoke** (compositae family; Cynara scolymus) seedlings in **house to greenhouse** when seedlings are 6 weeks old. Temperature should not go below 25 degrees. But it should be as cool as possible. Stays in greenhouse 6 weeks. So at 12 weeks (**mid May to early June**), it is planted outside.

(For **growing system and sowing,** see February Greenhouse. For **second possible sowing date**, see March Greenhouse. **To move from greenhouse to outside**, see June Garden-Plant. See August/September Garden-Harvest with details in August.)

 Start **sweet potato** slips (sprouts) early April or late March in house. (For **how to start slips**, see March Greenhouse. See June Garden-Plant and October Garden-Harvest.)

 Sow **tobacco** (solanaceae family; Nicotiana tabacum) in greenhouse **4-8 weeks before last frost** so that is mid March to mid April. Tobacco is a member of the **nightshade family** that includes eggplant, peppers, potatoes and tomato. Tobacco has similar growing needs as tomatoes. (For **tomato sowing and growing,** see March Greenhouse. See "Plant Families" in back of manual.)

An ounce has 300,000 seeds. Can mix 1 part seed to 2 parts sand to make it easier to spread seed. Soak mixture with water before sowing seeds. It is **legal to grow tobacco** if you only grow it for your own use.

To **germinate** seed, temperatures need to be at least 65. Seeds need light to germinate but not bright sun. So cover with a light cloth until seeds germinate. Or can cover with plastic to keep soil moist. Once seedlings just begin to emerge, remove the plastic because they need good air circulation. Germinates in 7-10 days.

After last frost plant outside 2-3 feet apart in rows 4 feet apart. Grows to 3-6 feet tall. Matures in 90-120 days.

Care: Likes full sun. Likes soil pH around 5.8. Likes well drained soil. Fertilize the same as you would for tomatoes, potatoes and pepper.

Cut off the **tops** (remove terminal bud) of the plant as soon as the flower forms. This lets the upper leaves get larger and thicker.

Usually, each tobacco plant produces about 3-4 ounces of dry, **cured tobacco**. A pound of tobacco will produce about 2 cartons of cigarettes. You need about 4 plants to produce 1 carton. Tobacco should be aged, so the first year you grow, double your normal tobacco needs for the year.

Cigarettes sold in stores can have up to 599 additives. Many of these are added for flavor, or used in the glue, bleach, and dyes on the paper.

The variety, how grown, and how dried/cured, all affect the taste of the tobacco. You can mix 2 or more varieties for different tastes. Tobacco is poisonous. Use with care. Consult your doctor and/or health care specialist.

Varieties:

1. Cigarette Tobacco: Argentinean, Big Gem, Black Sea Basma, Black Sea Samsun, Burley Original, Burley Hampton, Burley Variation, Walker's Broadleaf, Dark Virginia, Del Gold, Blue Tree Glaucia, Gold Seal Special, Lizard Tail Orinoco, Lizard Tail Turtlefoot, Monte Calme Blonde, Original Wild Rustica, Salamena Blue, Stag Horn, Turkish Izmir, Virginia Gold Types, and Maryland.

2. Cigar Tobacco: Black Sea Samsun, Black Stalk Mammoth, Burley Hampton, Burley Variation, Walker's Broadleaf, Cuban Habano 2000, Cuban Criollo 98, Dark Virginia, Gold Dollar, Greenwood, Isleta Pueblo, Little Crittenden, Lizard Tail Orinoco, Lizard Tail Turtlefoot, Monte Calme Blonde, Narrow Leaf Madole, Original Wild Rustica, Shirey, Stag Horn, and Turkish Izmir.

3. Pipe Tobacco: Argentinean, Big Gem, Black Sea Basma, Black Sea Samsun, Black Stalk Mammoth, Burley Hampton, Burley Original, Walkers Broadleaf, Cuban Habano 2000, Cuban Criollo 98, Dark Virginia, Del Gold, Glaucia Blue Tree, Gold Dollar, Greenwood, Isleta Pueblo, Little Crittenden, Lizard Tail Orinoco, Lizard Tail Turtlefoot, Original Wild Rustica, Narrow Leaf Madole, Perique, Salamena Blue, Shirey, Southern Beauty, Turkish Izmir, and Maryland.

4. Snuff Tobacco: Argentinean, Black Stalk Mammoth, Black Sea Samsun, Burley Hampton, Dark Virginia, Del Gold, Gold Seal Special, Greenwood, Shirey, Stag Horn, Turkish Izmir, and Virginia Gold Types

Saving Seed: It is usually **self pollinating** but can sometimes cross pollinate. If growing more than 1 variety, grow them as far apart as possible. If you are growing only 1 variety, then you do not have to be concerned about cross pollination. Let several plants form seed. The seed pods turn brown and dry in late summer.

(For **transplanting**, see June Garden-Plant but most details are in this section. See August Garden-Harvest. For **using tobacco to treat mites and lice in poultry**, see March Farm Animals.)

Read the book "How to Grow Your Own Tobacco: From Seed to Smoke" by Ray French.)

Sow **watercress** (brassica family; Rorippa nasturtium) in greenhouse. It can also be planted in **late fall**. A **perennial related to mustard**. (For mustard see February Greenhouse.) Grows easily from seed and propagated easily by stem cuttings put in wet soil. Watercress sold in a grocery store will root.

Sow seed on top of soil or cover with just a little soil. Keep moist. Germinates in 7-14 days. Matures in 60-150 days. Grows to 6-24 inches tall. Hardy to minus 10 degrees.

Can plant in stream. Or can plant in container put in dirt. Put holes in bottom of container and water.

Care: Prefers full sun but tolerates shade. Likes lime (alkaline soil). Likes soil pH of 6.5-7.5. It is one of the few plants that is watered through winter.

Harvest mid-fall to spring. It has a peppery taste. Eat raw or cooked. After flower buds appear, leaves become too bitter to eat. (See September Greenhouse but details are in this section.)

 Can sow **winter squash or gourds** in mid April in greenhouse for **transplant in May** but most people plant directly in May. (See May Garden-Plant. See October Garden-Harvest.)

Late April sow in Greenhouse:

Sow **basil** (labiatae family; Ocimum basilicum) in greenhouse late April or early May in house or greenhouse when 70 degrees day, 60 degrees night lowest. It is a **tender annual** that is a member of the **mint family**.

Sow 1/4 inch deep. Poor germination rates. Germinates in 7 days. When seedlings are 6 inches tall, pinch off the tops to get bushier growth. Thin to 6-12 inches apart.

Care: Needs 6-8 hours of sun. Likes soil pH of 5.5-6.5. Grows to 12-24 inches tall. Matures in 75-90 days.

Varieties: Most basil is dark green. Some such as Dark Purple Opal basil are red-purple.

(See May Greenhouse but details are in this section. For **transplanting** see June Garden-Plant. See July/September Garden-Harvest with **harvest details in July.** For **mint** see May Garden-Plant.)

Eggplant (solanaceae family; Solanum melongena) is a member of the **nightshade family.** (See "Plant Families" in back of manual.) Can start in greenhouse **8 weeks before last frost.** Transplant at least 2 weeks after last frost date. It is a **very frost sensitive annual.** In spring can grow under row cover.

Has growing needs similar to tomatoes. (For sowing/growing tomatoes, see March Greenhouse.)

Sow 1/4 inch deep. Germinates in 14 days. Germinates best at 75-90 degrees. Does not like being transplanted so sow in degradable pots not flats. Later transplant 24 inches apart in rows 12 inches apart. Matures in 56-76 days.

Care: Likes soil pH of 5.8-6.8. Needs full sun and well drained soil. Likes rich soil.

Plant outside **late May or early June after last frost** when temperatures are 65-70 degrees. Needs a lot of calcium so add lime. Tall varieties need to be staked. Grows on bush that can be 5 feet tall.

Colors: Green, lavender, pink, violet, white and yellow. Place a paper or metal collar around each stem to stop cutworms. (For **cutworm collar image,** see tomatoes in March Greenhouse.)

Harvest: Produces a lot. Harvest when 6-8 inches long and still glossy. They do not taste as good (may be bitter) when big. Harvest before seeds and flesh turn brown. Remove with a knife or pruning shears rather than breaking or twisting the stem. It is overmature if it is spongy. They do not store long.

(See June Garden-Plant but most details are in this section.)

Okra (malvaceae family; Abelmoschus esculentus) does not like being transplanted so sow in degradable pots not flats in greenhouse. (For **sowing details** see May Garden-Plant.)

Sow **marigold** (asteraceae family; Tagetas ercta and Tagetas patula) seeds in greenhouse in April. Marigold is an easy to grow **annual.**

Sow seed 1 inch apart and cover lightly with soil. Germinates in a few days. Grows 1-4 feet tall depending on variety. Blooms in about 8 weeks (56 days). Flower colors include orange, red, white, yellow and mixed. Flowers **bloom mid summer until frost.**

Care: Likes full sun or partial shade. Likes soil pH of 5.8-6.2. Likes somewhat rich, well drained soil but is tolerant of poor soils.

Transplant outside after last frost so that is **late May.** Plant small marigolds 4-6 inches apart, and large ones 1-2 feet apart.

Disease and Pests: Most insects avoid marigolds. Slugs eat the leaves.

Uses: An insect repellant is made from it. Plant to deter nematodes (microscopic worms in the soil) since marigolds release a chemical toxic to them.

The flower petals are edible for people and animals. Put in a salad. Or use in cooking as a substitute for saffron. If feed flowers to chickens, it makes their yolks more yellow.

During the Civil War and First World War, it was used to treat wounds. The flowers were made into a poultice or an infused oil for application on the wound. Consult your herbalist, doctor and/or health care specialist.

(See May Garden-Plant but details are in this section. For **using it as an insecticide**, see How to Make Organic Insect or Pest Sprays in "Plant Health" in back of this manual. For **how to make a tincture, decoction or infusion,** see April Garden-Harvest.)

Sow **melons and watermelons** in April (cucurbitaceae family) in greenhouse.

Place seeds in degradable pots since they are sensitive to transplanting. Start **2-4 weeks before last frost**. Plants are killed by frost. (See "Plant Families" in back of manual.)

Plant 2-3 seeds per pot in soil 1/2 inch deep. Melon seeds germinate in about a week. Watermelon germinates more slowly but do better if the seeds are nicked or soaked.

Thin to 2 plants once they are well up. Melon matures in 75-120 days but usually 74-90 days. Watermelon matures in up to 150 days.

Care: Prefers full sun and well-drained soil. Needs good air circulation. Watermelon likes soil pH of 5-7. Melon likes soil pH of 7-8. The roots are shallow and do not like compacted soil. They need a regular supply of water. Likes rich soil. Do not use a lot of nitrogen once flowering starts.

Transplant outside in late May or early June after last frost when soil is at least 60 degrees with 70 degrees being better when peonies bloom. Plant 12 inches apart in rows 4-6 feet apart. Or plant in hills 6-8 feet apart with 2 or 3 plants per hill. The vines take up a lot of space. Can be grown on a trellis 8 feet tall and 20 feet wide depending on variety.

Pollination:
There are male and female flowers on the same plant. Female flowers have a small swelling at the base. Male flowers open first. Insects need to **pollinate** them.

For plants in most families, cross pollination usually occurs only among members in the same species. However, **some cross pollination between species occurs in the genus Cucurbita** among pumpkins, squash and gourds. Usually even among Cucurbita the species will not cross but sometimes they do. However, **cross pollination does not occur between melons, cucumbers, squash or other species.**

(For more about **Cucurbita pollination**, see winter squash in May Garden-Plant. For **seed saving**, see "Plant Families" and "How to Save Seed" in back of this manual.)

Once fruit has formed, remove all defective fruit and leave 2-3 fruits per plant. Remove any blossoms that start to develop within 50 days of the first frost date since they will not have enough time to mature.

If you do not want a blemished spot on the melon from where it sits on the ground, you can rotate it occasionally or prop it up on a can, flower pot or similar object.

Harvest: They ripen **late summer to early/mid fall.** Melons range in size from a softball to 15 pounds. Ripe seeds are also edible. The more fruits that ripen at the same time, the less sweet they are because the vine has to divide the leaves' sugar production between fruits. You can remove a few melons very early if you want to.

Melons by Genus (They are a very variable species.):
- **Benincasa**
  - **Winter melon** (Benincasa hispida) is only member of genus Benincasa. Used in Asia & India.
- **Citrullus**
  - **Egusi** (Citrullus lanatus) looks like watermelon. The seeds are eaten instead of the flesh.
  - **Watermelon** (Citrullus lanatus) is a popular summer fruit. There are 120 varieties.
- **Cucumis**
  - **Horned melon** (Cucumis metuliferus) has distinctive spikes.
  - **Muskmelon** (Cucumis melo)
    - **C. melo cantalupensis** has warty skin, not netted. Removes easily from stem when ripe.
      - **European cantaloupe** is lightly ribbed with pale green skin. Varieties include French Charentais & Netted Gem. Yubari King is a Japanese cantaloupe.
      - **Persian melon** looks like cantaloupe with darker green rind and finer netting.
    - **C. melo inodorus (winter melon)**- late maturing. Not easily pulled from stem when ripe.
      - **Canary melon**-a large, bright-yellow melon with pale green to white inner flesh.
      - **Casaba**-bright yellow with smooth, furrowed skin. Less flavorful but keeps long.
      - **Hami melon** is from China. Flesh is sweet and crisp.
      - **Honeydew** has sweet, juicy, green-colored flesh.
      - **Kolkhoznitsa melon** has smooth, yellow skin and dense, white flesh.
      - **Navajo Yellow** is round or oval with ribbed, yellow skin, and orange flesh.
      - **Piel de Sapo** (toad skin) or Santa Claus has green skin and white sweet flesh.
      - **Sugar melon** has a smooth, white, round fruit.
      - **Tiger melon** is an orange, yellow and black striped melon with a soft pulp.
      - **Japanese melons** including the Sprite melon.
    - **C. melo reticulatus (true muskmelon)** has netted skin. Easily pull from stem when ripe.
      - **North American cantaloupe** distinct from European, has netted skin.
      - **Galia** is small and very juicy with either faint green or rosy pink flesh.
      - **Sharlyn melons** has netted skin, greenish-orange rind, and white flesh.
  - **Modern crossbred varieties** include **Crenshaw** (Casaba × Persian).

Melon Weight: American cantaloupe (muskmelon) are netted and weigh 2-7 pounds each. European cantaloupe (true cantaloupe) are not netted. They weigh 4-9 pounds. Casaba weigh about 4-8 pounds. Crenshaw weigh around 6 pounds. Honeydew weigh 4-8 pounds. Watermelon weighs from 1-200 pounds.

Diseases: (See "Plant Health" in the back of this manual.)
1. **Mosaic Virus** creates mottled, distorted leaves that are brittle. Yields are reduced. It is spread by

aphids and leafhoppers. Remove diseased plants and broadleaf weeds that serve as hosts.

2. Powdery Mildew is white powdery spots and coating on leaves caused by fungus. Spores germinate on dry leaf surfaces when the humidity is high. Common late summer or fall but does not kill plants. Remove infected leaves. (See apple tree diseases above in Garden-Plant.)

Pests: Can use row covers to keep out pests except remove so pollination can take place.

1. Aphids make leaves curl, become deformed and yellowish. Aphids are tiny, oval, and yellow-green insects that live on the undersides of leaves. They leave behind a sticky substance called honeydew. Use insecticidal soap. (For more about aphids, see "Brassicaceae Family Disease and Pests" in the back of this manual.)

2. Leafhoppers create white speckling on leaves and may turn leaves brown. Leafhoppers are green, brown, or yellow with wedge-shaped wings. Use floating row covers to exclude and spray with insecticidal soap. (See blueberry pests above in Garden-Plant. See Solanaceae Family in "Plant Families" in back of manual.)

3. Leafminer larvae make trails and tunnels in leaves. Remove infected leaves and cultivate the garden to destroy larvae and keep adult flies from laying eggs. Use floating row covers. (See Solanaceae Family in "Plant Families" in back of manual.)

4. Spotted Cucumber Beetle chews holes in leaves and scars fruit. The beetle is green-yellow, 1/4 inch long with black spots and black head. Pick off by hand. Mulch around plants. Plant resistant varieties. Dust with wood ashes. (See Cucurbitaceae Family in "Plant Families" in back of manual.)

(See May Garden-Plant but details are in this section. See August Garden-Harvest.)

Read "Melons for the Passionate Grower" by Amy Goldman.)

 Sow **Tarragon (Russian)** (asteraceae family; anthemideae tribe, Artemisia dracunculus. Also known as Dragons Wort.) in greenhouse.

Both tarragons are a **perennial related to wormwood**. They are a shrub that grows 1-2 feet wide and tall.

Care: Likes full sun or partial shade. Likes soil pH 6.6-7.5. They do not like acid soil. Likes well drained soil.

French tarragon (Artemisia dracunculus sativa, true tarragon) is weed-like in its appearance, with long, narrow, smooth-edged leaves which look like spears. Hardy to zone 5. Others say it can not take a hard freeze so keep in greenhouse or other protected area in winter. Seeds do not breed true. Can only be propagated by plant cuttings.

Russian tarragon (Artemisia dracunculus inodora, false tarragon) seed is covered with 1/2 inch of soil. Plant outside when 4 inches tall and when frost is over (**in May**). Transplant 1 1/2 to 2 feet apart. Has narrower leaves. It produces flowers but French tarragon rarely does. Bees and beneficial insects like the flowers. Russian tarragon is much hardier and vigorous. It prefers poor, rocky soils and tolerates drought.

Both tarragons are easily **propagated** by dividing mature plant every 3-4 years. This improves the health of the mother plant. Can propagate both by rooting small, vigorous cuttings in the spring. Russian tarragon readily self seeds.

Harvest: Russian tarragon produces many leaves starting early spring that are mild and good in salads and cooking. Young stems in early spring are cooked as an asparagus substitute.

To **harvest French tarragon** cut back to 2 inches tall in **July or early August.**

Uses: Drying both tarragons significantly reduces their flavor so preserve in vinegar or freeze.

French tarragon is preferred in the kitchen over the weaker flavored Russian tarragon. They both have a sweet anise flavor with Russian being a little bitter. (See May Garden-Plant but details are in this section.)

 Farm Animals (April)

Give **high magnesium minerals** (Sweetlix or other brand) to **goats and other ruminants April through June.** High magnesium makes it less likely that the animals will get **grass tetany or bloat**. This can happen when ruminants graze young, fast-growing grass and weeds during the spring. Do not feed copper to sheep. Consult your veterinarian or County Extension office.

Bloat is when too much gas from fermentation forms in their stomachs, particularly on the left side. It interferes with breathing. They can die. Make sure they have sodium bicarbonate (baking soda) available at all times. (For where to buy 50 lb bags **buffered sodium bicarbonate,** see fertilizers in "Garden Tips". See **hay** in June Garden-Harvest.)

Grass tetany is a serious, often deadly metabolic disorder of low levels of magnesium in the blood of ruminants. Water logged soils can have plants low in magnesium. Add dolomite lime (21% calcium and 12% magnesium) and rock dusts to the pasture. (For more about **fertilizers**, see "Garden Tips" in back of this manual.)

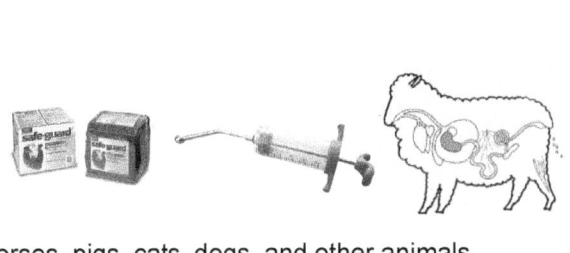 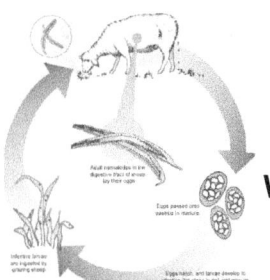 **Worm** cattle, goats, sheep, donkeys, horses, pigs, cats, dogs, and other animals.

Appalachian Folklore: **Worm animals for 2 days before the full moon, during the full moon, and 2 days after the full moon.** This is a total of 5 days. Around the full moon worms detach from the walls of the infected animal and then breed. They are more vulnerable at this time. Consult your veterinarian and/or **County Extension office**.

Worm female animals 1 month before giving birth.

Pastures, Worms, and Rotational Grazing

Worms in a pasture are worse in warm, humid, wet weather. Dry or cold weather decreases the number of worms. Larvae do not travel more than 12 inches from a manure pile.

In the last image above adult worms lay eggs in the digestive track of animals. (Eggs stay in the stomach at most for 1 year.) The eggs are dropped onto pasture in feces. The eggs hatch and larvae develop. (Larvae need moisture to crawl up plants.) Infected larvae crawl up plants about 1-3 inches and are eaten by animals, thereby infecting them. **You can break this cycle by not letting animals graze down pasture more than 4 inches.** (Animals can also be infected by eating feed that has manure in it.)

Do not put too many animals on a pasture. Around 4-5 goats or sheep per acre is best. For small breeds of cattle, 1 cow needs 1 or more acres. For larger breeds of cattle, 1 cow needs 2 or more acres. One horse needs about 1 acre. A donkey needs 1/2 to 1 acre. These are very rough estimates. It depends upon the quality of the pasture, the amount of rain, and other conditions. All need hay in the winter.

Grazing pasture with multiple species is good. It helps to reduce worm loads. For instance, graze ruminants with poultry at the same time or different times. The poultry break up the feces piles so it is harder for larvae to grow. They eat larvae too. Or graze cattle, and then remove them and put in goats or sheep. Cattle only share a few parasites with small ruminants. Or alternate with swine.

It is best if **pasture is left empty 2-3 months** so larvae die without infecting animals. At the very least a pasture should be left empty for 2 weeks on a regular basis. This is done successfully with rotational pasture management.

Rotation or strip grazing is called **Management Intensive Rotational Grazing** (MIRG). Animals are regularly and systematically moved to fresh pasture. This improves the quality and quantity of forage growth.

A herd grazes one part of a pasture or paddock, while other pastures are left empty. The length of time it is grazed depends on the size of the herd and the size of the pasture. **At least 8 different pastures is best, with around 1 week per pasture.**

The ungrazed land grows and rebuilds itself creating plants with deeper root systems which creates more biomass production in the long run. They withstand drought better.

If you have pasture with a wide variety of **grass, legumes and herbs,** it is more likely the animals will get the nutrients they need so will need less worming. The animals will be more resilient to worms. The more fertile the pasture, the higher quality the forage and the fewer unwanted weeds. **Tanin-rich forages** such as sericea lespedeza, plantain and chicory reduce worm egg counts.

Read "Greener Pasture on Your Side of the Fence: Better Farming Voisin Management-Intensive Grazing" by Bill Murphy. See **pasture** in "Garden Tips" in back of manual.)

Fecal (Feces) Testing

It is better to do **fecal testing** before giving a wormer so you know if and what treatment is needed. Then 10-14 days after you worm, do another fecal test. If the worm egg count is down 90% or more, then the wormer was effective.

Sometimes the person looking at the eggs under a microscope mis-identifies the eggs. So if one type of wormer does not work, you may want to try another wormer that kills different types of worms.

Your veterinarian can do this testing for you, or you can do it yourself. It is not hard. There are 2 types of tests: McMasters and Fecal Flotation. You need a microscope with 4X, 10X and 40X power with a light. You need slides, test tubes, and flotation solution. Contact your veterinarian and/or County Extension office.

FAMACHA Worm Load Testing in Animals

FAMACHA is a diagnostic tool to help goat and sheep owners determine the amount of parasite infection in their animals. **It uses a color chart and compares it to the inner eyelid color.** The lighter pink the color, the more anemia an animal has. If an animal has a healthy color, it does not need to be wormed. Animals with a very light pink color need to be wormed. If white, then the animal is near death. This is most effective with determining the infestation by Barber Pole worms.

It is possible that animals may only need to be wormed twice a year. The goal is to have healthy animals, not worm-free animals since totally worm-free animals may be impossible. Certain animals have a genetic predisposition to being wormy. For instance, within a herd or flock 80% of the worms may be in 20% of the animals.

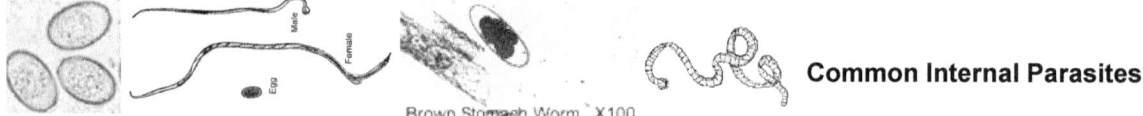
Brown Stomach Worm X100

Common Internal Parasites

Try to breed animals that are resistant to parasites. Worm and quarantine for 2 weeks new animals bought from other owners.

There are **3 classes or phylum of parasitic worms:** Cestoda (tapeworms), Nematoda (roundworms), and Trematoda (flukes). Some worms suck the blood of the animal and cause anemia. Others live off the nutrients eaten by the animal and cause weight loss. Young or pregnant animals have lowered immunity so are more prone to problems with parasites.

Symptoms of parasite problems are rough coat/fur, diarrhea, poor condition, pale mucous, and death.

1. Coccidia is a protozoa, not a worm. (See first image above.) It infects the intestines. It is spread by infected feces or other tissue. The main symptom is diarrhea. It is more common if animals (goats, sheep, horses, cattle, poultry, dogs, cats) are confined to a pen that is moist and warm. Overcrowded conditions make it worse. Young animals may die from it due to dehydration. For goats and sheep keep no more than 3-7 animals per acre. Sulfadimethoxine (Albon) can be used to treat sick animals. Contact your veterinarian and/or County Extension office.

2. Barber Pole Worm or Red Stomach Worm (Nematoda phylum, Haemonchus contortus) is the number 1 cause of death in goats and the number 1 disease in sheep. (See second image above.) It also infects cattle and other ruminants. It attaches to the true stomach and sucks blood. It is about 1 inch long.

It has a 21 day life cycle. The worms deposit eggs in the stomach, that later come out in the feces. Larvae crawl up vegetation up to about 2 to 3 inches high. The animal eats the vegetation with the larvae and is infected. So a significant way to prevent infection is to keep stocking rates low enough so that **pasture is not grazed below 3 inches**. To break the cycle **every pasture should spend 2 weeks with no animals on it.**

Barber Pole Worms are becoming resistant to wormers. Only worm animals that need it so you need to test feces for eggs. In the winter there are the fewest eggs. In spring and mid-summer it is the worst. Safeguard and other wormers kill Barber Pole adult worms.

Barber Pole worm eggs are very similar to **Liver Fluke** (Trematoda phylum, Fasciola hepatica) eggs. If a wormer designed to kill Barber Pole worms is not working, try one that kills Liver Flukes such as Clorsulon which is in Ivomec Plus, Alverin Plus, and Noromectin Plus.

Give wormer and leave animals on the same pasture for 2 weeks. Then move them to a new pasture. It takes about 14 days for larvae to grow. Worm again 10 days after the first worming Then worm again 10 days after the second worming.

Bottle Jaw is a symptom of severe Barber Pole Worm infection. There is fluid buildup at the chin.

3. Brown Stomach Worm (Nematoda phylum, Ostertagia species) is found in sheep, goats, cattle, and other ruminants. (See third image above.) It is about 1/2 inch long and red/brown. Symptoms are lethargy, weight loss, appetite loss, poor coat condition, and diarrhea. It is worst in fall and winter. Safeguard and other wormers kill it. Contact your veterinarian and/or County Extension office.

4. Tapeworms are found in goats, sheep, cattle, swine, poultry, horses, donkeys, cats, dogs and other animals. (See fourth image above.) They are white, flat and segmented. They can be over 60 feet long depending on the species. They need an intermediate host such as fleas or pasture mites to complete their life cycle. Tapeworms are killed only by drugs in the Benzimidazoles class (see number 4 below).

Common Internal Parasites of Poultry: Keep litter as dry as possible to reduce parasites.

Hair Worms are found in the crop, esophagus, proventriculus, and intestine.

Large Roundworms (Ascaridia galli) do more damage than other worms. They are found in digestive system.

Small Roundworms (Capillaria sp.) are found in the digestive system.

Gizzard Worm are found in gizzard, mainly in geese.
Tapeworms are found in the intestine.
Gape Worms are found in the trachea and lungs.
Cecal Worms (Heterakis gallinarum) do little damage but transmit Blackhead (Histomonas melegridis) to turkeys.
Poultry Chemical Wormers: Use once every 3-6 months if needed. Repeat 14 days after first treatment. See below for natural wormers that can be used every week.

Flubenvet can be used for chickens, geese and turkey. It is put in the food.

Solubenol is put in the water. It is used to kill roundworms, caecal worms and capillaria worms.

Ivermectin kills most internal worms (not tapeworm), external mites, and some lice such as scaly leg mite and northern fowl mite. Ivermectin Pour-on can be applied to the skin.

Panacur is used on cats, dogs and poultry. Contact your veterinarian and/or County Extension office.

Wormers for Ruminants and Other Animals:

1. Diatomaceous Earth (DE, diatoms) kills good and bad insects. It is an abrasive dust made from fossilized silica shells of algae. Add food grade (not pool grade) to animal feed to control internal parasites. Use 5% diatoms in feed once every 3 months or when you buy a new animal. There is controversy about whether this works. (For **where to buy**, see fertilizers in "Garden Tips".)

2. Natural Wormer Recipe for Goats by Pat Coleby from "Natural Goat Care"
1 teaspoon dolomite lime, 1/2 teaspoon copper sulfate, 1 teaspoon vitamin C (ascorbic acid preferred). Can be given to pregnant goats too. This formula may not be right for all goats. Consult your veterinarian.

Make sure your goat minerals contain copper. **Copper reduces or eliminates worms. Do not feed copper to sheep.** The book "Soil Fertility and Animal Health" states that goats getting enough copper do not have worms.

Make this **goat stock lick** for every day use and the above wormer may not be needed:
6 pounds dolomite lime, 1 pound yellow dusting sulphur, 1 pound copper sulfate, 1 pound seaweed (kelp). It must be kept dry. Otherwise, the dolomite deactivates the copper sulfate. For some soils 1/16 pound of cobalt sulphate may be needed but it can be toxic. It is safer to use seaweed (kelp) as a cobalt source. Get your soil tested.

3. Herbal Wormers: Herbal wormers kill and prevent parasites so can (should) be used all year.
Average Relative Dosages of Herbs:
Cat- 1/4 teaspoon, Chicken- 1/2 teaspoon, Cow- 3-4 tablespoons, Dog- 1/2 to 1 teaspoon (depending on size), Goat- 1 tablespoon, Horse- 2-3 tablespoons, Pig- 2-3 teaspoons, Rabbit- 1/4 teaspoon, Sheep- 2 teaspoons. Do not give black walnut to horses.

Dosages below are for one goat of weight 150-200 pounds. Most can be used once a week except wormwood. Consult your veterinarian.

Birch- Strong brew made with one handful of leaves to one cup of water.
Chives- a handful daily in feed.
Fennel- 2 handfuls of whole herb fed raw twice daily.
Garlic- 2 bulbs or whole plants twice daily though can impart taste to milk.
Honeysuckle- handful of leaves or flowers chopped and mixed in feed daily.
Hops- 5 handfuls of flowers once daily.
Horseradish- 1-2 roots grated into feed twice daily.
Hyssop- 2 handfuls of leaves given twice daily in feed.
Mulberry- several handfuls of fruit twice daily.
Nasturtium- 1 regular spoon of the seeds.
Nettle- seeds mixed into food.
Mustard seeds- 2 handfuls of the whole herb or the seeds fed raw twice daily.
Rue- 1/2 handful chopped small given in feed.
Tansy- 1 handful herb brewed into two pints of water plus 2 tablespoons honey.
Thyme- 1 handful brewed, finely cut and mixed in food morning and night.
Valerian- 4 roots finely sliced in 1-quart water, giving 1-pint morning and night.
Walnut- 2 handfuls leaves brewed in 2 pints water add honey. Do not give to horses.
Wormwood can be used safely on animals occasionally.

Brew 1 handful herb in 1 1/2 pints water + 1 tablespoon honey. Give 1 cap twice daily. Use 3 days in a row every 6-8 weeks. Do not use weekly since it is hard on the animal's kidneys and liver. Do not use when pregnant. (For growing wormwood, see May Garden-Plant.)

4. Chemical wormers are divided into 3 types or classes:
a. Benzimidazoles such as Fenbendazole sold under the brand names Panacur and SafeGuard. Others include Albendazole (Valbzen), Oxybendazole, and Thiabendazole. They are used against gastrointestinal parasites including roundworms (nematodes), hookworms, whipworms, tapeworms, pinworms, aelurostrongylus, paragonimiasis, strongyles and strongyloides. It can be given to sheep, cattle, horses, fish, dogs, cats, and rabbits. Restrict food 24 hours before giving wormer. Dose once and again 12-24 hours later. Many worms are resistant to these wormers.

b. Imidazothiazoles such as Levamisole (Tramisol), Pyrantel (Strongid), Tetramisole, and Morantel (Rumatel). It is used against gastrointestinal roundworms, lungworms, Ostertagia (brown stomach worm), and some eyeworms. They do not kill tapeworms (cestodes) or flukes (trematodes). Not many worms are resistant to these wormers.

c. Macrocytic Lactones such as Ivermectin, Doramectin, Quest, Cydectin, and Moxidectin. It is used against gastrointestinal roundworms, lungworms, eyeworms, grubs, sucking lice, mange mites, gastrointestinal/pulmonary nematodes, and ticks. It is used against: Haemonchus spp., Ostertagia spp., Trichostrongylus spp., Cooperia spp., Oesophagostomum spp., Dictyocaulus viviparus, Dermatobia hominis, Boophilus microplus, Psoroptes bovis, and other internal and external parasites. Many worms are resistant to Ivermectin.

"Ivermectin 1% Injectable" is one type. Use it orally (not injected) at about 1 ml per 25-50 pounds for goats. Cattle dosage is 1 ml per 110 pounds. Swine dosage is 1 ml per 75 pounds. Worms in goats have developed a lot of resistance to Ivermectin and related wormers.

Chemical worming is usually done **once every 3-4 months (3-4 times a year)** unless fecal or other testing indicates more or less is needed. Be sure to give enough wormer (do not underdose) so most worms are killed. Do not worm animals who have a very low worm count.

When giving chemical wormers to ruminants such as goats, sheep and cattle, use a **drenching gun** (see second image at beginning of this worming section). It is important to place the tip of the drenching gun at the back and to the left of the animal's tongue so that the medicine goes into the correct stomach.

When you use a chemical wormer, you are **poisoning the parasites** to kill them. You want to use enough poison to kill the worms but not the animal. However, a weak or ill animal may not be strong enough for a regular dose. Some wormers are not safe to give to pregnant animals. In healthy animals a dose 20% over the recommended amount is usually OK.

No new chemical wormers are in the process of being developed. **Parasites are developing resistance to all of the current wormers. If parasites have developed resistance to one brand in a particular family such as Macrocytic Lactones, then the parasites are resistant to all other brands in that family.**

In the past, people were told to **rotate wormers** to reduce the chance that parasites would become resistant to one type of medicine. But recently some experts are saying that using a different type/class of wormer each time is not that important. You may want to use one type/class of wormer for 6 months or a year, and then another type/class the next 6 months or year. There are still disagreements about which method is best.

It is possible to have healthy animals without using chemical wormers by using natural wormers combined with proper pasture rotation. Consult your veterinarian and/or County Extension office.

(For **breeding**, see September Farm Animals. For **breeds, care and lice/mites/mange**, see February Farm Animals.)

Read the books "Natural Goat Care" and "Natural Sheep Care" by Pat Coleby. Read "Soil Fertility & Animal Health: The Albrecht Papers, Vol II" by William A. Albrecht.)

Homing pigeons start hatching babies. Put nest bowls in loft. They have babies through August. (For **basic care**, see August Farm Animals.)

Peak **chicken, duck and turkey egg** production April through May.

In April and May have mail order **chicks, ducklings, and turkey poults** delivered.

Pullets hatched very early in spring or late winter usually start laying in 6 months. Pullets that hatched in late summer usually start laying in 9 months. The first eggs of pullets are small and then get larger over one month or so.

Each year put different colored **plastic leg bands** on chickens, ducks and turkeys to tell which year born. Kill chickens that are 3-4 years old since by that time egg production is way down.

Continue with **incubating eggs.** (For **doing your own incubating and brooding**, see March Farm Animals.)

 Set up **nesting area for broody hens**. Most commercial (not heirloom) hens have been bred to not be broody because it interferes with egg production. A broody hen is less productive because she does not lay eggs while incubating eggs and raising chicks. However, on the small farm **broody hens are very valuable**. It is much easier to let the hen incubate and brood chicks than it is for you to do it. ("Don't count your chickens before they hatch!")

Bantam hens can successfully hatch about 10 of her own eggs but only about 5 eggs from standard chickens.

Standard size hens can successfully hatch about 12-15 of her own eggs. Though 12 is a safer number.

Bantam chickens tend to be much more broody than standard size chickens. Among the bantams (banties) Old English Game, Cochin Bantams, Cornish Bantams, Kraienkoppes, Malays, Shamos, Asils, Madagascar Games, and Silkies are especially broody.

Standard size chickens most likely to go broody are Australorps, Cochins, Buff Orpingtons, Light Brahmas, Dark Cornish, Dominique and Buff Rocks. Breeds such as Leghorns, Minorcas, Rhode Island Red, and Barred Plymouth Rock almost never go broody.

If you want a hen to hatch eggs, set up an area with a nest that is somewhat secluded and quiet. She will start incubating once enough eggs have been laid. She will incubate her own eggs or other hens' eggs. If you need to move her or switch/add eggs, do it **at night**. (For instance, if you want to put some standard size eggs under a bantam hen.) She leaves the nest only once a day.

Once eggs hatch, she sits on them 24-48 hours. She then brings the chicks with her around the chicken yard. She is very protective. Put down chick waterers and feeders. In most cases they can stay with the main flock of birds.

Breaking Up a Broody Hen: If you do not want a hen to be broody, then take her away from her nest for a few days. You may have to put her in a wire floor cage. Give her food and water. Or put her in a pen with a cock and he may keep her from nesting.

(For **chicken breeds and care,** see February Farm Animals. For **incubating eggs**, see March Farm Animals.)

  **Honey Beekeeping in April**

The beehive is almost fully functioning. Early blossoms begin to appear. The bees **bring more nectar and pollen** into the hive. The queen is busily **laying eggs**, and the population is growing fast. The drones (male honey bees) begin to appear even more.

On a warm, still day do your first comprehensive **inspection**. Find the queen and make sure she is healthy. Make sure there are enough eggs and brood. Later in the month, on a mild and windless day, reverse the hive deeps (brood chambers). This allows for better distribution of the brood.

It may be necessary to **combine 2 hives** together if they did poorly over the winter.

Feed the hive medicated syrup. Also add menthol as mite control.

About 30 years ago most beekeepers used a standard hive configuration of **deep boxes for brood chambers** and **smaller boxes for heavy honey frames (supers).** The lower boxes were called brood boxes and everything above that was a super. Today many beekeepers use small boxes throughout the hive so may call all boxes supers.

If the bees have stopped eating the stored honey or sugar, you can **add supers for honey collection.**

(For **beekeeping overview,** see January Farm Animals. For **honey harvest,** see September Farm Animals.)

April p. 46

 # MAY
May 1 daylight is 13 hours 37 minutes. May 31 daylight is 14 hours 21 minutes.

Rain. Last frost usually early May or late April.

Early May: Comfrey, chives and horseradish bloom. Red and white clover blooms.
Poplar tree leaves are half of full size. Sourwood tree leaves are about size of a quarter.

Mid May: First strawberries from Everbearing plants. Blackberries and blueberries bloom.
Pokeweed is 1 foot tall. Walnut trees and cherry bushes bloom.

Late May: Sage, Mountain Laurel, Hardy Kiwi and Rosa Rugosa bloom. Oak leaves are full size.

 **Last Killing Frost in Spring: March 15 - April 15**

Last frost according to North Carolina Cooperative Extension Service: First line across N. Carolina map (on left)= April 15; Second line= April 1; Third line (far right)= March 15. Last frost is early May in mountains around 3000 feet.

 Garden- Maintenance

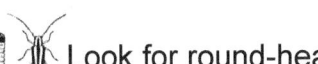 Look for round-headed **apple tree borers** in apple and other fruit or nut tree trunks.

Check again in **September**. Adult borer beetles (5/8-1 inch long) emerge in **May or June**. But it is their larvae that do the damage. The larvae feed under the bark as shown by patches of reddish-brown sawdust and sunken, dark bark that may ooze sap. Borer-infested trees grow slowly and have sparse foliage.

At the end of **June insecticide sprays** used to control plum curculio and codling moth are somewhat effective.

Remove infested crabapple, hawthorn, mountain ash, and shadbush (serviceberry) within at least 100 feet of apple and other fruit/nut trees. Remove vegetation and debris around the tree. Apply a 50:50 mixture of **white latex (not acrylic) paint** and water to the lower trunk. This deters egg laying. (For **whitewashing trunks**, see June Garden-Maintenance.)

Barrier: You can try to keep out beetles with mosquito screen or 1/4 inch mesh hardware cloth surrounding the lower two feet of the trunk. The barrier can be loose around the trunk but sealed at the top with a rope and at ground level by mounded soil.

If you **find beetles**, try to dig out shallow larvae by removing decayed tissue with a sharp knife. You may be able to kill the larva with a stiff wire. (For fruit tree planting, see April Garden-Plant. For tree borers, see September Garden-Maintenance but details are here. See "Plant Health" section in back of this manual.)

   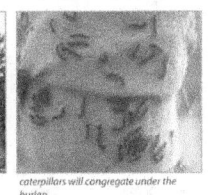 Put **burlap bag tree bands** on fruit and nut trees late **May through August.**

wrap the burlap around the stem — tie a rope in the middle — fold the burlap in half — caterpillars will congregate under the burlap

Trap and destroy caterpillars and other crawling insects by placing burlap cloth bands on fruit trees. Wrap burlap 12-18 inches wide around trunk at chest level. Tie it in the middle with a rope/string and fold it over so it hangs. Caterpillars feed at night and hide during the day in the burlap.

Check every day between mid-afternoon and 6 pm. Destroy caterpillars before they crawl back to the leaves to feed. Can put caterpillars in soapy water to drown. (See "Plant Health" section in back of this manual.)

 Cut **stinging nettle and other perennials** to encourage new growth that is tender to eat and to keep plants bushy. (See March Garden-Harvest, April Garden-Plant.)

 For **strawberries** set beer traps for slugs and snails.

You can also spread wood ash, sand, small gravel or diatomaceous earth around plants to discourage slugs and snails. They are worse in wet weather. Make sure there are no dead plants or leaves that give shelter from the sun.

(For **sowing and growing details**, see April Garden-Plant. See May/June/July Garden-Harvest with **harvest details in May**. See March Garden-Plant but details are in this section. For **maintaining beds**, see June Garden-Maintenance. For how to **make slug and snail traps**, see "Plant Health" section in back of this manual.)

 Clean driveway ditches, gutters, drain pipes, etc. Open all vents under house.

Clean and oil tools. Run outdoor pumps if they have not been run in a few months.

Garden- Plant or Transplant (May)
Early to Mid May:
Arnica, cauliflower, chamomile, chives, cotton, fenugreek, good king henry, mint (+ anise hyssop, catnip, lemon balm), oregano, potatoes, sage, saint johns wort, salad burnet, salsify (+ scorzonera), sorrel, Swiss chard, thyme, valerian and winter savory.

Late May or Early June:
Beetberry, calendula, celeriac, celery, chicory, cucumbers, feverfew, ground cherry, horehound, Jerusalem artichoke, leeks, marigold, marjoram, melon/watermelon, okra, pepper, purslane, rosemary, sea kale, sorghum, summer squash, soybean, sweet violets, tarragon, wormwood and yarrow.

Three Sisters:
Beans, corn, and winter squash / pumpkins / gourds.

 **Sow frost-sensitive, warm weather cover crops** after the spring frost-free date and at least **6-8 weeks before the fall first frost date**. It is best if soil has warmed to at least 60 degrees in the spring. This is usually **May** when corn is planted.

Warm weather cover crops include black eyed pea (cowpea), **buckwheat**, **garbanzo bean (chickpea)**, millet, sesame, **sorghum, soybean, and sudangrass** (Sorghum bicolor drummondii).

(For **beans, sorghum/sudangrass, and soybean** see May Garden-Plant. For **buckwheat** see June Garden-Plant. See **cover crops** in "Garden Tips" in back of manual.)

Early to Mid May:
 Sow or transplant in May **perennial herbs** such as chamomile (roman), chives, **good king henry, mint, oregano, parsley, sage, salad burnet, sorrel, thyme, valerian, and winter savory**. Plants planted previous year(s) usually harvested starting early spring. (See March Greenhouse for many.)

Plant tender annuals and perennials when new growth on green ash, maples, grapes and bur oaks.

Read the book "Rodale's Illustrated Encyclopedia of Herbs".

 Transplant **arnica** (asteraceae or compositeae family; Arnica montana or Arnica chamissonis) from **greenhouse to outside** in May after last frost. (See March Greenhouse.)

 Transplant **cauliflower** from greenhouse to outside. (For **early cauliflower and sow/grow details**, see April Greenhouse. For **late cauliflower,** see June Garden-Plant and September Garden-Harvest.)

 **Chamomile** has small daisy-like flowers with yellow centers and white petals. The cone in the center of the daisy is solid in Roman Chamomile and hollow in German Chamomile. Roman Chamomile has a tiny papery bract (leaflike or scalelike) between the florets (small flowers) that German Chamomile does not.

Roman chamomile (asteraceae family, Anthemis nobilis) is a low growing (about 3-6 inches tall) **perennial** hardy to minus 10 degrees. Roman chamomile germinates in 10 days at 70 degrees. Thin seedlings to 6-12 inches apart. Very aromatic. Likes cool summers. Flowers less than German chamomile. Divide and replant every 3-4 years.

German chamomile (asteraceae family, Matricaria recutita) is an extremely **hardy annual**. It needs light to germinate. Germinates in 7-14 days. Can plant seed in August for germination in spring. Or start in greenhouse in March. Or sow seed outside after last frost. Thin to 6 inches apart. Grows 1-2 foot tall. Likes pH 5.6-7.5. Poor to average soil.

Uses for both types: Harvest just as flowers bloom. Can use fresh or dried flowers. Consult your herbalist.

To make tea: Steep 1 cup fresh flowers in a pint of boiling water. Or use 2 tablespoons dried powdered flower heads to a cup of boiling water. Drink no more than 2-3 cups a day.

To use externally, as a poultice for skin problems or minor wounds, use 3-4 cups of fresh flowers to a pint of water or 6-8 tablespoons dried herb to a cup of water.

Use as **sedative tea**. Good to use with calming herbs valerian and catnip.

It is antibacterial, antiviral and antifungal that is good on minor wounds, skin conditions and as a gargle for sore throats and inflamed gums. It relaxes the smooth muscles, particularly the intestines, and is used to calm upset stomach. It relieves gas pain, cramping, and minor diarrhea.

Used in beauty products.

Chamomile tea is sprayed on propagation flats to prevent **damping off of seedlings**. (See "Garden Tips" in back of this manual.)

 **Chives** (amaryllidaceae family; Allium sativum or Allium tuberosum) are a **perennial in the onion family.** (See "Plant Families" in back of this manual.)

Seeds store only 1-2 years. Seeds germinate in 7-15 days. Grows slowly. Thin to 8-9 inches apart. Grows 12-18 inches tall. It takes about a year before they are big enough to harvest. Each year the clump grows bigger.

Care: Likes full sun but tolerates partial shade. Likes soil pH 6.1-7.8. Likes moist well drained soil.

Harvest: All edible including flowers. Stays green most of the year. Starts growing very early spring.

Can also be **propagated** by dividing clumps of bulbs. The best time to divide is in spring after the leaves have emerged and grown 3-4 inches tall. (See March Garden-Maintenance.)

 **Cotton** (Gossypium hirsutum, upland cotton, 90% of world production, white fiber) is a frost-sensitive perennial that is grown as an annual. Other varieties have fibers that are naturally brown, pink or green. Cotton is grown in Alabama, Arkansas, Arizona, California, Florida, Georgia, Kansas, Louisiana, Mississippi, Missouri, New Mexico, North Carolina, Oklahoma, South Carolina, Tennessee, Texas and Virginia.

Most cotton grown in North Carolina is grown in the eastern part of the state (Piedmont, Coastal Plains) because the frost-free period is longer. **In eastern North Carolina it is planted mid to late April. In western North Carolina it is planted early May. It is harvested September through November.**

Care: Cotton plants require a long frost-free period, lots of sunshine, and moderate rainfall (24-48 inches). It tolerates hot weather well. Soil fertility can be average. Likes soil pH of 5.8-6.5.

Sow 1 inch deep. Plant 3 seeds together with each set of 3 about 6 inches apart. Plant in rows 3 feet apart. It germinates in 7-10 days. It grows 3-7 feet tall. It is shrub-like.

In 45 days large yellow flowers appear. Later the flowers fall off and form bolls (rounded seed capsule). About 16-18 weeks after sowing, stop watering. Later the bolls split open showing white fluff.

Harvest: It is ready to harvest in 160-180 days. **Remove the bolls from the plants by hand.** Then use a cotton gin to remove the seeds from the fiber. You can spin cotton directly from a boll but there will be some seeds stuck to the fiber.

You can **make your own cotton gin** by attaching wire mesh to a frame with a backing board. (See second image above.) Then use a flicker to draw the cotton though the mesh, leaving seeds on the bottom. (A **flicker card** is used to prepare long fibers for handspinning. It looks like a dog brush, with wire bristles that separate and lift the fibers. See third image above.)

Uses: Cotton fiber is used to manufacture textiles (cloth or woven fabric). The fiber is spun. The cotton seed, which is extracted during the ginning process, is crushed to produce cooking oil and animal feed.

Sow **fenugreek** (leguminosae or fabaceae family; Trigonella foenum-graecum) **4-6 weeks before last frost in spring. Sow April or early May.** Sow 1/4 inch deep, 4-6 inches apart. Germinates in 7-10 days. Thin seedlings to 9 inches apart when they are 1-2 inches tall.

Grows 2 feet tall. Matures in 3-5 months (90-150 days). Hardy **annual legume**. Does not like being transplanted.

Care: Likes full sun. Likes well drained soil. Likes neutral to slightly acid fertile soil such as 5.3-8.2.

It is a **cool weather cover crop.** (See cover crops and grains in March Garden-Plant.)

Harvest: Pick seed pods when they turn yellow. Lay the pods in a sunny spot until completely dry. Then break open pods to get seeds. Smells like maple syrup.

Uses: The seeds are used in curry. Seeds can be sprouted and eaten. The seeds **increase milk production** in women and animals. The seeds are used to treat sore throats, inflammation, fungal infections, and stomach upsets. It improves circulation. It is an aphrodisiac. Young leaves are edible though slightly bitter. Do not use if pregnant. Consult your herbalist, doctor and/or health care specialist.

Good King Henry (chenopodiaceae family; Chenopodium bonus-henricus, goosefoot or lincolnshire spinach) is a **hardy perennial**.

Stratify seeds before sowing. (For **how to stratify** see "Garden Tips" in the back of this manual.) Sow seeds 1/4 inch deep. Needs light to germinate. Thin seedlings to 12 inches between plants. Grows to 2 feet tall. (If sowing in greenhouse, grow in biodegradable pot since it does not like being transplanted.)

Care: It takes a full year for the plant to develop. Prefers full sun but will tolerate partial shade. Likes soil pH of 6.5-7.0. Likes rich soil. After established, self seeds.

Harvest: One of the first greens in **early spring**, and one of the last in **fall**. Flower buds, leaves, shoots and stems are edible raw or cooked. Shoots and stems are better cooked. Grown mainly for the shoots like asparagus. Harvest shoots when 5 inches tall, then peel and boil. Chickens love it.

Mint is a **perennial** in the lamiaceae family and genus Mentha. Plants in the **lamiaceae family** include basil, fennel, lavender, rosemary, sage, and thyme. Plants in this family are rich with nectar, and repel pests and disease. Mints are a cooling herb. (See "Plant Families" in back of manual.) Harvest in the morning.

Propagation: Mint is best grown from a cutting from another mint plant or from root divisions in **spring or fall.** Seeds do not usually grow true to type. Plant 12-18 inches apart. Grows up to 2 feet tall depending on variety.

Most mints are invasive, some are very invasive. The safest way to plant them is in a pot.

Care: Likes slightly acid soil. Prefers partial shade but will grow anywhere. It is invasive so plant them in bottomless containers sunk in the soil. (Some mints are more invasive than others.) Keep mints cut back so they stay bushy. Harvest **spring through fall.**

Uses: Most mints can be eaten raw or cooked. Exceptions are listed below. Leaves are best harvested just as it starts to flower but can be harvested at any time. Dry to use all year. Mint contains vitamins A and C.

Varieties:

    1. Most varieties of mint are edible except pennyroyal (lamiaceae family; Mentha pulegium). It is used as an **insect repellant**. Rub leaves on your skin. Some people get a rash so start with a small area. Or make an oil (such as olive oil) infusion with the leaves and other insect repelling herbs. (For **making oil infusion,** see April Garden-Harvest.)

    2. Mints with underlying flavors include banana (Mentha arvensis), lavender (Mentha piperita), lemon (Mentha piperita citrata), and spearmint (Mentha spicata).

Use **spearmint leaves** in mint sauces, teas and sweets. Good for some stomach disorders, nausea, fevers, headaches, colic and gas. Good as poultice (soft, moist mass applied to body to relieve soreness) on bruises.

3. Menthol containing mints used for upset stomach include basil (Mentha aquatica spicata), chocolate (Mentha piperita), lavender (Mentha piperita lavendula), lime (Mentha piperita citrata), moroccan (Mentha spicata), orange bergamot (Mentha piperita citrata), and peppermint (Mentha piperita). (Basil mint is not the same as basil. Though basil is in the mint family.)

Peppermint flowers are lilac colored. The tea is good for muscle aches, gas, headaches, colds, nausea and indigestion. Used in toothpaste, mouthwash and soap.

4. Fuzzy mints used for cooking such as apple (Mentha suavelens), Egyptian (Mentha niliaca) and pineapple (Mentha suaveolens). Apple mint is very invasive.

5. Anise Hyssop (lamiaceae family; Agastache foeniculum, licorice mint, hummingbird mint) is a **biennial or short-lived perennial** (lives up to 4 years) in the mint family. Start indoors a few weeks before last frost or sow outside after last frost. Stratify seeds for 2 months so start in March. Seeds need light to germinate. Germinates in 5-30 days.

Transplant seedlings to 1 foot apart. Grows 2-5 foot tall, and 2 foot wide. Likes sandy soil. Likes full sun or partial shade. Will do well in shade.

Self seeds and spreads by runners. Bees and other insects love the flowers. Repels cabbage moth.

Anise Hyssop Uses: Flowers (pinkish-purple, lavender-blue or white) and leaves have a sweet licorice aroma and flavor. It is best to harvest leaves for drying before it blooms. Harvest flowers when 3/4 open.

The leaves and flowers make a good tea. Can mix with other herbs in tea. Use 1 teaspoon dried or 3 teaspoons fresh leaves/flowers to 1 cup boiling water.

The leaves and flowers are good in salads, baked goods, and cooking with fish or chicken. Used by native Americans to relieve depression. Leaves are good for fevers, colds, coughs, to induce sweating, and to strengthen a weak heart. Consult your herbalist, doctor and/or health care specialist.

(For **when to stratify anise hyssop,** see March Garden-Plant. **For information about stratifying,** see "Garden Tips" in back of this manual.)

6. Catnip / Catmint (lamiaceae family; Nepeta catariais) is in the mint family. Thin seedlings to 20 inches apart. Grows to 2-3 feet tall. Plant in full sun or partial shade. Cut back regularly for bushy growth unless want seeds to form. It is a **strong smelling perennial that cats and bees love.** Has clusters of white or purplish-blue flowers in mid summer. Make tea out of leaves for colds and for calming.

7. Lemon balm (lamiaceae family; Melissa officinalis) is in the mint family. **It is not invasive.** Plant seedlings 2 feet apart. It is aromatic and sweet, growing to 2 feet tall. Grows rapidly. Likes full sun or partial shade. A tea made with its leaves **calms the nerves** and eases digestive problems. Native Americans rubbed lemon balm leaves on their skin as an insect repellant. (Pennyroyal above is also an insect repellant.)

(For **dividing mature plants,** see April Garden-Maintenance.)

 **Oregano** (lamiaceae family; Origanum vulgareis, pot marjoram) is a **perennial** sown in May outside. Sprinkle oregano seeds on top of soil. Germinates in 5-7 days. Thin seedlings to 10-12 inches apart. Grows to 2 feet tall and 20 inches wide. Grows wider every year.

Care: Likes full sun. Likes soil pH of 6-9. Likes moderately rich soil that is well drained.

Pinch off flower buds unless saving seed. When gets fairly tall, cut back to 6 inches tall to encourage new growth.

Harvest: Best flavor is when flowers first appear.

Uses: It aids digestion. Good for respiratory diseases and sore throat. It is antibacterial, antiviral and antifungal. (See April Garden-Harvest, July Garden-Maintenance.)

Can continue to plant **potatoes** (white, red or blue) (solanaceae family). **Potatoes are planted April through June, and harvested June through October.** (For when to plant early/mid/late season potatoes and planting details, see April Garden-Plant .)

In **May or April** plant mid season and late/storage potatoes for harvest in August through October.

(For **sprouting to plant in April,** see March Greenhouse.)

(**Potato Harvest:** For early/short season potatoes, see June/July Garden-Harvest. For mid-season potatoes, see August Garden-Harvest. For late/storage potatoes, see September/October Garden-Harvest. **October Garden-Harvest has most of the details about harvesting.**)

 **Sage** (lamiaceae family; Salvia officinalis) is a hardy, **evergreen, short-lived perennial** that grows to 2 feet or more tall. Lives about 5 years. There are **over 750 varieties** such as Common Sage, Berggarten Sage, Golden Sage, Golden Pineapple Sage, Pineapple Sage, Purple Sage, and Tricolor Sage.

Seeds germinate in 21 days. Thin to 20 inches apart. They are slow growing when young. Grows 18-24 inches tall and 24-36 inches wide.

Care: Likes full sun. Likes soil pH 6-7. Likes well drained, slightly dry soil. Likes bonemeal (phosphorus). Easy to grow. **Most sage is hardy to 0 degrees** but white sage is hardy to only 20 degrees so may want to cover it in the winter.

Trim to about half its size or about 12 inches tall **after the flowers fall off.**

Propagation: After 3-4 years it loses some of its vitality, so take cuttings and start new plants. Cut stem 6 inches long on growth from this or previous year. Best done in spring. Remove lower leaves. Can dip in rooting hormone. Put stem in potting mix about half way in. Keep moist. Should be well established by next spring. (For how to make rooting hormone, see "Garden Tips" in back of manual.)

Can propagate by dividing plant but the above method is preferred.

Most **harvesting** is done in the second year and after. Best harvested before it flowers for maximum taste. But can harvest at any time.

Uses: High in antioxidants, flavonoids and other beneficial compounds. It aids digestion. Gargle for sore throat. It is a warming herb. Good for night sweats. It is antibacterial. Consult herbalist, doctor and/or health care specialist.

 **Saint John's Wort** (Hypericaceae family; Hypericum perforatum) is a **hardy perennial** that grows 2-5 feet tall. Has over **300 varieties.**

Sow seed in spring. Soak seeds overnight. Can stratify them. Germinates in up to 3 months. When 2 inches tall transplant seedlings to 1 foot apart.

Care: Likes full sun or partial shade. Likes moist soil with pH of 5.5-7.0. Blooms second year and thereafter. Flowers are yellow or white and bloom in **July and August.**

Can be invasive. Can propagate by cuttings. Grows wild in most of the United States.

Harvest in July by cutting off top 1/3 of plant. Dry by hanging upside down in a cool, dry place.

Uses: For **medicinal** purposes, use plant varieties that have black dots on the flower petals, and clear dots when leaves are held up to light. Used to help treat depression and anxiety. Leaves are used as a tea. Consult your herbalist, doctor and/or health care specialist.

To check your variety of St John's Wort for the **active ingredient,** hypericin, make a tincture. The tincture should be bright to dark red.

(For **how to make a tincture, decoction or infusion,** see April Garden-Harvest.) (For details about herbaceous/softwood/semi-hardwood/ hardwood cuttings and how to stratify seed, see "Garden Tips" in the back of this manual.)

 Salad Burnet or Garden Burnet (rosaceae family; Sanguisorba minor) is a short-lived **hardy perennial, about 3-4 years.** It is one of the first perennials to grow in spring and one of the last to stop growth in winter. It is **evergreen** and can be harvested in the winter. Good for erosion control.

Sow seed 1/2 inch deep several inches apart after last frost. Germinates in 7-21 days. Thin to 12-15 inches apart.

The first year it grows to 6-8 inches tall. The second year it grows to about 18 inches tall. It grows about 2 feet wide. It grows in a circular mound.

Trim back regularly to encourage new growth. Cut off pink/purple flowers unless saving for seed. It will self sow.

Care: It likes full sun but can take partial shade. Likes soil pH 6-8. It does OK in poor soils.

Harvest: To use, strip leaves off of the stalks. Leaves and stalks can be eaten raw or cooked. Flowers are edible.

(For **dividing mature plants**, see April Garden-Maintenance. See June Garden-Maintenance.)

 Sow **salsify** (compositae family; Tragopogon porrifolius, oyster plant, vegetable oyster) **starting 2 weeks before first frost in spring. Best sown in April or May.** Sow 1/2 inch deep about 1-2 inches apart. Poor germination if seeds older than 1-2 years. Germinates in 7-21 days. Leaves look like grass. Slow growing. Flowers are purple and look like a daisy.

When 2 inches high, thin to 2-4 inches apart. Later thin 6 inches apart in rows 18-24 inches apart. Grows to 2 feet tall with 18 inch flower stalk above that. Matures in 100-120 days (14-17 weeks, 3-4 months).

Can sow seed in **late fall** for growth starting in early spring. Best sown in spring.

Care: Likes full sun. Prefers soil pH 6.0-6.8. Likes loose, well drained soil.

A **biennial.** Looks like a white carrot. Root can get tough if temperatures go above 85 degrees. No serious diseases or pests. (For how to make **large roots**, see carrots in June Garden-Plant.)

Spring and Summer Harvest: Young shoots, flower buds and flowering shoots are edible.

Black salsify (compositae family; scorzonera hispanica) is called **scorzonera**. Scorzonera is a hardy perennial. (See September Garden-Harvest.)

Sorrel (polygonaceae family; Rumex acetosa or Rumex alpinus or Rumex scutatus) is a **cool season perennial** that is easy to grow from seed. Sow 1/4 to 1/2 inch deep, 2-3 inches apart. Thin seedlings to 6-8 inches apart in rows 15-18 inches apart. Matures in 60 days.

Care: Likes soil pH of 5.5-6.8. Prefers full sun but can do well in partial shade. Remove seed stalks unless saving seed. Can **propagate** by dividing mature plants every 3-4 years.

Uses: Has sour leaves that contain oxalic acid (also in rhubarb and spinach). Eat in moderation. The older the leaves, the more acid they contain. Cooking reduces the amount of acid. It is a laxative.

Varieties:

Garden sorrel (Rumex acestosa) grows about 3 feet tall. Leaves are good for salads.
French sorrel (Rumex scutatus) grows 6-12 inches tall. Leaves are good for salads.
Patience or spinach dock (Rumex patientia) grows to 4 feet tall. Leaves are eaten fresh or cooked.
Spinach rhubarb (Rumex abyssinicus) grows up to 8 feet tall. Leaves are eaten like spinach. The stalks are eaten like rhubarb.
Common or sheep sorrel (Rumex acetosella) is a wild plant whose leaves are eaten when very small.
(For **dividing mature plants**, see April Garden-Maintenance. See August/October Greenhouse.)

Sow **Swiss chard** (chenopodiaceae family; Beta vulgaris) in May or when maple trees are in bloom or **2 weeks before last frost date in spring** when soil is 50+ degrees. **In fall plant 8-10 weeks before first frost date. Hardy to 29-32 degrees.**

Soak seeds for 15 minutes before sowing. Sow seeds 1/2 inch deep, every 2-3 inches in rows 18-24 inches apart. Thin to 8-10 inches apart. But only thin when gets crowded since can eat thinnings. Matures in 50-60 days.

Care: Likes full sun but tolerates some shade. Likes soil pH of 6.2-7.0. Easy to grow. It is a **biennial.** Has few pests. Related to beets. (See "Plant Families" in back of manual.)

Harvest all summer and into fall (tolerates light frost down to 29-32 degrees). Pick 3-4 leaves per plant except not from center/crown leaves. New leaves will grow. Eat raw or cooked.

There are 3 main types:

White-stemmed Swiss chard is more productive and bolt (flowering) resistant than colored-stem (red, pink, orange, yellow) varieties. White-stemmed include Fordhook Giant Lucullus, and Silverado.
Colored-stemmed include Pink Passion, Burgundy, Orange Fantasia, Golden Sunrise, Bright Lights.
Perpetual varieties (perpetual spinach) are short, stocky with thinner stems and smaller, smoother leaves.

(For **late Swiss chard**, see August Garden-Plant but details are here. See September Greenhouse.)

Thyme (lamiaceae family; common thyme: Thymus vulgaris) is sown in **early May**. A hardy **perennial** with over 100 varieties that vary in height from 2-12 inches.

Sow seed 1/4 inch deep. Germinates in 21-28 days. Thin to 8-12 inches apart. Grows slowly. Seeds may not grow true to type. Better to grow from cuttings or mature plant divisions. Flowers June or July.

Care: Likes soil pH of 6.0-8.0. Needs full sun and good soil drainage. Cut back half or more of growth when gets too leggy. Few pests or diseases.

Varieties: English thyme is the most common. Lemon thyme smells of lemons. Variegated lemon thyme has bi-color leaves. Orange thyme is very low-growing, smells like orange. Creeping thyme is the lowest-growing of the widely used thyme. Silver thyme is white/cream. Summer thyme has unusually strong flavor. Caribbean thyme has same flavor as English thyme but 10 times stronger.

German Winter thyme has lavender, white or pink flowers. Not as flavorful as other thymes. It deters cabbage worms. It is hardy to minus 20 degrees.

Propagation: Can divide mature plants (about 3-4 years old). Can grow from tip cuttings.

Harvest: Harvest lightly the first year. Best harvested in June or July just before flowering.

Uses: Can eat leaves fresh or dry. Good in salad, vinegar, soup, gravy, stew, sauces, sausage, dressing and other dishes. It is used as an antiseptic and stimulant in herb lotions and baths. Consult your herbalist, doctor and/or health care specialist.

(For details about herbaceous, softwood, semi-hardwood and hardwood cuttings, see "Garden Tips" in the back of this manual.)

 **Valerian** (Valerianaceae family; Valeriana officinalis, garden heliotrope) is not the same as red valerian (Centranthus ruber). Consult your herbalist.

Sow in greenhouse in March or outside in May. Needs light to germinate. Germination rate is variable and usually poor. They do self seed. Seedlings need a lot of moisture and are frost hardy. A **perennial** that grows to 3-5 feet tall, and 1 foot or more wide.

Can also divide mature plants in **spring or fall** every 3 years. Or at the **end of summer** replant daughter plants from runners from mature plants. Plant 1 foot or more apart.

Care: Likes rich soil. Likes full sun or partial shade. Likes soil pH of 5.5-7.0.

Blooms (red, pink or white) in July/August but does not bloom first year. If not saving seed, cut flower buds off.

(For **transplanting daughter plants**, see August Garden-Maintenance. For **harvesting roots**, see October Garden-Harvest. For **dividing mature plants,** see April Garden-Maintenance.)

 Winter savory (lamiaceae family; Satureja montana) is a 2 foot tall **perennial**, semi-ever-green, semi-woody herb. Sow 1/8 inch deep, 3 inches apart. Thin to 10-12 inches apart. Grows 2 feet tall.

Care: Easy to grow. Likes full sun or partial shade. Likes soil pH 6.5-7.3. Likes well drained soil. Drought tolerant.

Replace with new plant every 2-3 years. Can propagate plant by cuttings or dividing plant.

White or lavender flowers bloom July to September. Best **harvested** before it flowers. For harvesting hang in bundles upside down in an airy place to dry.

Uses: Leaves are stiffer and thicker than summer savory. It is less sweet with a stronger flavor than summer savory. Used in liqueurs, vinegar, herb butters, bean dishes, creamy soups, and tea.

Late May or early June (after last frost) or when apple blossoms begin to fall, and flowering dogwood is at peak bloom, or when daylilies start to bloom, **sow or transplant heat loving plants:**

Sow **beetberry or strawberry spinach** (chenopodiaceae family; Chenpodium capitatum, strawberry blite). A **hardy annual and sometimes biennial.** Sometimes survives through winter. It is 400 years old.

Needs light to germinate. Sow 1/4 inch deep or less. Germinates in 3-6 days. Germination is poor. Matures in 40-60 days. Grows to 1-3 feet tall and 12 or more inches wide. Bushy, sprawling growth. Self seeds readily. Easy to grow.

Flowers in summer. Berries ready late summer and fall. Pick when dark, crimson red.

Care: Likes full sun but will tolerate partial shade. Likes soil pH of 5.5-7.5. Drought tolerant.

Uses: Grown for their triangular, toothed leaves. Also for their sweet, juicy but somewhat bland red berries that are 1/2 inch across. Flowers are edible. Eat raw, cooked or make jelly. Leaves are high in vitamins A and C.

 **Calendula** (asteraceae family; Calendula officinalis) is also known as pot marigold but is not in the same family. It is in the **daisy family.** It is an **annual or short-lived perennial.**

Sow after last frost in spring. Sow about 1/4 inch deep. Germinates in 7-14 days. Thin 6-12 inches apart. Grows 1-3 feet tall.

Care: Likes full sun or partial shade. They like a good soil but will do OK in a somewhat poor soil. Likes soil pH of 5.5-7.0. Keep watered during drought.

Blooms 40-50 days after planting. They produce an abundance of yellow, gold, orange and lemon (even some pink and cream) colored flowers that begin blooming **early summer** and continue until frost. They easily reseed. The seeds germinate better after being stored for 6 months.

If you want bushier growth, trim the plants back. Remove dead flowers to encourage development of more blooms.

Uses: Harvest late morning. They make great cut flowers. Calendula petals are edible. They are slightly bitter. To dry petals, break apart flowers on screen so all get good air flow. Stir occasionally.

Medicinal: The flowers are used externally for its **antiseptic and healing properties** in skin infections, cuts, punctures, scrapes, burns, varicose veins and chapped/chafed skin. The tea or the tincture in water can be swished and swallowed to help heal oral lesions, sore throat, or gastric ulcer. Contact your herbalist and/or health care specialist.

Mash flowers into a paste with a little water in a blender or mortar/pestle. Then rub directly onto affected areas. Dried flowers are made into an aromatic infused oil, tea or tincture. To **test the tincture**, put one drop on the surface of a plate. When the alcohol dries off, there should be a raised drop of sticky, golden resin. (For **how to make a tincture, decoction or infusion,** see April Garden-Harvest.)

 Celeriac (umbelliferea family; Apium graveolens, celery root) **is sown outside after last spring frost. Sow late May**. Seeds need light to germinate. Best to soak seeds overnight. Germinates in 2-3 weeks. Thin to 6-8 inches apart. Plant in rows 24-30 inches apart. Matures in 90-120 days (3-4 months) so that is **late September or October.**

Care: It is easier to grow than celery. Likes soil pH 5.5-6.5. Likes moist, rich soil. As it grows remove any lower stems that are growing sideways so root is smoother. It is a **biennial.**

Harvest: Best harvested after a light frost. The knobby roots are eaten. It stores 4-6 months in a root cellar or similar storage. (See June Greenhouse but details are in this section. See October Garden-Harvest.)

Sow **celery** (umbelliferea family; Apium graveolens) in May for harvest in **September.** Or transplant celery from greenhouse to garden **early June.** Better started in greenhouse.

Plant 8-10 inches apart. Keep well watered. (For **sow/grow information,** see March Greenhouse. For **transplanting,** see May Garden-Plant. See August Garden-Harvest.)

 **Chicory** (compositae family; Chrysanthemum intybus, blue dandelion) is a **hardy biennial or perennial** with blue flowers. It is not the same as Belgian Endive or Radicchio. Flowers are blue.

Sow 2-3 weeks before last frost in spring. If growing for roots, time the sowing so comes to maturity in cool weather. Sow seed 1/2 inch deep, 1-2 inches apart in rows 24 inches apart. Seeds germinate in 7-21 days. Thin plants to 1 foot apart. Can eat the thinnings.

Leaves are about 18 inches tall. Looks like dandelion leaves. Grows 3-4 feet tall if include flowerstalk.

Care: Likes soil pH 5.0-6.8. Likes well drained soil. Keep soil free of rocks so roots do not split. Likes full sun but will tolerate some shade. **Roots** mature in 85-100 days (2 1/2 to 3 1/2 months).

Harvest and Uses: You can also forage for chicory.

Young leaves eaten raw or cooked. Older leaves need boiling and water changes to remove bitterness.

Roots store for 4-5 months. Mature root is bitter. Roasted roots are made into a caffeine-free coffee substitute. It is a general herbal tonic. It is diuretic (increases urination) and laxative. Consult your herbalist.

(See **related plants endive/escarole** in August Garden-Plant. For **radicchio,** see February Greenhouse.)

 Late May sow or transplant **cucumbers** (cucurbitaceae family; Cucumis sativus) when lilac flowers have faded. **Sow several weeks after sow summer squash.** Do not thin too soon. Wait to see if cucumber beetle kills any plants first. Can sow again directly in **July**. (For **sowing details and types**, see mid-April Greenhouse. See July Garden-Plant.)

 Sow **feverfew** (asteraceae family; Tanacetum parthenium also known as Chrysanthemum parthenium or Pyrethrum parthenium) after last frost in spring. **Can sow in greenhouse 6-8 weeks before last frost in spring** to plant outside after last frost. Needs light to germinate. Germinates in 7-14 days.

Thin to 6-8 inches apart in rows 18-20 inches apart. Grows up to 2 feet tall.

Care: Likes full sun. Likes soil pH 5.5-6.5. Drought tolerant. Can grow in poor soil. Flowers mid summer through fall with small daisy-like flowers. Leaves smell like citrus. **A short-lived perennial**. Can be invasive.

Propagation: Can propagate by dividing mature plants in the fall. Can propagate by cuttings. Easily self sows.

Medicinal: Harvest **leaves and flowers** when it is in bloom. Has been used for more than 2,000 years. Do not use if pregnant or taking blood thinners. It is anti-inflammatory so is used to treat migraine headaches, and reduce fevers. In Europe, it is more popular than aspirin for treating arthritis. Take 2-3 leaves a day. Do not use all the time. Contact your herbalist and/or health care specialist.

 Late May or early June transplant **ground cherry** from greenhouse to outside after last frost. Grow similar to tomatoes. (For **sowing/growing details,** see March Greenhouse. See July Garden-Harvest.)

Sow **horehound** (lamiaceae family; Marrubium vulgare; common horehound, white horehound) **after last frost in spring** when temperatures are consistently above 50 degrees. Or sow **3 weeks before last frost in fall.**

Sow 1/4 inch deep. Germinates slowly. Thin seedlings to 10-15 inches apart. It grows 1 1/2 feet to 3 feet tall.

Care: A drought tolerant **perennial**. It does not like heavy soil. Tolerates poor soil. Likes neutral to somewhat alkaline pH around 7. Likes well drained soil. Likes full sun.

Member of the **mint family**. (See "Plant Families" in back of manual.) Can be invasive. The white flowers **bloom** the second year after planting. They bloom June through September. They attract beneficial insects. A good companion plant for peppers and tomatoes. Can **propagate** by cuttings or by dividing plants.

Harvest leaves while plant is flowering. The first year it can be harvested even though it has not flowered. You can cut up to 1/3 of the plant. Cut off stalks with leaves and flowers, and hang to dry out of direct sun.

Medicinal: It has been used medicinally for centuries. The bitter leaves are used to make tea, ale, cough drops, cough syrup and candy. Good for bronchitis, flu, colds, and sinus infections. Do not use if you have low blood pressure or heart conditions, are using insulin, or are pregnant. Contact your herbalist and/or health care specialist.

Cough syrup: 1/4 to 1/2 cup dried leaves and/or flowers, 1 cup water, 2 cups honey, 1 tablespoon lemon juice or cider vinegar. Boil horehound in water for about 10 minutes. Remove from heat and let sit for 5 minutes. Strain out the horehound. Add honey and lemon. Stir. Use one tablespoon as needed.

(For **how to make a tincture, decoction or infusion,** see April Garden-Harvest.)

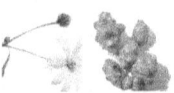 Plant **Jerusalem artichokes** (compositae family; Helianthus tuberosus, sunchokes) **after last frost in spring or when soil can be worked. Can plant in fall.**

Cut tubers in sections with at least 2 eyes each that weigh at least 2 ounces. Do not let them dry out before planting. Plant 6 inches deep, 12-18 inches apart in rows 2-4 feet apart.

Care: Does well in soil that is good for corn and potatoes. Likes well drained soil with a pH of 7. Likes full sun. Do not let soil dry out too much.

Grows 3-12 feet high with large leaves. Yellow flowers are 1 1/2 to 3 inches in diameter. Tubers begin to form in **August.** Tubers are up to 4 inches long and 2-3 inches wide.

Be careful where you plant it because it is almost impossible to get rid of it once it has grown someplace. So consider it's location as permanent.

Jerusalem artichokes are a very hardy, prolific **perennial of the sunflower family.** (See "Plant Families" in back of manual.) They are native to central North America. The tubers are good for humans and farm animals.

Varieties:

White-skinned varieties include early-maturing Stampede that has crisp, round roots. It is good to grow if summers are short. Slower-growing Clearwater and White Fuseau varieties are longer so are easier to scrub and peel.

Red-skinned ones include Red Fuseau that has red skin on cone shaped roots that have few attached nodules so they are easier to clean. Red Rover and Waldspinel are very long so are called fingerling sunchokes.

(See March/April/October Garden-Harvest with **harvest details in October.** See July Garden-Maintenance.)

Transplant **leeks** from greenhouse to outside 1 week after last frost. (For **sowing and planting details**, see February Greenhouse.)

Transplant **marigold** (asteraceae family; Tagetas ercta and Tagetas patula) from greenhouse to outside. (For **sowing and planting details,** see April Greenhouse.)

Sow **marjoram** (lamiaceae family; Origanum majorana), a **frost sensitive annual** that grows 1 foot tall. Do not cover seeds with soil, it needs light. **Can sow seeds in the fall** for growth in the spring. Can start in greenhouse **6 weeks before last frost.**

Germinates in 8-14 days. Transplant seedlings 6 inches apart. Grows 24-36 inches tall and 15-18 inches wide. Can propagate by seeds or softwood/hardwood cuttings.

Care: Likes full sun. Likes pH 6.5-7.5. Likes rich, well drained soil.

Uses: Has milder flavor than its cousin oregano.

Sow or transplant **melons and watermelons** late May or early June.

Types of melon: Cantaloupe, crenshaw, honeydew, muskmelon, watermelon and many more. Sow seeds or transplants in hills 2 feet apart. (For **sowing and planting details,** see April Greenhouse. See August Garden-Harvest.)

Soak **okra** (malvaceae family; Abelmoschus esculentus, gumbo) seeds for 12 hours before planting. Seeds do not stay viable for longer than a year or two. Sow seeds 1 inch deep in hills 12-24 inches apart. When seedlings are 3 inches tall, thin all but the strongest plant per hill. Or in a row thin plants to 2 feet apart. (See Malvaceae Family in "Plant Families" in back of manual.)

Frost will kill it. Does not like being transplanted. Okra needs 50-60 days to mature. An **annual** that grows to 4 feet tall (or taller) and 3 feet wide. Can trim plant to height you want. Can use trellis. Popular in southern gardens. Flowers are yellow or pink.

Care: Needs full sun. Likes soil pH 6.0-8.0. Likes well drained soil. Likes rich soil. Sensitive to day length. (It flowers based on the amount of day light.)

Harvest: Pick soft pods every other day when pods are 2-3 inches long. This is about 4 days after flowers fade. Large pods are too tough to eat. Keeps producing pods until frost.

(See April Greenhouse but details are here.)

Sow or transplant **peppers** (solanaceae family; Capsicum annuum) after dogwood blossoms fall. Frost sensitive. Member of **nightshade family**. (See "Plant Families" in back of manual.)

Plant seedlings 12-15 inches apart in rows 24-36 inches apart. Plant deeper than was sitting in pot.

Care: Likes hot weather and full sun. Tolerates acid soil. Do not add too much nitrogen or will have a lot of leaves and little fruit. Protect from cutworm. (See tomatoes in March Greenhouse.)

(For **sowing/growing details,** see March Greenhouse. See June Garden-Harvest.)

 **Sow purslane** (portulacaceae family; Portulaca oleracea) after last frost. Better if stratify seeds. (For how to stratify see "Garden Tips" in back of this manual.) Needs light to germinate. Germinates in 7-21 days. When 2-3 inches long, they start to creep along the ground.

Care: Prefers full sun. Likes hot weather. Likes sandy soil with pH 5.5-7.0.

It is an **annual** that readily self seeds and can be invasive. It is a low-growing, succulent plant with small, fleshy leaves and tiny yellow flowers. Commercial varieties are larger than wild ones.

Harvest: Leaves, stems and seeds are edible. Can harvest what you sow, or **forage June through September.** Considered medicinal. Contains oxalic acid like spinach does so eat in moderation. Good in salads.

 **Rosemary** (lamiaceae family; Rosmarinus officinalis) a hardy, woody, **evergreen perennial that is a member of the mint family**. (For mint see May Garden-Plant. See "Plant Families" in back of this manual.) Seeds are difficult to germinate and usually do not grow true to the parent. So propagation by cuttings is better.

To **propagate** cut a 2 inch piece from the soft, new growth of an established plant. Remove bottom leaves and dip in rooting hormone. Then place in damp mixture of peat moss, vermiculite and perlite. Place in indirect light. (For details about herbaceous/softwood/semi-hardwood/hardwood cuttings, potting mixes, and how to make rooting hormone, see "Garden Tips" in back of this manual.)

Care: Likes full sun but will tolerate partial shade. Likes soil pH 6.5-7.5 Likes well drained soil. It is drought tolerant.

It grows to a small shrub up to 5 feet tall and 3 feet wide. Keep pruned to stay bushy. Blue flowers in the **spring** (April or May) so do not trim early spring. Either trim late fall or after flowering.

It is somewhat sensitive to cold temperatures so cover it with a sheet or row cover in the winter. It is hardy to 15-20 degrees.

Harvest: In greenhouse or other protected location, can harvest all year. Can use in cooking or make tea to reduce congestion.

 Sow **sea kale** (brassicaceae family; Crambe maritima), a very **hardy perennial,** in May after last frost. Seakale has a corky covering over the seed. Remove the covering to speed up germination that can take up to 3 years! May germinate quickly if remove covering. Sow 1 inch deep, 12-18 inches apart.

Thin seedlings when they have 3-4 leaves. Grows 2 feet tall and 3 feet across.

Care: Likes full sun. Grow in sandy soil with pH around 7.0. Must have good drainage. Drought tolerant.

Has thick blue-green-gray leaves. Plants live about 7 years. Very few diseases or pests except for clubroot.

Edible parts are usually blanched. Cut stems when 8 inches tall. Stems, young flowerheads and very young leaves are eaten raw or cooked. Leaf midribs are cooked like asparagus.

Can also **propagate sea kale** by root cuttings or thongs in November or December. Nantahala Farm (www.nantahala-farm.com) sells Sea Kale. (See March Garden-Plant but details are here. See November Garden-Maintenance.)

 Sow **sorghum** (poaceae/gramineae/grass family; sorghum genus with over 30 species).

Sow seed mid-May for long-season varieties, and **early June** for short season varieties. Plant 5-10 pounds of seeds per acre. The sweet sorghum plant looks like corn, but it has no tassels or ears. It grows 6-12 feet tall. The stalk is about 2 inches in diameter.

Sow when soil is between 65 to 70 degrees. Germinates in 3-5 days and looks like blades of grass. Sow 3/4 to 1 inch deep, 4-6 inches apart, in rows are 3 to 3 1/2 feet apart. Thin to 2-3 plants per foot. If using a small garden seeder, use the carrot plate. (See **Garden Seeders and Broadcast Spreaders** in "Garden Tips" in back of manual.) Or sow in hills with 7-12 seeds per hill, with hills 1 foot apart. Then thin to 5 seedlings. Matures in 90-120 days.

Care: Likes soil pH 6.0-7.0. Needs full sun. **Most varieties are annuals; a few are perennials. Likes warm weather. Very drought tolerant.** Tolerates heat and drought better than most grains. Self pollinating but will cross pollinate by wind. It is a little more frost hardy than corn.

Varieties:

1. Broomcorn (Sorghum vulgare var. technicum) is an upright grass with seeds on the end of long, straight branches. It grows 6-10 feet tall. After harvest, the stiff bristles are bound to make broom heads and brushes. Its seeds are used to feed livestock.

2. Forage Sorghum (Sorghum bicolor) is a large, warm-season annual grass that is closely related to the grain variety. When temperatures range between 75 and 80 degrees, forage sorghum produces its highest yields. Grows 6-12 feet tall. Used to feed animals and for biomass such as biodiesel.

3. Grain Sorghum (Sorghum bicolor or Sorghum japonicum) is also called **milo**. Similar to corn in appearance and growth. Leaves are 2 1/2 feet long. Prior to the 1940s, grain sorghums were 5-7 feet tall. Today they are 2-4 feet but can be 15 feet tall. In the United States it is fed to livestock. In other countries it is also eaten by people. It has more protein and fat than corn, but is lower in vitamin A. Better to crack or roll grain before feeding to livestock.

Sudan grass (Sorghum bicolor subspecies drummondii) is usually used as feed grain for livestock. Can be used for silage or pasture.

4. Sorghum-Sudangrass is a cross between sorghum and sudan grass. Used for grazing (pasture), silage and hay. It can grow up to 15 feet tall. Uses 1/3 less water than corn. Varieties with the brown midrib gene are easier to digest. Silage made with sorghum is usually this variety.

5. Sweet sorghums (Sorghum bicolor) is also called **sorgos**. They are cultivars primarily grown for forage, silage, and syrup production. Has high sugar content. **White sorghum** is sweeter and used as a grain crop. Outside the United States it is used for making bread and porridge. **Red sorghum** is less sweet (even bitter) and is not eaten by birds as much. Used for fodder and making beer. It grows 6-15 feet tall.

Sugar Drip (Sorghum vulgare), a sweet sorghum hybrid, was the most widely grown variety in the **mountains of western North Carolina**. Produces 168 gallons of syrup per acre. It grows 9-12 feet tall. Matures in 110 days. This variety has relatively low yields and is susceptible to most sorghum diseases, particularly stalk red rot and maize dwarf mosaic. It is prone to lodging (falling over in wind or heavy rain). Too much nitrogen can increase lodging. (See "Plant Health" in back of manual.) (See October Garden-Harvest.)

 In May sow **squash (summer)** (cucurbitaceae family; Cucurbita pepo) **3 weeks after last frost** when soil is 60 degrees or warmer. (Can start earlier in greenhouse in biodegradable pots.) It is a member of the **cucumber family** Cucurbita which includes gourds, pumpkins and summer/winter squashes. (See "Plant Families" in back of manual.) One or two plants per person is usually enough. Can plant again **mid-June**.

Can presprout seeds indoors. Sow seed 2 inches deep. Sow 6-8 seeds in hills 3-4 feet apart, or every 4-8 inches in rows 3-4 feet apart. Thin by cutting (not pulling) to best 2-3 plants in hill. Does not like being transplanted so sow in degradable pot. Can trellis or let grow along ground. Matures in 43-75 days.

Care: Easy to grow. Likes soil pH 5.8-6.8. Needs full sun. Do not let soil dry out too much.

It is a bush or weak-stemmed vine. It has male and female flowers. The first flowers to appear are male and they do not produce fruit. Insects pollinate the female flowers that produce fruit.

Uses: It is eaten immature when skins are soft and thin. They stay good in the refrigerator for about 1 week.

Varieties include patty pan/scallop, yellow crookneck, yellow straightneck, and zucchini. Harvest yellow crookneck when 4-5 inches long, 4-8 inches for zucchini, and the size of a silver dollar for patty pan. (For **pests and diseases**, see Cucurbitaceae Family in "Plant Families" in back of manual.)

 Sow **soybeans** (leguminosae or fabaceae family; Glycine max) **in May and June about 2-3 weeks after last spring frost** or when soil is at least 60 degrees. Seeds are planted soon after wheat. (For **map of when to plant wheat,** see October Garden-Plant.) Do not soak seeds before planting since they may crack.

Sow 1-2 inches deep, 2-4 inches apart in rows about 24-30 inches apart. They can be cultivated/tilled in with a tractor/planter or drilled into the ground with a no-till planter. Germinates in 4-7 days. Thin 4-6 inches apart. Cut seedlings rather than pull so as not to disturb other plants. Matures in 5-6 months.

It is an **annual bushy legume** that grows 12-36 inches tall. Stems and leaves are hairy. Pods grow 1 to 4 inches long in clusters of three to five. Each pod has 2 to 4 seeds.

Care: Likes full sun and warm weather with best temperatures 70-80 degrees. Likes soil pH of 6.0-6.8. Likes well drained soil. Will grow in poor soil. They are more drought tolerant than most other crops. It is best to water at ground level rather than from the top. Usually not bothered by many pests or diseases.

Plants bloom in **July, August and September**. The flowers are small and white with some violet or purple. In **late September** the soybeans begin to mature. The leaves begin to turn yellow. By **mid-October and November,** the leaves turn brown and fall off leaving pods of soybeans.

Harvest: For **green beans**, harvest when pods are green (half mature), full and plump about 2 to 3 inches long. This is about 45 to 65 days after sowing. **Dry soybeans** require 100 or more days to reach maturity. Soybeans mature at the same time. Pull up the whole plant and hang it upside down to dry. Shell when pods are fully dry.

Uses: Do not eat raw since they contain a protein inhibitor. Soybeans are high in protein and used in animal feed. Oil is extracted from the beans. People eat soybeans roasted in forms such as soy sauce, tofu, crunchy beans, soy milk and imitation meats. Soybeans are also used in various industrial products.

Varieties: There are more than 10,000 soybean cultivars. Colors include black, gray, brown, green, yellow, white, and striped. Seeds can be smaller than a pea or as large as a kidney bean. Green bean cultivars are most tender and best flavored. Black beans are used for drying. Yellow beans are used to make soy milk and flour.

Soybeans and corn are the most genetically modified (GMO) plants. The long-term effects of this on human/animal health and the environment have not been determined. (For more about GMO plants, see corn in May Garden-Plant. For **open and hybrid pollination and about GMO,** see "How to Save Seeds" in the back of this book.)

 Sow **sweet violets** (violaceae family; Viola odorata, wood violet, garden violet) **after last frost.** Seedlings grow slowly. Flowers are dark violet or white with a little violet.

Care: Likes a slightly acid soil around pH 6.0-6.5. Prefers partial shade. A hardy clumping **perennial** that blooms early spring. Grows 4-6 inches tall and 8-24 inches wide. Spreads easily. Has few diseases or pests.

Harvest: Leaves and flowers are edible. Leaves are tart, flowers are sweet. Uses include making tea or garnishing salads. Eat raw or cooked. Used for sore throat. Consult your health care specialist.

Transplant **Tarragon (Russian)** (Asteraceae family; Artemisia dracunculus. Also known as Dragons Wort.) seedlings from greenhouse to outside in May. It can take hard freezes.

Transplant **French Tarragon** outside or to greenhouse or other protected area. It can not take hard freezes. (For **sowing and planting details,** see April Greenhouse.)

 Sow **wormwood** (asteraceae family; Artemisia absintium) seeds on top of soil. (**Mugwort** is Artemisia vulgaris also known as Saint Johns plant. See last paragraph.)

Needs light to germinate. Thin seedlings to 18 inches apart. Grows 2-5 feet tall, 2-3 feet wide. Grow several feet from other plants because it's roots give out a chemical that is toxic to other plants.

Care: Likes full sun but does OK in partial shade. Likes soil pH of 5.5-7.0. A **hardy perennial.**

Cut back late fall or early spring. **Fall pruning** is preferred. Few pests or diseases.

Can be **propagated** by root divisions or runners. (For details about herbaceous, softwood, semi-hardwood and hardwood cuttings, see "Garden Tips" in back of this manual.)

Harvest: Parts used are leaves, and fresh or dried flowers harvested before or during the blossoming season.

Uses: The leaves are used as an insect repellant. Crush a small amount of leaves into a moist mash, then blend with a little apple cider vinegar. Place some onto a small cloth and wipe on skin.

Used to rid the body of worms but is toxic in large amounts. Used to make absinthe (green, aromatic liqueur that is 68% alcohol; has a bitter, licorice flavor) and beer.

To make wormwood tea: Boil 2 cups water, remove from heat, then add 4 teaspoons of dried wormwood leaves or flowers. Or twice as much fresh leaves or flowers. Let soak. Drink the lukewarm tea on an empty stomach 3 times daily. It is very bitter so can add sweetener. Only use occasionally since it is toxic in large quantities or if used frequently. Consult your herbalist and/or health care specialist.

(For **how to make a tincture, decoction or infusion,** see April Garden-Harvest.)

Mugwort is Artemisia vulgaris. Stratify seeds (See "Garden Tips" in back of this manual.) Seeds need light to germinate. Can grow to 10 feet tall, 2-3 feet wide. Easy to grow. Likes full sun or partial shade. A drought tolerant **hardy perennial.** Can be invasive but if remove flower buds then not a problem.

Mugwort Harvest: The **leaves** are picked in August and dried. Poultice made from leaves is said to relieve poison ivy rash.

The **root** is dug in fall and dried. Make a tea to ease PMS and cramps. Relieves fatigue.

(For **using wormwood as an animal wormer,** see April Farm Animals.)

 Sow **yarrow** (asteraceae family; Achillea millefolium) seeds outdoors after last frost. Needs light to germinate. Germinates in 15-20 days. Thin seedlings to 1-2 feet apart.

Care: Likes full sun. Likes well drained soil. Likes soil pH of 6.0-6.5. Is drought tolerant. Tolerates poor soil.

A **perennial.** Flowers are white, yellow, gold, pink or red. Can **propagate** by dividing roots or hardwood cuttings in spring or fall. (For details about herbaceous, softwood, semi-hardwood and hardwood cuttings, see "Garden Tips" in back of manual.)

Harvest: Do not use during pregnancy. **Use flowers, leaves and stems.** Used to stop bleeding. A leaf held against a wound stops bleeding. Yarrow tea is good for colds and flu especially if add peppermint leaves and elderberry flowers/berries. Yarrow acts as an activator to speed decomposition of **compost.** Consult your health care specialist.

For soil and compost information, read the book "Secrets of the Soil: New Solutions for Restoring Our Planet" by Peter Tompkins and Christopher Bird.)

In May after danger of frost is over, or when lilacs are in full bloom, or when soil is 60+ degrees, or when oak leaves are big as a squirrel's ear:

 Sow / Transplant the Three Sisters: beans, corn, and gourds / **pumpkins / winter squash.** Interplant corn with beans and winter squash/pumpkins. **(See corn below for details about planting the Three Sisters together.)**

 First of Three Sisters. Beans are in the leguminosae or fabaceae family.

17 Types (Genus) of beans: (See "Plant Families" in back of this manual.)
    1. Cajanus: cajan or pigeon pea.
    2. Canavalia: Ensiformis or jack bean.
        Gladiata or sword bean.
    3. Cicer: arietinum or chickpea (garbanzo bean). **Garbanzo/chick peas** mature in 63 days. Young leaves, shoots and pods are eaten cooked. Roasted roots used as coffee substitute.
    4. Cyamopsis: tetragonoloba or guar.
    5. Erythrina: E. herbacea or Coral bean.
    6. Glycine: max or soybean. **Soybeans/edamame** are high in protein and oil. Can eat young leaves and pods.
        Grows as a bush. Likes slightly acid soil. Matures in 77-100 days. (See above in May Garden-Plant.)
    7. Lathyrus sativus (Indian pea).
    8. Lathyrus tuberosus (Tuberous pea).
    9. Lens: culinaris or **lentil.** Matures in 84 days.
    10. Lablab: purpureus or hyacinth bean.
    11. Lupinus or Lupin: L. mutabilis or tarwi, Lupinus albus or lupini bean.
    12. Macrotyloma: M. uniflorum or horse gram
    13. **Phaseolus: Most beans that Americans eat are in this group.**
        Acutifolius or tepary bean.
        Coccineus or **runner bean.** Can eat young leaves, pods and flowers. Does well in cool weather.
        Lunatus or **lima bean.** (See June Garden-Plant.)
        Vulgaris or common bean includes anasazi, black turtle bean, caparrones, green bush/pole,
            kidney bean, pinto bean, Taylor horticultural, wax and many others.
    14. Psophocarpus: tetragonolobus or winged bean.
    15. Stizolobium: several species, velvet bean.

16. Vicia: **Faba (fava)** or broad/bell bean. Also used as cover crop. (See March Garden-Plant.)
17. Vigna: Aconitifolia or **Moth bean**. Matures in 84-98 days. Very drought tolerant.
     Angularis or **azuki/adzuki** bean. Need at least 120 frost free days.
     Mungo or urad bean.
     Radiata or **mung bean**. Grows as bush. Needs 90-120 frost free days.
     Umbellatta or ricebean.
     Unguiculata or **cowpea/ cow pea**. Includes the **black-eyed pea, yardlong bean and others.**
       Cowpeas mature in 63-84 days. Drought tolerant. Good for forage.

Sow All Beans: Soak beans for 1-2 hours. Sow 1 inch deep, 2-3 inches apart, in rows 3-4 feet apart. Germinates in about 10 days. Beans do not transplant well. (See **Garden Seeders and Broadcast Spreaders** in "Garden Tips" in back of manual.)

You can **inoculate beans** if you have poor soil. You buy a powder that you coat the bean in. It has bacteria that improves the plant's ability to draw nitrogen from the air and ground. You may already have enough beneficial bacteria if you have fertile soil.

Thin bush varieties to 4-6 inches apart, in rows 18-24 inches apart.

Thin pole varieties to 6-8 inches apart in rows 3-36 inches apart. Pole beans need a stake or trellis.

Care: Easy to grow. Likes soil pH of 5.5-6.5. Needs full sun. Does not need rich soil. Beans have shallow roots.

Sow **succession plantings every 2-3 weeks until late July** of bush and pole beans. **Lima beans** are planted later. (For lima beans see June Garden-Plant.) All beans mature in 40-100 days.

Bush beans mature in 50-60 days with a large harvest for a short time. **Frost sensitive annual.**

Pole beans are continuously productive until first frost so 1 or 2 plantings is enough.

There are 2 varieties of beans:

1. Snap/green beans are tender enough to eat right off the vine. The bean and pod are eaten. They lack the strong, fibrous pods that many dry beans have. Some popular varieties are Blue Lake, Kentucky Blue, Kentucky Wonder, and French Haricort Vert. Wax beans (butter beans) are yellow.

Green bean harvest: Snap/green beans are harvested before the pods are fully mature. Pods should be full sized with small seeds. The pod is firm and crisp. Pods are ready for harvest 7-14 days after flowering. Pick regularly unless you are saving seed for planting.

2. Dry or shell beans dry in their pods and are stored for eating later in the year or in a few years. (When young some varieties can be eaten like green beans.) Mature beans must be shelled because pods are too tough to eat.

Popular shell beans are anasazi, black-eyed peas (cowpea, southern pea), black turtle, cream peas, field peas, garbanzo (chick pea), kidney (red), lentils, pintos, and purple hulls. Cowpeas and field peas are also used as warm weather cover crops. (For **warm weather cover crops,** see June Garden-Plant.)

Shell bean harvest: Dry/shell beans are allowed to mature and dry before being picked. In fall let beans dry on vine to save for planting next year and eating over winter. Shell beans are **easiest to shell** when they are dry but not over dry (where the pod is rock hard). So it is better to shell soon after the pod dries rather than waiting to do it months later.

Favorite beans in Appalachia:

Cut-Short beans are a tender, squared-off bean. They are a very old type of bean which most people, except for gardeners who have saved their own seeds, have never heard of.

Half-Runner beans have recently had poor quality control among commercial growers that has made more than half the beans too tough to eat. If you still have the old fashioned beans, you are lucky. They are productive, flavorful and tender.

Greasy beans have been grown in Southern Appalachia for many generations. They are particularly found in parts of southeastern Kentucky and western North Carolina. Greasy beans do not have the tight knit fuzz like that on the hulls of other beans. They look shiny and greasy instead. People who know them usually think they are the best of all beans. They usually cost several times more than commercial beans at farmers markets.

Greasy Cut-short beans combine the best characteristics of both greasy beans and cut-short beans. They are very slick, have tender hulls with tightly packed beans.

Cornfield beans also known as pole beans, stick beans and trellis beans. They are climbing beans that need vertical support. Historically grown in corn with the stalks providing the support. Hybrid corns are usually too weak to provide support. Use heirloom open-pollinated corn varieties to give good support.

October beans or fall beans used to be grown by most people in Appalachia. A few are bush beans but most are climbing beans. Some are stringless and some are string beans. Most Appalachian heirloom beans are climbing beans and have strings, with exceptions being some October beans. Nantahala Farm sells October beans (www.nantahala-farm.com).

Other good beans include Mountain Speckled beans that are dried. Very good for cooking. They are various shades of purple. Grows 13-15 feet tall. Plant 1 1/2 to 2 inches deep.

(For **fava beans** see March Garden-Plant. For **green beans** see August Garden-Harvest. For **shell beans,** see September Garden-Harvest. For **late beans** see July Garden-Plant but details are in this section.)

 **Second of Three Sisters.** Sow **corn (maize)** (poaceae/gramineae/grass family; Zea mays) in May when the soil is 65-70 degrees or more. Be sure to wait until the soil is warm enough. Sow when oak leaves are as big as squirrel's ears. It is **frost sensitive annual**. Plant when the **moon** is in the second quarter (waxing) and in one of these zodiac water signs: Cancer, Scorpio, or Pisces.

If you plant more than 1/4 acre, a **corn planter** really helps (see above image of manual planter). There are planters that attach to garden tractors. Earthway makes a manual Precision Garden Planter that has 2 wheels. It plants corn and most other seeds except really small seeds. (See **Garden Seeders and Broadcast Spreaders** in "Garden Tips" in back of manual.)

Soak seeds overnight (optional) plus not practical if sowing acres. Sow every 4 inches, or sow 3 seeds together every 7-15 inches. **Sweet corn** rows should be 30-36 inches apart. **Field corn** rows should be 40 inches apart. For early plantings, sow seeds 1 to 1 1/2 inches deep. In midsummer, plant seed 2-3 inches deep because soil is drier. Field corn is planted deeper than sweet corn. You may need row covers so birds and other animals do not eat seed.

Germination is about 75 percent. Germinates in 7-10 days. When 3-4 inches tall, thin to one plant every 10-12 inches for sweet corn and every 15 inches for field corn. **Hill soil** around the base of stalk when they are 6 inches high.

Care: Likes soil pH of 6.0 to 6.5. Likes full sun. Needs at least 8 hours of sun. Likes well drained soil. It does better if provided with a wind break. Needs a rich soil especially high in nitrogen. It is good to have previously planted the area with soil-enriching crops such as beans, hairy vetch or clover.

Can be **successively planted** especially for sweet corn so can harvest a long time. You can sow an early maturing variety every 2 weeks for 6 weeks. Or plant early/mid/late varieties at same time. Or plant an early variety as soon as the soil warms up, a mid season variety 5-10 days later, and a late variety in another week.

Early varieties have smaller ears and are of less quality than mid/late varieties. Early corn grows 4-6 feet tall and is planted 9-12 inches apart. It matures in about 60-75 days.

Mid-season corn grows 6-8 feet tall and is planted 12 inches apart. It matures in about 76-84 days.

Late corn grows 7-10 feet tall, and is planted 12-18 inches apart. It matures in about 85-100 days. Late corn **stores best over the winter.**

Cultivation: Keep the soil well cultivated because weeds reduce productivity. Be careful not to damage the corn roots. When the plants are about **knee height**, you should have cultivated them 3 times. At this point you can stop, so you do not damage the roots.

Corn needs about **1 inch of water per week** especially when the stalks begin to tassel (top of plant). Water stress during pollination creates ears with lots of missing kernels. It is best to apply water at the soil surface by using a soaker hose or drip irrigation. Avoid spraying plants from above since it could wash pollen off the flowering tops.

When stalks are 6 inches tall, side-dress them with blood meal, diluted fish-based fertilizer, or other **nitrogen.** Repeat when about knee-high. Corn does better if interplanted with a nitrogen-fixing legume like beans or a compost crop. Do not remove any side shoots or suckers since they do not harm production, and cutting them might damage roots.

Pollination: Corn is pollinated by the **wind.** Pollen from the tassels (plant tops, male part) fall onto each of the silks on the ears (female part). If a silk is missed, then the ear will be missing one kernel. Corn must be planted in blocks of at least 5 square feet (5-6 rows) to ensure cross-pollination by the wind.

You can also **pollinate the ears yourself.** Collect pollen when the silks emerge from the ears and the tassels look loose and open. It is best to do this when there is little or no breeze. Shake the tassels over a dry bucket to collect the pollen. Shake pollen from several plants. Transfer the pollen into a small paper bag. Sprinkle the powder onto the silks of each ear. Repeat 1 or 2 times over the next few days.

Different varieties **cross-pollinate** so if saving seed for planting, keep far apart (at least 400 yards) or plant so they tassel at least 2 weeks part. (See "How to Save Seed" in back of this manual.)

Pests: (See "Plant Health" and "Plant Families" in back of this manual.)

1. Cutworms sometimes attack corn seedlings. They feed on stems mostly on cloudy days and at night. Cutworms overwinter in plant debris so remove it or till under. Weeds can be burned about 3 or more weeks before planting. Pyrethroid (permethrin) insecticides are sometimes effective. (See Brassicaceae Family in "Plant Families".)

2. Deer and raccoons love corn. May need fence around it. (See **Animal Pests** in "Plant Health".)

3. Earworm (silkworm) moths lay eggs on corn silks. The larvae crawl inside the husks and feed on the tips of developing ears. Earworms are yellow-headed and grow to about 2 inches long with yellow, green or brown stripes. To prevent earworm damage, take an eyedropper or spray bottle and apply a mixture of vegetable oil, Bt (Bacillus thuringiensis), water and a few drops of dishwashing liquid to the tip of each ear a few days after silks emerge. Or use mineral oil. (See Gramineae Family in "Plant Families".)

4. Flea beetles sometimes chew holes in the leaves of young plants. They are small, shiny and black about 1/16 inch long. Cold winters kill off many of the beetles. Plant resistant hybrids. Plow under crop residue and maintain good weed control to eliminate overwintering sites. (See Brassicaceae Family in "Plant Families".)

Three Sisters: Corn has been grown in North America for over 4,000 years. Corn can be grown along with pole **beans** and **winter squash/pumpkins/gourds**. They are called the 3 sisters by Native Americans. **Sunflower** is sometimes called the fourth sister. (For sunflower, see June Garden-Plant.)

One way is to create flat-topped mounds of soil about 12 inches high and 20 inches wide. A few **corn (maize)** seeds are planted close together in the center of the mound. (Later thin to 1 plant.) You can bury rotten fish or other high nitrogen products in the mound. When the corn plant is 6 inches tall, plant **beans** (thin to 2 plants) and **squash or pumpkin or gourd** (thin to 1 plant) around the corn. Corn plants should be about 3 feet apart.

The corn gives a place for the beans to climb. (Hybrid corns are usually too weak to provide support. Use heirloom open-pollinated corn varieties to give good support.) The beans provide nitrogen for the other plants. Squash/pumpkin/gourd spreads along the ground keeping weeds under control. And the leaves help keep soil moisture. The prickly hairs of the vine deter pests. Corn does not have the amino acids lysine and tryptophan (needed by people to make proteins and niacin), but beans have both so the diet is balanced.

There are 5 types of corn:

1. Dent/field corn has a depression in the top of the kernels. It is starchy. Dent corn is used for roasting ears, corn bread and hominy. (For **how to make hominy,** see sweet corn in August Garden-Harvest.) It is more insect resistant than other types because of the tight husks. Used as food for animals. Kernels are yellow or white. Has 4% sugar at same stage when sweet corn is picked (sweet corn is 10%).

2. Flint or Indian corn is a larger-kernelled, pop-type corn that is very hard. Can be blue, black, red, white, yellow or multi-colored. Indian corn is often used in autumn decorations.

3. Flour corn is used in baked goods. It has a soft, starch-filled kernel that is easy to grind. It is usually white, but can be blue.

4. Popcorn is a type of flint corn. It has a soft starchy center surrounded by a very hard exterior shell.

5. Sweet corn is eaten on the cob, canned or frozen. Rarely used for feed or flour. Extra sweet because it has more natural sugars (10%) than other types. About 50% of sugar is converted to starch 24 hours after being picked.

Sweet corn is divided into several types:

a. **Normal sweet corn (su)** kernels contain moderate but varying levels of sugar. Sugars convert to starches rapidly after harvest. These should be picked and eaten within a very short time. Most are hybrid. Some are open-pollinated.

b. **Sugar-enhanced (se, se+, or EH)** have increased tenderness and sweetness. Conversion of sugar to starch is slowed. It is a hybrid. Gourmet quality sweet corn.

c. **Super-sweet or Xtra-sweet (sh2, sh)** has greatly increased sweetness and slow conversion of starch. The dry kernels are smaller and shriveled. It is a hybrid. It is not that good when canned or frozen but is good fresh. Their pollen is weak.

Colors of all corn: blue, red, white, yellow, bi-color, or multi-color.

Open-pollinated / old-fashioned corn is not as uniform in size and maturity as man-made hybrid corn. **Hybrid corn** usually matures all at once but this is a disadvantage if you want to eat fresh sweet corn for many weeks. Hybrids have weak seedling vigor, are more prone to ear damage by insects, and seeds rot in cool soil.

When chickens, pigs, cows and horses are given a choice between open-pollinated and man-made hybrid corn, the animals prefer open-pollinated. Most commercial varieties of corn are hybrid and GMO.

Genetically modified organisms (GMO) have had specific changes made to their DNA by genetic engineering in the laboratory such as adding pesticide production to the plant's properties. Most are transgenic meaning genes from other species have been inserted into the DNA. **Corn and soybeans** are the most genetically altered (GMO) plants. The long-term effects of this on human/animal health and the environment have not been determined yet. However, the Committee for Independent Research and Information on Genetic Engineering have done studies that show GMO food may cause kidney, liver, heart, adrenal and spleen problems. (For **open and hybrid pollination and about GMO**, see "How to Save Seeds" in the back of this manual.)

(For **sweet corn** see August Garden-Harvest. For **field and flint corn** see September Garden-Harvest. For **general grain harvesting,** see May Garden-Harvest.)

 Third of Three Sisters. Gourds (cucurbitaceae family which include pumpkins and squash; and Lagenaria siceraria). (See "Plant Families" in back of this manual.)

It is a **frost sensitive annual.** Soak seeds 2-3 days before planting. (Plants can be started in the greenhouse early.) Sow 4-5 seeds per hill, 4-5 inches apart with rows 5-10 feet apart. Thin to 2-3 seedlings per hill.

Gourd plants can take up as much space as pumpkin plants. Space small ornamental gourds 18-24 inches apart if they are trellised vertically 6-8 feet. Larger gourds require wider spacing and a strong trellis to hold the fruit. (For **trellis system**, see grapes in February Garden-Maintenance.)

Care: Needs full sun. Likes soil pH of 5.8-6.8. Keep well watered. Likes rich soil especially early in growing season. Care is **similar to squash.** (For squash see below.)

For plants in most families, cross pollination usually occurs only among members in the same species. However, **some cross pollination between species occurs in the genus Cucurbita** among pumpkins, squash and gourds (in the Cucurbita genus). Usually even among Cucurbita the species will not cross but sometimes they do. **Cross pollination does not occur between melons, cucumbers, squash or other species.** This is only important if you are saving seed.

Gourds are very susceptible to pests and diseases. (See "Plant Health" in back of this manual.)

Uses: Gourds are used as bird houses, containers, dippers, scoops, and sponges. They can be shaped with string or cloth when growing.

Types of gourds: Ornamental (mature in 90-100 days, about 10 varieties). Hard-shelled (larger and longer gourds, mature in 110-130 days, about 30 varieties). Luffa (sponge squash, mature in 110-180 days).

1. Cucurbita Gourds (Yellow Flowered) (Cucurbitaceae Family, Cucurgita genus): Aladdin, Egg, Orange, Warted, Apple, Flat Striped, Pear (Bi-color), Bell, Malahar Melon, Spoon, Crown of Thorns, Miniature, and Turks Turbin. (**If saving seed,** be careful about cross pollination with pumpkins and squash.)

2. Lagenaria Gourds (White Flowered) (Cucurbitaceae Family, Lagenaria genus): Bottle (Giant, Miniature, Siphon), Dipper, Longissima, Calabash (Penguin, Powderhorn), Dolphin (Maranka), Martin (Birdhouse), Caneman's Club, Hercules Bulb, Carsican Flat, and Italian Edible (Cucuzzi).

3. Miscellaneous Gourds (Cucurbitaceae Family): Luffa genus (Dishrag), Serpent Gourd (Trichosanthes genus), Momordica genus, and Teasel (genus cucumis, Squirting cucumber).

(See April Greenhouse but most growing information is in this section. See September Garden-Harvest for **harvest and curing.** See Cucurbitaceae Family in "Plant Families" in back of manual.)

 **Third of Three Sisters. Pumpkin** (cucurbitaceae family; Cucurbita pepo, Cucurbita mixta, Cucurbita maxima, and Cucurbita moschata). Matures in 100-115 days. **Grown the same as winter squash** (see below).

Types are listed in the **Winter Squash** section. (See October Garden-Harvest.)

 **Third of Three Sisters. Winter squash** (cucurbitaceae family; genus Cucurbita) is sown outside in May when soil temperature is 70 or higher about when lima beans are sown. It usually is a vine, some are bush. (Historically squash was planted for its seeds not its flesh.)

Soak seeds overnight. Germinates in 3-5 days. Plant seeds 1 inch deep, 4-5 seeds per hill. When young plants are well-established, thin each hill to 2 or 3 plants. (Extra plants are insurance if insects kill or damage vines.)

Vining squash needs 50 to 100 square feet per hill. Have 10-12 feet in all directions for vining squash and 5-6 feet for bush squash. Be sure plants are not crowded since this will reduce yield.

Care: Likes full sun. It likes soil slightly acid at 6.0-6.8. They do not like compacted, clay soil. It helps to plant a fall cover crop that is dug into the soil in late spring. They like a lot of **compost and manure** but do not overfeed nitrogen or phosphorus. Too much nitrogen reduces the number of female flowers.

They do not like cold weather. The ideal is days in the 80s and nights in the 60s. **Frost kills the plants.**

They grow better if **black plastic** is placed on the ground around the plants. It reduces weeds and retains water. The disadvantage is that roots along the vine can not grow into the dirt. Or you can mulch with straw.

Keep weeds under control until vines are well developed and can suppress weeds on their own. Put a few shovelfuls of dirt over sections of vine to encourage more rooting in case squash borer kills other parts of vine.

Pollination and Seeds

For plants in most families, cross pollination usually occurs only among members in the same species. However, **some cross pollination between species occurs in the genus Cucurbita** among pumpkins, squash and gourds. Usually even among Cucurbita the species will not cross but sometimes they do. Only important if you are saving seed.

If you want to save pure seed, plant far from other squash (minimum 500 feet with 1/2 mile apart better). There are male and female flowers. Bees frequently pollinate squash. If pollinated by a different variety, the fruit will still look the same. However, when the seeds are planted, a hybrid plant will grow. This is only an issue if you are saving seed. If you are not saving seed and just eating them, cross pollination does not matter.

The safest way to save pure seed is to hand pollinate. There is disagreement about which species will cross with each other but most say:

C. pepo will cross with **C. mixta** and **C. moschata**. **C. pepo** will not cross with **C. maxima**.
C. maxima will cross with **C. moschata**. **C. maxima** will not cross with **C. mixta**.
Cross pollination does not occur between melons, cucumbers, squash or other species.

There are 4 major domesticated squash species: (See "Plant Families" in back of this manual.)

1. **Cucurbita maxima** includes varieties such as Australian blue, banana, buttercup (very sweet), hubbard, mammoth, turban, zapallito, some varieties of prize pumpkins such as Big Max. Maximas have mild flavor, high in starch/sugar, low fiber/strings, and brilliant orange flesh. Great for canning. Many are large. Tolerant of cool temperatures.

2. **Cucurbita mixta (argyrosperma)** includes varieties such as green-striped cushaw, and Tennessee Sweet Potato. Has coarse, thin, pale flesh. Originally grown for seeds. Has large, easy-to-hull seeds. Likes hot, dry climates.

3. **Cucurbita moschata** includes varieties such as butternut, calabaza, winter neck/crookneck, Golden Cushaw, Seminole and Cheese pumpkins. They have buff (pale, yellow-brown) colored skin and have a flat, rounded shape like a wheel of cheese. They are thin skinned. The flesh is brilliant orange and high in carotenoids. They tolerate heat and humidity better than other squash and are more resistant to insects and disease. They do not like cold weather.

4. **Cucurbita pepo** includes **most pumpkins** such as oil and summer pumpkins of Europe, grooved pumpkins of North America, and ribbed Mexican pumpkins. All are hard skinned. Also includes **summer squash:** pattypan (scallop), yellow summer crookneck, yellow summer, and zucchini. (For **summer squash**, see May Garden-Plant.) **Winter squash:** acorn, delicata, spaghetti, and sweet dumpling. Also includes some **ornamental gourds** that are not edible. (See gourds above.) All do well in temperate zones with cool climates. They have good texture but are not as flavorful as other squash except for delicata that is very flavorful.

Other Tips

Store winter squash for 1 month after mature before **saving the seed**. Then wash seeds in a strainer and blot dry. Dry at room temperature in a well ventilated place out of direct sun for several weeks. They last 6 or more years.

Remove defective squash fruit from plants while small. In **fall** remove all young squash fruit that will not have time to mature so energy goes to those that are almost mature. Squash fruit can be moved a little if in a bad position. Leaves are edible cooked.

Pests and Diseases: The main diseases are vine borer, squash bug, and cucumber beetles (striped and spotted). Rotate crops to reduce soil borne and insect diseases. Floating row covers can be used to protect plants from insects but need to be removed for pollination to take place. Aluminum foil wrapped around the stem at the base of plants helps prevent squash maggots and vine borer from tunneling into the stem. (See "Plant Health" and "Plant Families".)

Vining Winter Squash: Acorn-Table Queen, Buttercup-Burgess, Buttercup-Sweet Mama, Butternut-Puritan, Butternut-Zenith, Hubbard-Blue, Hubbard-Golden, Hubbard-Green Warted, Spaghetti-Vegetable, Sweet Potato-Delicata, and Turban-Turk.

Bush Winter Squash: Acorn-Cream of the Crop, Acorn-Table King, Butternut-Waltham, and Spaghetti-Tivoli.

Popular Varieties: They range in size from 1 to 66 pounds.

1. Acorn squash (cucurbita pepo) is 80-100 days to harvest. Usually blue-green but can be orange, yellow, white, tan or multi-colored. Sweetest when turns bronze. Mild tasting. **Weighs 1-2 pounds. Stores 1-2 months.** Table Queen acorn squash matures in 58-90 days. Vine grows 6-8 feet long. Squash is 5-6 inches long. Sweet flesh is light yellow to deep orange. Thin seedlings 12-18 inches apart. Germinates in 6-10 days.

2. Banana squash (cucurbita maxima) is 105-120 days to harvest. Shaped like a cucumber or banana. Grows up to 2 feet long and 6 inches diameter. The skin is a creamy peach, blue or pink. Bright orange flesh. Sweet tasting. **Weighs 10-30 pounds sometimes 50 pounds. Stores about 2-3 months.**

3. Buttercup squash (cucurbita maxima) is 95-105 days to harvest. It is in the Turban group of squash. It has a large navel like a navel orange. It is turban shaped, dark blue/green with lighter green stripes. Creamy, brilliant orange flesh, not stringy. Very sweet (**one of the sweetest squash**). Tastes like a sweet potato. It is the driest, sweetest, finest-grained, and richest flavored of all fall squashes. Good for baking, steaming or mashing. **Weighs 3-4 pounds. Stores about 2-3 months, sometimes up to 4-6 months.**

4. Butternut squash (cucurbita moschata) is 80-110 days to harvest. Skin is beige and flesh is deep orange. Least hard rind. The darker the rind, the sweeter. Tastes like sweet potato with a nutty flavor. It is a little watery when cooked. Good for soup because it is not stringy. Good for baking. **Weighs 1-3 pounds. Stores about 2-3 months.**

5. Cushaw squash/pumpkin (argyrosperma/mixta) is 100-110 days to harvest. It looks like a large, striped butternut squash with a curved neck. The light yellow flesh is slightly sweet, medium-coarse, and fibrous. Good for baking or canning. Reliable producer. Resistant to squash vine borer. An heirloom squash that can grow as large as 30 pounds. **Usually weighs 10-12 pounds. They keep over the winter in storage extremely well, up to 6 months.**

6. Delicata known as sweet potato squash, (cucurbita pepo) is 90-100 days to harvest and is easy to grow. Heirloom squash with thin, yellow with green striped, ridged skin. Skin is edible. Creamy orange flesh tastes like corn and sweet potato. Very tasty. Good baked or steamed. It is 5-10 inches long. **Weighs 1-2 pounds. Stores 2-3 months.**

7. Hubbard squash (cucurbita maxima) is 110-120 days to harvest. Blue-gray skin, warty and thick skinned. Or orange, red, blue, green or white skin. Very large and irregularly shaped that tapers at ends. Sweet, dense orange flesh that takes a long time to cook. Cooks faster if cut up in smaller pieces. The skin, stringy center, and seeds are not edible. Good baked or steamed. Good to make pumpkin pie. **Weighs 10-15 pounds or more. Stores 5-6 months** at 50 degrees, 70% humidity. Remove stem before storing.

8. Spaghetti squash (cucurbita pepo) is 70-115 (usually 90-100) days to harvest. Small, watermelon/oval shape with golden-yellow skin. Flesh has a mild nut-like flavor that is stringy. The more yellow or tan the skin, the sweeter. Good baked. Low calorie. **Weighs 2-5 pounds. Stores about 2 months or less.** The strands can be dried in the sun.

9. Turban or Turk's Cap squash (cucurbita maxima) is 90-110 days to harvest. Has bright orange, green or white skin. Bright orange-red flesh. Has a mild hazelnut taste. Flesh is not very sweet. Good for soup. Ornamental too. **Weighs 3-10 pounds. Stores about 2-3 months.**

(See April Greenhouse but most details are in this section. See October Garden-Harvest.)

Read the book "The Compleat Squash" by Amy Goldman.

🧺 Garden- Harvest 🧺 (May)

Vegetables and Fruit: Carrots, flowers, rhubarb, spinach, and strawberry.
General Hay and Grain: Hay making, grain harvesting, and testing / drying / storing grain.
Grain: Rye-winter and wheat-winter.

 For how to make a **tincture, decoction or infusion,** see April Garden-Harvest.

Harvest early **carrots** sown in March. (For **early carrots** see March Garden-Plant and June Garden-Harvest with no details in either. See August/December Greenhouse. For **late carrots and sowing details**, see June Garden-Plant. See September/October Garden-Harvest with **harvest details in September.**)

Edible flowers: Anise hyssop, apple blossoms, arugula, borage, broccoli, broccoli raab, calendula, chamomile, chives, cilantro, daylily, dill, elderberry, fennel, honeysuckle, lavender, lilac, marigold, mustard, nasturtium, pea, pansies, pumpkin, radish, rosemary, rose, sage, squash, strawberry, thyme, viola, violet and others. Consult your health care specialist or a book specializing in edible flowers.
(For **how to make tincture, decoction or infusion**, see April Garden-Harvest.)

Continue harvesting **rhubarb** outer stalks if 1+ inch in diameter. Do not harvest all the stalks. The plant needs some left to grow well next year. (For **dividing and planting crowns**, see March Garden-Plant. For **how to harvest,** see April Garden-Harvest. For **removing flowers,** see April Garden-Maintenance.)

Harvest early **spinach** sown in March. Also harvest in **June.** (For **early spinach and sowing/growing details**, see March Garden-Plant. See May/June/October Garden-Harvest with **harvest details in October.** For **late spinach** see August Garden-Plant but no details. See August/September/October Greenhouse.)

First **strawberry** harvest mid to late May. **Pick June-bearing** strawberries every day for 2-3 weeks. Best picked in morning. Berries picked unripe will not ripen later. Do not leave picked strawberries sitting in sun.

1 quart = 2 pints = 4 cups strawberries weighs 1 1/4 pounds. About 25 plants started in a matted row system can yield more than **50 pounds of strawberries** after one year.

Handle gently. Remove all disease or damaged berries so disease does not spread.

Uses: Do not wash until ready to eat or process. Can eat fresh, dry or freeze. Good if frozen in milk. Cook some with rhubarb since they are in season at the same time

Pests: Chipmunks like to eat strawberries so set traps in burrows or next to holes. Or get cats.

(For **sowing/growing details**, see April Garden-Plant. See June Garden-Harvest but details are here. See June Garden-Maintenance. For **how to make trap boxes,** see September Farm Animals.)

 **Hay Making, Grain Harvesting, Testing/Drying/Storing Grain**

*Above images: First 3 are **sickles**. Then a regular **scythe**, a cradle scythe and a man using a scythe. The next image is **grain being harvested** in a field. The stalks are being gathered into **sheaves and shocks**. The next image is a **hay rake**. The last image is stalks/leaves/grain in **windrows for hay making**. (The light rows are hay.)*

*Above images: The first is a man with a **flail for threshing**. The next 3 images are **winnowing**. You bind and tie cut grasses or grain stalks into **sheaves**. Sheaves are bundled together upright to dry. These larger bundles are called **shocks**. The 5th and 6th images are sheaves. The last image is a bicycle-powered **drum thresher**.*

 **Hay Making** (without large equipment) ("Make hay while the sun shines.")

The best hay comes from pasture that is high in minerals and is grown without pesticides or chemicals. It should have many different types of plants in it for variety in nutrition. (For **fertilizers including rock dusts**, see "Garden Tips" in back of this manual. For **when to harvest hay**, see June Garden-Harvest. For **cool weather cover crops and grasses**, see March/September Garden-Plant. For **pasture**, see "Garden Tips".)

Use a **scythe** or brush mower to mow the field. (For **scythe information**, see General Grain Harvesting below.)

Then with a **hay rake**, move the leaves and stalks (grass) into windrows. **Windrows** are long rows where the stalks/leaves are gathered on the ground to dry. Turn the windrows every day for several days to dry all parts. Usually ready to store in several sunny days. There should still be green color to the hay when it is baled.

Or spread the grass evenly around with a three-pronged pitch fork instead of putting into windrows.

The most common way to **store the dried hay** is to stack the fodder in inverted cones or big rolls that shed water quickly and keep the inner layers of hay dry. Or can tie into bales with jute/nylon twine or baling wire. You can make a box to put the hay in to **create the bale** before putting on the twine. Hay can also be stored loose in a barn or shed. Do not store wet or damp hay, because it may mold or start a fire.

You can make your own **hay rake** with hazelnut for the handle, aspen or ash for the head, and ash or rowan for the teeth. You can use all ash or if that is not available, then use spruce. (See hay rake image above.)

   **Grain Harvest**

Basic Process for Grain Harvesting: Cut grain stalks and stack upright by tying them into **bundles** (sheaves and shocks) to dry in the field for several weeks. Then remove grain/seed heads (top of plant that flowered) from stalks by **threshing**. Then **winnow** (separate) grain from chaff (seed coverings and other debris). **Dehull** (remove hull or outer coating) from grain. **Winnow** again to get rid of leftover hulls and debris. Grain may have to be **dried** more. Then **store**.

Ripening grain goes through 3 stages: milk, dough and hard/ripe. In the **milk stage** the grain is a little liquidy and the leaves/stalks start turning from green to a little yellow. In the **dough stage** there is less liquid but the grain is still easy to dent. The leaves/stalks are changing to yellow and tan. In the **hard/ripe stage** the grain is mature. The leaves/stalks turn golden brown. Seed heads usually bend over.

Test the grain by biting on it or pressing between your fingernails. (See image below after solar dryers.) If it is soft, it is not ready to harvest. Though some people harvest at the late dough stage and then cure the grain in the field or a shed. If you wait too long, birds and other animals will eat the grain.

To harvest a small plot up to 150 square feet, cut off the seedheads with a **sickle** or pruners. Spread out to dry for a few days. Then thresh and winnow. (See below for threshing and winnowing details.)

To harvest larger areas, use a **scthye** or electric/gas blade trimmer. A scthye has a long handle and a 20-36 inch curved blade. It has 3 types of blades: grass, ditch and bush. Use a grass blade. Cut with the first third of the blade.

Grab the lower grip of the scthye with your right hand, palm down. Grab the upper grip with your left hand, palm up. Then stand with your legs apart to balance the swing and do a whole body pivot with knees bending. Shift your body from right to left and back. Keep the blade parallel to the ground. Sharpen the blade frequently. (Read "The Scythe Book" by David Tresemer.)

Bind the cut grain stalks into sheaves about 12-14 inches around. Bind the same day you cut. Use string or twine. Then take 8-12 **sheaves** and form a **shock/shook/stook** by piling shocks against each other. You can cover with a loose weave cloth to keep birds and rain away. Or cover with a bent sheaf to keep rain off. Keep in shocks until dry.

Threshing is removing grain from the seed heads and straw/chaff. There are many ways to thresh. Amaranth and quinoa are easy to thresh by rubbing seedheads between your hands. Thresh other grain by banging grain in a trash can, flailing with chains in a plastic bucket or trash can, flailing with 2 pieces of wood on a tarp, walking on the grain (animals can be used), hitting with a plastic baseball bat, or make a drum thresher (see above image of bicycle thresher). A **flail** is made from 2 or more sticks attached by a short chain. One stick is held while the other is swung.

Winnowing is separating the grain from the straw/chaff. Pour grain from 1 container to another in front of a fan, hair dryer, or strong breeze. It gets rid of light stuff not heavy stuff like rocks. (See above images.)

Some grains need to be hulled such as barley (hard to hull, there are hulless varieties), buckwheat (hard to hull), emmer wheat (Triticum dicoccum, hard to hull), millet (difficulty depends on variety, some are easy to hull), oats (hard to hull, but there are hulless varieties), rice (hard to hull), spelt (hard to hull), triticale (a wheat/rye hybrid that breeds true, easy to hull), and wheat (easy to hull, dehulls during threshing). Other grains **do not need hulling** such as amaranth, quinoa and rye. (For field and flint **corn harvesting and processing**, see September Garden-Harvest.)

Some grains such as buckwheat need to be sifted. To sift use a **seed/soil screen/sieve** to separate out the different sizes of seed. (Bountifulgardens.org sells a soil sieve with 3 interchangeable metal screens of 3, 5 and 10 mesh per inch.) (For **buckwheat** see June Garden-Plant.)

You can modify a home electric grain grinder or flour mill to **hull difficult grains** such as millet. A primitive method is using a blender. You can also use any machine that creates a hard striking action against the grain such as a modified hammer mill or grain blower.

Use a manual or electric grain mill to make **flour.** Or whole grains can be soaked in water overnight and then cooked. Or soak for a few days until they sprout, and then cook. This makes it easier to digest.

(Read "Homegrown Whole Grains" by Sara Pitzer, "Small Scale Grain Raising" by Gene Logsdon, and "Nourishing Traditions" by Sally Fallon.)

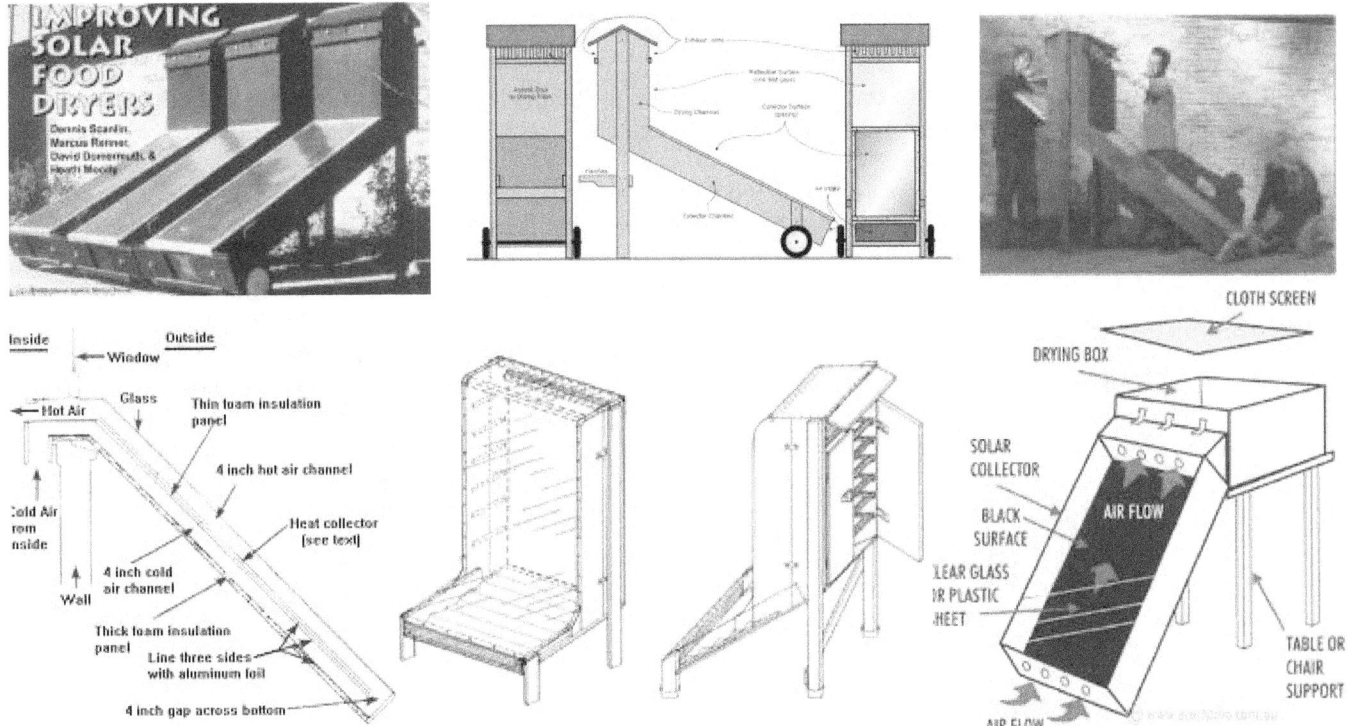

(Top 3 images are for the Homepower solar dehydrator. The second 3 images are other solar dehydrator designs.)

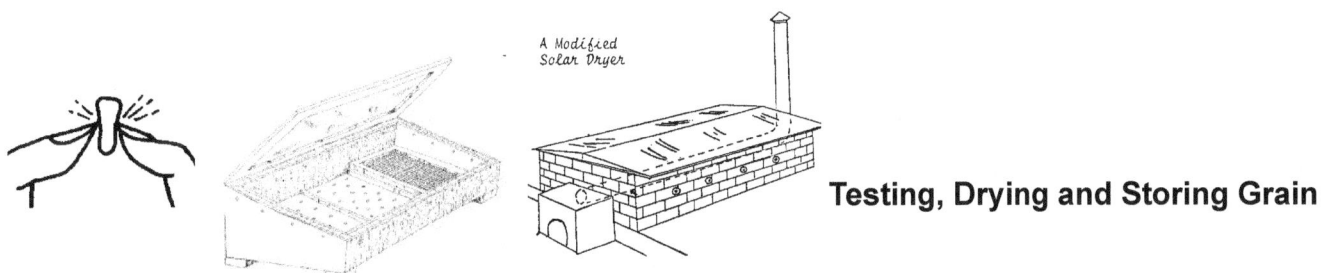

Testing, Drying and Storing Grain

Moisture Content: Most small grains are mature at **35 percent kernel moisture**. At this stage, wheat kernels are yellow without any trace of green and can be crushed between the fingers without finding any milk.

The moisture content of grain can be ballparked using **in-the-field methods**. You can nibble the grains. Or score grain with a thumbnail or crush between your fingers. Or smell a handful of grain. Or put a few grains in a metal box and see if there is a dull or sharp rattle when you shake it.

Of all these methods, the **best is the salt test.** Mix some grain in a glass with some dry ordinary kitchen salt. After shaking a few times, look at the walls to see if salt has stuck to them. If the salt sticks to the walls, it means the moisture content is higher than about **15 percent.**

Ways to Dry: Grain can be dried in the field on the stalk but it risks being eaten by birds and rodents or being rained on. If there are many rainy days, it could mold.

Grain can be separated from the hull and then dried in the sun on plastic or other surface. Or it can be dried in **a solar dryer.** A perforated floor or vent system is recommended for all in-bin or in-box drying. You can build a solar dryer with a sheet of glass/plexiglass and some wood.

(See above images from **Home Power magazine** 1997 and 1999. It can be used to dry vegetables or fruit too. Go to http://www.homepower.com. One article is called "Indirect, Through-Pass, Solar Food Dryer" by Dennis Scanlin, HP 57, February/March 1997. Another article is called "Improving Solar Food Dehydrators" by Dennis Scanlin, Marcus Renner, David Domermuth and Heath Moody, HP 69, February/March 1999.)

Grain should be **raked or moved** every few hours so new parts of the grain are exposed to the sun and air flow. Supplemental heat can be added such as from a propane heater. Drying may take several weeks depending on the airflow, weather, and the amount of water to be removed. (See above image.)

Storage: Grain that is too moist will heat up while stored and may mold. Store so rodents can not get to it. A metal/plastic bucket or barrel with a lid is good. The cooler the storage temperature, the longer the grain can be stored.

Put **diatomaceous earth (DE, diatoms) in stored grain to kill insects**. Use food grade not pool grade diatoms. Fifty pounds of diatoms protects 7 tons of grain from insects. That is 1 cup to 50 pounds of grain, or 1 pound for 300 pounds of grain.

Diatomaceous earth is a nonselective (kills good and bad insects), abrasive dust (powder) made from fossilized silica shells of algae called diatoms. It kills by projecting little needles into insects and by dehydration. (For **where to buy**, see fertilizers in "Garden Tips".)

(For **amaranth** see June Garden-Plant, September Garden-Harvest. For **buckwheat** see June Garden-Plant. For **corn** see May Garden-Plant, September Garden-Harvest. For **flax and oats**, see March Garden-Plant, July Garden-Harvest. For **sorghum** see May Garden-Plant. For **wheat, barley and rye**, see below and October Garden-Plant.)

Maximum Recommended Moisture Content of Grain, Bean and Seeds for Storage:

	Short term % (less than 6 months)	Long term % (more than 6 months)
Barley	14 %	12 %
Corn	15	13
Edible Beans	16	13
Flax seed	9	7
Millet	10	9
Oats	14	12
Rye	13	12
Sorghum	13	13
Soybeans	13	11
Sunflower- NonOil	11	10
Sunflower- Oil	10	8
Wheat	14	13

 **Harvest fall planted barley in May or June.** Spring-planted barley ripens in 60-100 days while **fall-planted** ripens 60-90 days (usually 60) after spring growth starts. **Flowers and matures about a week earlier than wheat.**

It is **mature** when it is golden and brittle. It moves easily in wind just like wheat. It is more prone to shattering (seeds falling out of seedhead) so harvest before it is fully mature and in the morning. A hulled seed should be hard to dent with your fingernail.

Bearded barley can be dehulled somewhat successfully with a blender. It dehulls easier if it has first been roasted.

Beardless (naked) barley does not have a hull so is better for the small farmer. (See October Garden-Plant.)

Seeds should be 14% or less moisture for good storage. Stalks and leaves can be fed to animals (ruminants).

Saving Seeds: It is outcrossing (outbreeding) and wind pollinated. You may need around 500 plants to get good genetic variety. (See "How to Save Seeds" in back of this manual.)

Using: Hulled barley (covered, bearded) is eaten after removing the inedible, fibrous outer hull. Once removed, it is called **dehulled barley (pot barley, scotch barley)**. **Pearl barley (pearled barley)** is dehulled barley which has been steam processed further to remove the bran. Pearl barley is processed into flour, flakes and grits.

(See October Garden-Plant. For more details, see **General Grain Harvesting** above.)

Read "Homegrown Whole Grains" by Sara Pitzer, "Small Scale Grain Raising" by Gene Logsdon, and "Nourishing Traditions" by Sally Fallon.)

Harvest fall planted rye (winter) in May or June. Winter rye **ripens about 1 week earlier than winter wheat.** It is ready in the spring 40-50 days after temperatures reach the 40s. Harvest when seed heads are full, almost dry and have changed from green to golden. Test grain by biting on it. Harvest at 22% moisture and dry to

15%.

Cut the grain with a **sickle or scythe**, putting it in windrows or shocks to dry. Thresh and winnow. Rye does not have a hull. Yield is 4 to 24 pounds from 100 square feet with the average being 8-10 pounds. Stalks and leaves can be fed to ruminant animals (sheep, goats, cattle).

(See October Garden-Plant. For more details, see **General Grain Harvesting** above.)

(Read "Homegrown Whole Grains" by Sara Pitzer, "Small Scale Grain Raising" by Gene Logsdon, and "Nourishing Traditions" by Sally Fallon.)

 Harvest fall planted **wheat in May or June.** Matures about 17-19 weeks (4-5 months) from date of planting. **Warm spring weather** causes rapid growth. Harvest is within 2 months. Stop watering when 85% of seed heads are golden.

Harvest when almost all **seedheads** are golden and the heads are tipping towards the ground. Stalks turn from green to yellow, brown or gold. Test every day by biting on it. If you wait too long, grain can shatter and fall to the ground.

Cut off seedheads and put in paper bag to dry. Or can use **sickle or scythe** to cut down entire plant. Then bind into sheaves. About 8-12 sheaves make a shock. Leave in field or bring to open shed to dry for 7-10 days. Dry fully before threshing and winnowing. Yield is up to 20 pounds per 100 square feet. Stalks and leaves can be fed to ruminant animals.

(See September/October Garden-Plant with **harvest details in October**. See **General Grain Harvesting** above.)

(Read "Homegrown Whole Grains" by Sara Pitzer, "Small Scale Grain Raising" by Gene Logsdon, and "Nourishing Traditions" by Sally Fallon.)

Greenhouse, Hoop House or Cold Frame  (May)

 Freezing temperatures are over so turn on greenhouse **water** and other unprotected water.

 Early May or late April sow tender annual herbs such as basil to **transplant after last frost**. (**Basil:** For sowing/growing details, see April Greenhouse. For transplanting see June Garden-Plant. See July/September Garden-Harvest.)

Move **tomato and tomatillo** seedlings from **house to greenhouse**.

(**Tomato:** For sow/grow details see March Greenhouse. See May Greenhouse and June Garden-Plant. See August/October Garden-Harvest with harvest details in August.)

(**Tomatillo:** See March/May Greenhouse and June Garden-Plant but sowing/care details are in March Greenhouse. See August Garden-Harvest.)

Farm Animals  (May)

 Turn automatic **waterers** back on in coop and barn. Turn off water heat tapes and heat bulbs.

 **Chickens, ducks and turkey** are still broody. (For **broody hens,** see April Farm Animals.)

Best **pasture growth** for animals is in May and June. A **spring flush takes place in cool season grasses and most legumes** because it is the best temperature (70-85 degrees) for growth and water is readily available. This grass is high in nutrients. Rotate livestock through paddocks or fields so they have just enough time to graze the tops off of plants but do not overgraze. Feed ruminants **high magnesium mineral salts** April through June. (For high magnesium minerals, see April Farm Animals. For **pasture** see "Garden Tips" in back of manual.)

 **Honey Beekeeping in May**

The beehive is fully functioning. **Nectar and pollen** begin to come into the hive very fast. Strong hives do not need any extra feeding from you. By May 1st **add extra honey supers** (boxes) to each hive on top of the top deep, anywhere from 2-4 may be needed. Add a queen excluder.

The queen is reaching her **greatest rate of egg laying** with brood chambers being filled up fast.

Bees swarm in May. Have an extra hive ready just in case you need to capture a swarm. Inspect the hive weekly.

Remove Apistan strips (treats varroa mite infestation) if they have been in the hive for 45 days. Remove menthol.

(For **beekeeping overview,** see January Farm Animals. For **harvesting honey,** see September Farm Animals.)

 # JUNE
June 1 daylight is 14 hours 22 minutes. June 30 daylight is 14 hours 29 minutes.

June 20 or 21 is **Summer Solstice- longest day of the year.** Somewhat rainy.
Early-Mid June: Honeysuckle, day lilies, lavender, rosa rugosa and elderberry bloom. Currants are ready to harvest.
Late June: Second-year parsley blooms. Raspberries are ready to harvest.
Self-seeded dill blooms. Marshmallow and anise hyssop bloom. Sourwood trees bloom.
June-bearing strawberries have stopped production for the year.

Garden- Maintenance

 Let **salad burnet** go to seed if want it to self sow or want to save seeds. If do not want seeds, cut back for new growth. Divide plants every 3-4 years in spring or fall. (See May Garden-Plant. For **dividing mature plants**, see "Garden Tips" in back of manual.)

 Late June or early July **weed eat and mow** along driveway, around apple trees, berries, etc.

General Fruit in June

  Set **apple maggot fly traps** on apple and other fruit trees.

Hang **red plastic or wooden balls** (see above image) that look like apples. Apply a sticky coating that catches apple maggot flies. (For **sticky paste formulas**, see How to Make Traps for Flying Insects in "Plant Health".)

Use 1 trap per 100 apples which is about 2-6 traps per standard tree and one trap per dwarf tree. Set out **3-4 weeks after petal fall (mid-June).** Continue through **September**. Make sure trap can not hit tree.

Apple maggots riddle fruit with holes while codling moths make one entry and one exit hole. Remove any apples that fall to the ground.

(For **pruning** see February Garden-Maintenance. See April Garden-Plant. See August/September Garden-Harvest with **harvest details in August.** See **How to Make Traps for Flying Insects** in "Plant Health" in back of manual.) (Read "The Apple Grower: A Guide for the Organic Orchardist" by Michael Phillips.)

Bagging Fruit to Protect from Insect Pests

In a small orchard individual fruits can be protected from insects by putting bags over each **3 weeks after petal fall** (all petals have fallen off). The bag stays on until **2-3 weeks before harvest**. It controls coddling moth, curculio beetle and other insect pest damage on apples, pears and other fruit. No pesticide sprays are needed once bags are put on fruit.

Before bagging, use **dormant oil spray** on trees in late winter or early spring. (For dormant oil spray, see February Garden-Maintenance.) May need to apply pesticide to kill plum curculio and coddling moth at the petal fall stage before bagging. Early in the season before bagging, also treat trees for **fungus disease** such as apple scab, cedar apple rust and fire blight. (Some sources say fire blight is a fungus, other sources say it is a bacteria.) (For **fire blight see pears** in April Garden-Plant. For **copper/sulfur sprays**, see February Garden-Maintenance. See **Sprays and Dusts** in "Garden Tips".)

After thinning fruit, place bags over remaining fruit. (For thinning fruit, see below.) Or put bags on fruit when 1/2 to 3/4 inches across.

Can buy **Japanese apple bags** designed for this. Or buy 3-pound paper bags cut to 6 inches long. Or use no. 2 paper bags (standard lunch bag size measuring 7 1/4 by 4 inches). Cut a 2-inch slit in the bottom of the bag and slip this opening over the fruit to form a seal around the stem. Staple the open end shut.

Some say use only paper or Japanese bags because **plastic bags** lead to overheating. Others say plastic bags such as regular baggies are OK. However, plastic bags do need 1-2 small holes cut in the bottom of the bag so water can drain. Tie on with string or twisty ties or staple shut.

(For **more about sprays,** see "Plant Health" in back of manual.)

Propagate **blueberries step 3.** Transplant 10 week old blueberry plant. (For **step 1** see December Garden-Maintenance. For **step 2** see March Garden-Maintenance. For **how to plant**, see April Garden-Plant.)

Look for **fire blight in fruit trees** now and throughout summer. Some sources say fire blight is a fungus, other sources say it is a bacteria. It is found in apple, Asian pear, pear and other trees. It destroys fruit, branches and even entire trees. Remove sick limbs and fruit. (May affect raspberries.)

Symptoms: Shortly after bloom there is **blossom blight**. In the early stages, blossoms look watersoaked and gray-green but then turn brown or black. Then there is **shoot blight** occurring 1 to several weeks after petal fall. The leaves and stem on young shoot tips turn brown or black and bend over into a shape like the top of a candy cane. Small droplets of sticky bacterial ooze may be seen on these shoots. Then it can travel down **stems** causing the tree to look scorched by fire. **Older wood** can be infected causing dark sunken cankers. **Fruit** may look small, dark and shriveled if infected when young, or have expanding red, brown or black lesions when infected later. They may have droplets of sticky bacterial ooze.

Control: Plant disease resistant varieties. Keep properly pruned. Do not over-apply nitrogen. Control insects with piercing and sucking mouthparts such as aphids, leafhoppers, pear psylla since they can spread the disease.

Apply a **copper-containing fungicide/bactericide** at or shortly after green tips appear in spring. Certain antibiotics can effectively protect against blossom infections when applied shortly before or immediately after they occur in spring.

Prune out infected blossoms, shoots, stems, branches or parts of trunk.

(See **pears** in April Garden-Plant. See **Apply Dormant Fungicide Sprays to Fruit and Nut Trees**, in February Garden-Maintenance. See **Fungi Disease** and **How to Make Organic Fungicide Sprays/Solutions** in "Plant Health".)

Prune hardy kiwi 2-3 times in summer by cutting non-flowering side shoots that go outside of trellis. Cut flowering shoots back to 4-6 leaves beyond last flower. Remove watersprouts (vigorous upright shoots) and shoots from trunk. A lot can be removed doing this.

(For **pruning** see February Garden-Maintenance. See April Garden-Plant. For **propagation** see July Garden-Maintenance. See September Garden-Harvest.)

 **Strawberry (June-bearing) Upkeep**

Keep well **watered** to encourage berry growth.

Late June or early July renovate (clean and renew) June-bearing strawberries bed after last harvest. Remove mother plants (plants 3 years old or older). Clip off old leaves or mow off all leaves but be careful crown is not cut. (If renovation is done in August or later, do not cut off leaves.) (For **how to create beds**, see April Garden-Plant.)

Then remove some plants so rows are 8-12 inches apart again. Mound dirt around crowns of plants. **Runner production** will start. Use runners to fill in bare spots. With this system the strawberries should maintain themselves for 4-5 years. At that time replace with new plants.

Fertilize now. This is better than fertilizing in spring which may cause soft fruit. Do not fertilize after August.

Set traps for slugs/snails in strawberries and spread wood ash. (For **how to make traps for snails and slugs,** see "Plant Health" in back of manual.)

(For **sow/grow details**, see April Garden-Plant. See May/June Garden-Harvest with **harvest details in June**.)

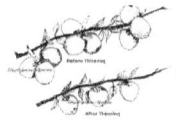 **Thin fruit and nut trees early June or late May.** (For fruit trees see April Garden-Plant.)

All fruit trees need thinning. The rule of thumb for **very small fruit** such as mulberries is: if you can touch 2 with one thumb, they are too close. For **larger fruit**, if you can touch 2 with 1 fist, they need to be thinned.

Thin apple trees before **June drop**. June drop occurs 6-8 weeks after full bloom. Thin when fruit is 1/2 inch diameter (size of a dime). This should be done sooner than 35 days after full bloom. First thin diseased or misshapen fruit. Then remove all but the largest fruit in a cluster. Be careful not to damage the stubby fruit spur branch.

Cut off fruit so remaining apples are **spaced 4-6 inches apart for small varieties, and 6-9 inches apart for large varieties.** Leave only 1 fruit per cluster.

 Thin Asian pears to 6 inches apart when fruit is size of a cherry.

Thin apricots so fruits are about 2 inches apart.

 Thin peaches and nectarines to 1 fruit per cluster when they are the size of hazelnuts (about 1 inch across). After the natural drop, thin again to 8-10 inches apart.

Thin pears midsummer to 1-2 per cluster or 6 inches apart. The natural fruit drop for pears occurs later in the summer than for most other fruits.

Thin plums 2-3 inches apart for small varieties, and 3-4 inches apart for large varieties.

Nut Thinning: Each nut may need 8-10 or more leaves to develop properly. If there are fewer nuts, the ones remaining develop bigger and healthier. Sometimes 20-30% of nuts need to be thinned off the tree. Trees with health problems need heavier thinning to give the tree a year to rest.

Thinning is best done in early morning or late evening **when nuts are the size of a dime or nickel.** By the **third or fourth week of June** most natural nut drop is finished. Natural nut drop occurs from low vigor, lack of pollination, lack of water, pests and disease.

You can remove nuts by hand. Or for larger trees, you can shake the tree. If leaves or branches fall, you are shaking too hard. Do not shake for more than 6 seconds. You can shake lightly 2-3 times with short bursts (2-3 seconds each). If shaking does not remove nuts, then **wait a few weeks or until mid July and try again**.

If the tree is too big to shake, then use a **pole pruner** and snip out some nut clusters. Or you can bump or jar limbs or nut clusters using a **padded pole, plastic bat, or stick** with a short piece of garden hose on the end.

If nuts are thinned properly, then it is less likely for there to be off years where few nuts are produced. If too many nuts are left on, then the tree may have alternate year bearing to recover from the abundant crop the prior year.

 **Whitewash trunks of fruit and nut trees** with a paint brush. Put on a light coat. If the bark is thick, then add extra water to the formula. May need to apply 2 coats if did not get it into all the crevices. Go **2 feet up trunk.** Can be done once a year or more. Can be done **April, June, and/or August**. If only once, do in April.

Whitewash Formulas:
1. Use **white water-based latex paint** mixed with an equal amount of water or with twice as much water as paint. If the bark is rough, use the formula with more water. Do not use oil-based paint.
2. Dissolve 3 pounds **table salt** in 12 quarts water. Stir in 10 pounds **hydrated lime.**
3. Or use pottery grade **Kaolin clay** (Surround brand, a chalky clay) mixed to a consistency of overly wet plaster.
4. Mix diatomaceous earth (DE, diatoms) powder with water and a little liquid dish soap until it is a thick slurry. (For **diatomaceous earth** and where to buy it, see fertilizers in "Garden Tips".) Do not use antibacterial soap since it may cause damage to plants. The best to use are **Ivory Liquid or Shaklee Basic H.**

Why Whitewash: Whitewashing stops winter sunshine from damaging the tree (sun scald) while it comes out of dormancy in the spring. Whitewashing makes the trunk a smooth surface that makes it hard for some insects to crawl up. And it leaves a powder residue that attaches to insects and they do not like biting it. **Kaolin clay** is particularly good at repelling insects. Kaolin can also be sprayed on the underside of leaves.

(See April/August Garden-Maintenance but details are here.)

Garden- Plant or Transplant 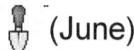 (June)

Early to Mid June:
　　Amaranth, basil, buckwheat, cabbage, eggplant, globe artichoke, lima bean, marsh mallow, nasturtium, peanuts, potatoes, sunflower, tobacco, tomato and tomatillo.

Late June or Early July:
　　Broccoli, carrots, cauliflower, rutabaga, and sweet potato.

Early or Mid June outside or when soil is 65+ degrees:
　　Plant summer cover crops (green manure) such as buckwheat, cowpeas (black eyed peas), millet, sorghum, sweet clover, soybean and **summer grains** such as amaranth. Green manure crops are cover crops grown and then plowed under to enrich the soil. Also used to prevent soil erosion.
　　(For **cowpeas, soybeans and sorghum** see May Garden-Plant. For **vetch and ryegrass** see September Garden-Plant. For **clover, fava bean and mustard** see March Garden-Plant. For **rye and wheat** see October Garden-Plant. For **more about cover crops**, see March/September Garden-Plant. For more about **cool and warm weather cover crops and also winter annuals**, see "Garden Tips" in the back of this manual.)

 In early June or late May sow **amaranth** (amaranthaceae family; Amaranthus cruentus, Amaranthus hybridus, Amaranthus hypochondriacus, Amaranthus tricolor; related to pigweed) **in spring after risk of frost** is over, or in **summer/fall** up to 2 months before cool weather. Best time to sow is **late May to early June** when soil temperature is 65 degrees or more. (See Amaranthaceae Family in "Plant Families" in back of manual.)
　　Seeds are very small. Germination may be poor. Plant 1/2 to 2 pounds of seed per acre. Sow 1/4 inch deep in rows 12-18 inches apart. Thin plants to 6-18 inches apart. Thinned plants can be eaten. Grows 4-8 feet tall and 3 feet wide.
　　Care: Likes soil pH 6.5-7.5. It is a **tender annual** that likes full sun. Needs a moderately rich soil. Water sparingly after established. Drought tolerant.
　　Some varieties are grown more for their grain and other for their leaves. Grain amaranth matures in 100-120 days. Leaf amaranth matures a little sooner. Will **cross pollinate** with pigweed and lambs quarters. Self seeds.
　　Uses: Leaves can be eaten raw or cooked. Leaves are high in oxalic acid (same as spinach) so eat in moderation. Cooking reduces oxalic acid. Harvest grain in September. (See September Garden-Harvest.)

Transplant **basil** from greenhouse to outside. (For **sowing and planting details,** see April Greenhouse. See May Greenhouse. See July/September Garden-Harvest with **harvest details in July**.)

 Sow **buckwheat** (polygonaceae family; Fagopyrum esculentum). Use as **cover or smother crop** and then dig under. Use as cover crop if ground will be bare for over 5 weeks. Or grow for **grain**. **A frost sensitive annual** that does not like excessive heat. Fast growing. Grows 1-3 feet tall.
　　Sow 3 or 4 months before first frost in fall so this June to July. There should be no hot weather (greater than 90 degrees) after flowering begins. **It needs at least 10 weeks (2 1/2 months) of frost-free growing weather.**
　　Broadcast seed at 1 cup per 100 square feet, 2-4 pounds per 1000 square feet, 1 pound per 300-500 square feet, or 60-80 pounds per acre. Rake in. Germinates in 3-4 days. (See **Garden Seeders and Broadcast Spreaders** in "Garden Tips" in back of manual.)
　　Starts **flowering** 4-6 weeks after sowing. Continues to flower for 4 weeks. New flowers form throughout the season so there are **ripe and unripe seeds** on the same plant. It takes 9-12 weeks for grain to mature.
　　Care: Grows OK in acid soils. Likes soil pH of 6.0-6.5. Easy to grow and does OK on poor soil. If given lots of nitrogen, it will grow more leaves and less seed. Has few pest or disease problems.
　　Bees love it. Beekeepers like it because of it's extended flowering period. The honey is dark with a different

taste from the more common clover honey.

Harvest when 3/4 of grain has turned brown. It is easier to harvest if it has not been killed by frost. Cut the stems gently with grass shears (mature seeds shatter easily), lay on a clean sheet, and beat with a broom or stick. Unripe seeds stay attached to the plant. Winnow by pouring seeds and chaff in front of a fan or in wind several times.

Hulling: Sift the buckwheat through a **seed/soil screen/sieve** to separate out the different sizes of seed. (Bountifulgardens.org sells a soil sieve with 3 interchangeable metal screens of 3, 5 and 10 mesh per inch.)

When you run it through a grain mill (a stone-burr mill is preferred), vary the setting for each size of seed. You will have to experiment. Try for a setting that breaks the hulls but leaves the interior buckwheat whole. It will evenly break the hulls apart if they are all the same size. If you do not sift it first, there will be small, intact hulls in the ground buckwheat.

Then resift the buckwheat through a seed screen with smaller holes. With the right screen size, all of the broken hulls stay on top, and all the hulled buckwheat falls through. Spread the buckwheat on a tray outdoors and blow across it to get rid of any leftover hull pieces. Or use a fan.

Uses: Once seeds are dehulled, the rest (groats) can be ground into flour or cooked whole (kasha). Buckwheat is high in lysine, which wheat and corn are low in. It is a high quality protein (12%) close to animal protein in content.

(See July Garden-Plant but most information is in this section. For **general grain harvesting**, see May-Harvest.)

(📖 Read "Homegrown Whole Grains" by Sara Pitzer, and "Small Scale Grain Raising" by Gene Logsdon.)

 Sow late/storage **cabbage** (brassicaceae family; Brassica oleracea). **Sow 10-12 weeks (2 1/2 to 3 months) before first fall frost. That is June or early July. Red cabbage** stores better than green cabbage.

(For **early cabbage**, see March Greenhouse but it has little information. For **details about sowing and growing,** see April Garden-Plant. For early cabbage, see June Garden-Harvest. For **late cabbage and harvest details,** see October Garden-Harvest.)

 Plant **eggplant** seedlings outside **early June** when temperatures are 65-70 degrees or when oak leaves are fully developed. It is sensitive to transplanting. (For **sowing and growing information,** see April Greenhouse.)

Early June or late May, transplant 12-week-old **globe artichoke** seedlings from **greenhouse to outside** when soil is 60+ degrees and there is no more frost. Plant 24-72 inches apart in 30 inch wide bed.

(For **growing system and sowing,** see February Greenhouse. For **second possible sowing date,** see March Greenhouse. To **move from house to greenhouse,** see April Greenhouse. See August/September Garden-Harvest with **harvest details in August**.)

Sow **lima beans** (leguminosae family; Phaseolus lunatus) **3-4 weeks after last frost in spring** and soil temperature is 65 degrees or more. **A frost sensitive annual.**

Sow seeds 1 inch deep, 2-3 inches apart. Sow with the eye side down. Thin seedlings to 4 inches apart in rows 2 feet apart. There are bush and vine varieties. Bush types grow 2 feet tall and mature in 60-80 days. Vine types grow 10-12 feet tall, so need a trellis or pole for support. They mature in 80-90 days.

Butter beans and Carolina beans are small varieties of lima bean.

Pods are pale green and 3-8 inches long. The outside pod is not eaten. Harvest when lima bean pod is round and firm and beans are visible. Pick throughout the season because mature beans on vine discourage new ones from growing. (For **other beans and more planting details,** see May Garden-Plant.)

 Sow **marsh mallow (marshmallow)** (malvaceae family; Althaea officinalis) on the soil surface. Germinates in 10-15 days. Thin seedlings 1 foot apart. Grows 3-6 feet tall. If sown in greenhouse, plant in degradable pots since they do not like being transplanted.

It is a hardy **perennial** with pink or white flowers in late summer. In fall harvest **root** in plants 2 years or older.

Care: Will grow in partial shade but prefers full sun. Likes soil pH 6.0-7.5.

Uses: Young leaves, flowers, seeds and roots are edible. **Leaves** are used for bronchitis, catarrh (nose or throat mucous), and coughs. Used to treat cystitis (bladder infection) by making a tea from the flowers and leaves.

The **root** has a soothing and protective mucilage (sticky substance) (11%). It is used in syrups, teas, poultices, lozenges and ointments. It is good for the treatment of coughs, minor skin wounds/irritations, and sore throats. Mild laxative and diuretic (increases urination).

Cook root to eat it. During the Middle Ages, people boiled marsh mallow roots, then fried them with onions in butter. Water from cooking any part of the plant can be used as an egg white substitute. Root is used to make marshmallow confectionery. Mild laxative and diuretic (increases urination).

Marshmallow Root Cough Syrup: 2 teaspoons chopped dried root, 2 cups water, 2 cups sugar, and 1/4 cup orange juice. Put the root in the water and bring to a boil. Lower the heat and simmer for 20 minutes. Strain into another pan. Over low heat, gradually stir in the sugar so a thick syrup forms. Simmer 5 minutes. Let cool slightly, then gradually mix in the orange juice. Consult your doctor or health care specialist.

(For **how to make tincture, decoction or infusion,** see April Garden-Harvest.)

Sow **nasturtium** (tropaeolaceae family; Tropaeolum majus) seeds. Rub seeds with nail file, then soak seeds overnight. Sow 1/2 to 1 inch deep, 10-12 inches apart. Needs darkness to germinate. Germinates in 7-14 days.

Care: Easy to grow. Likes soil pH of 6.5-7. Prefers full sun. Will grow in partial shade, but will not bloom. Drought resistant. An **annual** in zone 5-6. Grows bushy and 1 foot tall. Or grows as a vine 5 feet long. Few pests.

The fragrant flowers range in color from yellow to burgundy. They **bloom spring through fall**.

Uses: Flowers, seeds and leaves are edible. Older leaves are bitter. Seeds can be used like pepper.

Helps deter aphids, cucumber beetles, squash bugs, whiteflies and other pests from other plants. (For **insect repellant recipe** made with nasturtium, see "Plant Health" section in the back of this manual.)

Plant **peanuts** (leguminosae or fabaceae family; Arachis hypogaea, groundnut, earth nut, or goober peas). It is a legume in the **bean family**, not a nut. (For "Plant Families" see back of this manual.)

Sow 2 inches deep, 7-10 inches apart. Shell before sowing. Do not sow split kernels. Germinates in about 3-7 days. Grows 1 to 1 1/2 feet tall. The yellow flowers are similar to pea flowers. Can start in greenhouse earlier in season.

Care: Likes full sun. Likes soil pH of 5.9-6.3. Likes a moderately rich, well drained soil. Fertilizer with calcium helps the peanuts grow. Not many pests or diseases.

How peanuts form: The plant flowers at the end of runners. The runner (called a peg) bends to the ground where the peanut grows underground. This is a continuous process with nuts maturing at different times.

Popular varieties:

Spanish: Have a higher oil content than other types of peanuts. Prior to 1940 it was the most popular type grown in Georgia. Smaller but usually have 3 nuts in each shell. Matures in 140-145 days.

Runner: It has good flavor with better roasting quality and higher yields than the Spanish type. Harder to harvest since spread out. Matures in 160-170 days.

Virginia: Most common peanut sold. Produces large peanuts with 2 kernels in each shell. Grown in Virginia, North Carolina, Tennessee, Texas, New Mexico, Oklahoma, and Georgia. Can be either bunch (bush) that grows 18-22 inches tall and 28 to 31 inches wide. Or it can be running. Bush type is easier to harvest. Matures in 130-150 days.

Valencia: It is coarse with heavy reddish stems and large foliage. Matures before other types in only 100 days. Grown in Texas and New Mexico. Grows 49 inches tall. Matures in 90-110 days.

Tennessee Red and White: Similar to Valencia except stems are green to greenish brown. The pods are rough, irregular, and have fewer kernels. Matures in 110 days.

Harvest: Each plant produces about 30-50 peanuts in pods (shells) that are 1-3 inches long. When leaves turn yellow, pull out whole plant or dig out with a spading fork. The peanuts are attached to the roots. Let the nuts dry for 2-4 weeks before storing. For storage leave in the shell. Unrefrigerated but in a cool spot they will last about 3 months.

Uses: Can eat raw or roasted. They are high in protein. Some people are deathly allergic to peanuts.

How to Roast Peanuts: Freeze shelled peanuts overnight. Then rub skins off with your fingers. Or can roast with shells on. Spread in a shallow layer on a cookie sheet. Roast at 350 degrees for 15-20 minutes for shelled peanuts and 20-25 minutes for unshelled peanuts.

June is last month to plant **potatoes** (white/red/blue). Potatoes are planted April through June, and harvested June through October.

(For **when and how to plant early/mid/late season potatoes**, see April Garden-Plant. For **sprouting to plant in April,** see March Greenhouse.)

(**Potato Harvest:** For early/short season potatoes, see June/July Garden-Harvest. For mid-season potatoes, see August Garden-Harvest. For late/storage potatoes, see September/October Garden-Harvest. **October Garden-Harvest has details about harvesting.**)

Sow **sunflower** (compositae family; Helianthus annuus) in early June. (See "Plant Families".)

Sometimes called the **fourth sister** by Native Americans. They sowed them at the end of March as soon as the soil could be worked. You can experiment with this but they are **frost sensitive annual.** (The **three sisters** are corn, beans and winter squash. For more about three sisters, see corn in May Garden-Plant.)

Soil needs to be at least 46 degrees. Then cover with cold frame or row cover if planting before last frost. Matures in 70-120 days.

May need to cover seeds with screen or hardware cloth to prevent **animals** from digging them out. They do not like to be transplanted so use degradable pot if transplanting (start **2-3 weeks before last frost in spring**) from greenhouse.

Sow seeds 1 inch deep, 6 inches apart in rows 3-4 feet apart. Germinates in 5-10 days.

Spacing depends on variety: Giants are thinned to 3 feet apart in rows 3-4 feet apart. Regular or intermediates are thinned to 2 feet apart in rows 3 feet apart. Miniatures or dwarfs are thinned to 1 feet apart in rows 3 feet apart.

Care: Likes soil pH of 5.7-8.0. Prefers full sun (needs at least 6-8 hours a day). Needs well drained soil. Needs a rich soil. Keep well watered. Sunflower plants usually do not require staking but some do.

Apply extra phosphorus and potassium when the flower bud begins to develop for bigger flowers. They have few pests or diseases.

Types: Giants grow 10-20 feet tall with 20-24 inch wide flowers. They include Mammoth, American Giant hybrid, and Skyscraper. Regulars grow 6-10 feet tall. Miniatures or dwarfs grow 2-4 feet tall and are used ornamentally.

Varieties:

Sunflower varieties grown for their seeds are called confectionery sunflowers and are usually related to Russian Mammoth or Israeli varieties. They have black-and-white or gray-striped hulls.

Seed varieties include Russian Mammoth, Paul Bunyan, Giant, and Russian Giant that grow rapidly with the diameter of the head being 14 inches or more. About a third of the seed is oil and 1/5 is protein. There are 1,000-5,000 seeds in each head.

Sunflowers grown for oil have smaller seeds with black, thinner hulls. These are the type found in many bird feeder mixtures. After the oil is extracted from the seeds, the remaining seed cake is fed to animals.

Use: Sunflower is good feed for chickens, ducks, turkeys, goats and other farm animals. They are good for ruminants who are having skin problems (they eat them with the shells on). The leftover stalks and leaves are good for compost, mulch, feeding to ruminants, and other uses.

(See September Garden-Harvest.)

 Transplant **tobacco** (solanaceae family; Nicotiana tabacum) from **greenhouse to outside.** Plant similar to tomatoes. Final planting outside for tobacco is 2 feet apart in rows 3 feet apart in **early June.**

Needs full sun. In the **summer** the tops should be removed so upper leaves get larger. After the tops are removed, remove suckers or axillary buds if they grow to an inch or more.

Remove flower buds unless saving seed. (For **sowing and care details,** see April Greenhouse. See August Garden-Harvest. For tomatoes see March Greenhouse.)

Transplant **tomatoes and tomatillos** (solanaceae family) from **greenhouse to outside** when dogwoods are in full bloom. Can **sow directly in June** especially for cherry tomatoes or other early (short season) varieties. They are **frost sensitive.**

(See March and May Greenhouse with many **sowing/transplanting/care details in March Greenhouse.** See August/October Garden-Harvest.)

Late June or early July outside:

Sow late **broccoli** for fall eating. **Sow 12-14 weeks before first frost in fall.** Can also sow in **March** for early broccoli. (For **sowing/growing details,** see March Greenhouse. See October Garden-Harvest.)

Sow late **carrots** (umbelliferae family; Daucus carota) from **mid June to early July** for fall harvest. Sow **10-12 weeks before first frost in fall** for late carrots. Early carrots are sown in March, **2-3 weeks before last frost in spring**.

Carrots have slow **germination** usually 10-14 days but can be up to 1 month in cold soil. To make them easier to plant, mix seed with sand or vermiculite since the seeds are so small. Cover with 1/4 inch soil, never more than 1/2 inch.

Keep soil moist. You can cover seeds with a board to help keep soil moisture. You can plant radishes inbetween carrot rows in the same area since the carrots take so long to germinate. (See "Plant Families" in back of manual.)

When the seedlings are 2 inches tall, thin to 2-4 inches apart. Cut the seedling, do not pull it out since it will disturb nearby seedlings if pulled. Succession plant **March through June.** Matures in 60-76 days.

Care: Likes full sun. Likes soil pH of 6.0-6.8. Needs soil with few rocks and not too hard. A **biennial.**

Keep well weeded. Likes moderately rich soil but not too much nitrogen since it causes roots to branch. Put wood ashes around the plants to keep away rust flies, carrot weevils, wireworms, and other carrot/parsnip pests. (See "Plant Health" in back of manual.)

To get big roots: Drive a crowbar into the soil 2 feet. Rotate the bar in a circular motion until the hole is about 6 inches across the top. Fill hole with a mixture of sand, peat moss and sifted soil, leaving a small depression at the top of the hole. Place 2 or 3 pre-sprouted seeds in the depression, then cover with 1/2 inch of sifted sphagnum moss and water. Space holes 3-4 inches apart. Later thin to 1 strong plant. Can mulch with straw or leaves.

Most carrots are orange but some are black, red, purple, white or yellow.

Carrots are divided according to **how deep they grow**:
Amsterdam carrots are small and thin, growing up to 3 inches long.
Chantenay carrots grow around 5-6 inches long. They are a wide carrot.
Danvers carrots are thin and grow up to 7 inches long. The tops are thick and the flavor is strong.
Imperator carrots are thin and grow up to 10 inches long.
Nantes carrots are sweet and round that grow 6 inches long.
Paris Market carrots are very short carrots that grow to about 1 1/2 inches in diameter.

Carrots good for clay soil: Amsterdam Forcing 2, Amice, Autumn King, Chantenay Red Cored, Danvers Half-Long, Five-Star Baby, Little Finger, Mini Round, Nantes Half-Long, Parmex, Pioneer, Spartan Bonus, and Thumbelina.

Carrots good for fall planting: Artemis, Autumn King, Barwon, and Carotene 200.

Varieties recommended for fall planting in North Carolina: Danvers Half Long, Little Finger, Scarlet Nantes, Spartan Bonus, and Thumbelina.

Carrot Family: Anise, caraway, carrots, celery, chervil, coriander/cilantro, cumin, dill, fennel, parsnips, parsley and queen anne's lace are members of the carrot family (apiaceae or umbelliferae). (See "Plant Families" in back of manual.)

(For **early carrots** see March Garden-Plant but not much information. See June Garden-Harvest but not much information. See August/December Greenhouse but not much information in either. See September/October Garden-Harvest with **harvest and storage details in September**.)

Sow late **cauliflower** (brassicaceae family; Brassica oleracea) in late June. **Fall planting is better. Sow outdoors 10-14 weeks (2 1/2 to 3 1/2 months) before first fall frost. Will withstand light frost down to 29-32 degrees.** Cauliflower and chard are more sensitive to cold than broccoli, collards, kale, kohlrabi, or mustard.

(For **sowing/growing details and early cauliflower**, see April Greenhouse. See September Garden-Harvest. For **transplanting**, see May Garden-Plant but no details.)

Sow late/storage **rutabaga** (brassicaceae family; Brassica napus, swede turnip, winter turnip, yellow turnip). **Sow 12 weeks before first frost in fall.** Plant when nights are 50-60 degrees. Sow 1/2 inch deep, 1-2 inches apart in rows 15-24 inches apart. Germinates in 7-10 days. **Hardy to 25 degrees.**

When seedlings are 3 inches tall, thin to 4-6 inches apart. Thin a little over time. Eventually thin 12-24 inches apart. Can start eating when roots are 3 inches across so thin as they become crowded. Matures in 60-100 days. (Needs about 4 weeks longer than turnips to mature.)

Care: Easy to grow. Not usually transplanted from greenhouse. Likes soil pH of 6.2-6.8. Does well in heavy soil but it is good to remove rocks. Does OK in average soil but does better in somewhat richer soil. Likes well drained soil. Drought tolerant. But make sure gets enough water, otherwise root will be tough.

Description: A **biennial in the mustard family**. (See "Plant Families" in back of this manual.) Rutabaga has a large swollen root with skin that is purple, creamy brown or a combination of both. The inside is yellow or white. It is larger and sweeter than a turnip. Root is 3-5 pounds. It is a cross between a cabbage and a turnip that was developed in the Middle Ages. The greens are edible but not as good as turnip greens.

The most common **variety** is the American Purple Top (matures in 90 days). It has yellowish skin with purple shoulders and firm, light yellow flesh. It stores well.

Uses: Best harvested after frost. Good for animal fodder. Good survival farming food.

(See October Garden-Harvest. For **turnips** see July Garden-Plant. For **rutabagas as forage**, see Pasture in "Garden Tips" in back of manual.)

Transplant **sweet potato slips** (convolvulacae family; Ipomoea batatas) **from house or greenhouse to outside** as soon as **leaf cluster** forms above roots on slip and then slips have been rooted (roots soaked in water to form roots). Handle gently. If too cold outside, plant in greenhouse. They are **frost sensitive.**

Half of the slip should be in the dirt. Soil should be 70-85 degrees before planting outside, probably **June**. Plant 12-18 inches apart, in rows 3-4 feet apart. Soil should have few rocks and not be too hard. Water every day for the first week. Mound soil around plants as they grow.

Care: Likes full sun. They like slightly acid soil with pH of 5.5-6.5. Clay soil is OK. Does OK in moderately rich soil. Do not overfertilize because there will be too much growth of leaves and not enough of sweet potatoes.

It is sensitive to drought at the tuber initiation stage which is 50–60 days after planting. Keep well watered except do not overwater the last 3-4 weeks since it will cause the roots to split.

Maturity is 90-120 days. Georgia Jet has fewest days to maturity- 90. Vines grow slowly at first but by midsummer are growing rapidly. A vine can grow up to 8-20 feet long with tubers the entire length of the vine. Have few disease or pest problems.

Varieties:
Beauregard (100 days to maturity, light purple skin, dark orange flesh, extremely high yielder)
Bush Porto Rico (110 days, compact vines, copper skin, orange flesh, heavy yield)

Centennial (100 days, orange skin/flesh; good keeper; resistant to internal cork and wilt)
Georgia Jet (90-100 days, red skin, orange flesh, somewhat cold tolerant)
Jewell (100 days, orange flesh, good yield, excellent keeper)
Sumor (ivory to very light yellow flesh, good in warm areas of the country)
Vardaman (110 days, golden skin, orange flesh, compact bush type, young foliage is purple)

Uses: Young leaves and shoots are edible. Can eat a few early by digging some up in **late summer**.
Voles and gophers eat potatoes. (For **making trap boxes**, see September Farm Animals. See "Plant Health" in the back of this manual.)
(For **starting slips,** see March Greenhouse. See October Garden-Harvest.)

 Garden- Harvest 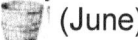 (June)

Vegetables and Fruit: Burdock, comfrey, currants/gooseberries, daylily, elderberry, hay, mushrooms, potatoes, strawberries and valerian.

 In June **harvest early planted broccoli, cabbage, carrots, fava beans, kale, peas, peppers, spinach, and turnips**. Or ready next month.
(For **broccoli** see March Greenhouse.)
(For **early cabbage** see March Greenhouse, April Garden-Plant, June Garden-Harvest. For **late cabbage** see June Garden-Plant.)
(For **carrots** see June Garden-Plant.)
(For **fava beans and peas**, see March Garden-Plant.)
(**Peppers:** For sowing details, see March Garden-Greenhouse. For transplanting see May Garden-Plant.)
(**Spinach:** See March/August Garden-Plant with details in March. See October Garden-Harvest.)
(For **turnips** see Mach/August Garden-Plant.)

 Dig up **burdock** root (asteraceae or compositae family; Arctium genus, Species: A. lappa, A. minus, A. minus nemorosum, A. pubens, A. tomentosum). It is a **biennial thistle.** (See "Plant Families" in back of manual.) Dark green leaves grow up to 28 inches long. Flowers July to October.

Harvest what you planted or forage for it in **June through September of the first year**. Can harvest in winter any time the ground is not frozen. The root is too tough to eat the second year.

Roots are 1-2 feet long and hard to dig up. The outer and inner root are both edible with a somewhat sweet, nutty taste. Eat raw or cooked. Cook root 30-45 minutes.

Leaves and stems (stalks) are edible. Peel stalks before eating. Leaves and stems are usually cooked like spinach. Stems can be eaten all year. Leaves are best eaten in spring.

Medicinal: Good for your liver, kidneys and skin. It is a blood cleanser. Consult your health care specialist.
(For **sowing/growing,** see March Garden-Plant. For **eating leaves**, see March Garden-Harvest.)

For **foraging guidelines** read "Nature's Garden: A Guide to Identifying, Harvesting and Preparing Edible Wild Plants" by Samuel Thayer.)

 Cut **comfrey leaves** (boraginaceae family; **Common Comfrey**: Symphytum officinale; **Russian Comfrey such as Bocking #4 and #14**: Symphytum x uplandicum).

Feed to horses, pigs, cattle, goats, sheep, turkey, chickens, rabbits and other animals. Comfrey is high in protein: 22-33%. It is high in vitamins A, C and B12. It has very high food value for animals.

Start harvesting leaves when the plants are about 2 feet tall. Cut back to within a few inches of crown (where plant stem meets the roots). **Late fall** stop harvesting so plant can build up energy to get through winter. (Stop harvesting end of September.) Should get about **4-8 cuttings per year** depending on location, fertilization and weather. Harvest of Russian Comfrey can be as much as **100-120 tons an acre.**

If using comfrey as **mulch/fertilizer** such as around fruit and nut trees, only cut off flower buds. Let the old leaves rot. It has deep roots and brings up minerals and other nutrients to its leaves and then into the soil.

How to Use Medicinally: (Consult your doctor and/or health care specialist.)

Comfrey root and leaves help wounds heal faster and reduces inflammation. It contains allantoin, a substance that speeds the production of new cells and aids in wound healing. It is safe to use externally.

Comfrey is not an antiseptic so should be used with herbs such as echinacea or calendula.

Comfrey root is boiled to make tea good for colds, stomach problems, bruises, dislocations, muscle pulls, swelling, wounds and sprains. The root is a good mouthwash and gargle.

Comfrey root was once used to treat broken bones. The pounded root forms a mucilaginous (having the sticky properties of an adhesive) mass, which is bound around a fracture. It dries and holds the bone in place.

Comfrey leaves: You can make a **poultice** with comfrey. A poultice is put on the skin to treat conditions such as a wound, bruise, boil, sprained ankle, burn or broken bone. Go to a doctor first but you can put this on after. Mash/grind dry or fresh comfrey leaves with mortar and pestle. Or cook fresh leaves in a little water for 2 minutes. Or put in a blender. Add warm water to make a paste. Depending on whether it is an open wound or not, you can put it directly on the skin or put a piece of gauze first and put the paste on the gauze. Cover with a clean cloth. Keep on 2-4 hours. Then if needed create a new poultice.

(For **planting details** see March Garden-Plant. See July/August/September Garden-Harvest but not much information there. For **re-planting,** see April Garden-Maintenance. For **how to make a liquid fertilizer / compost tea**, see March Garden-Plant.)

Currants and gooseberries (grossulariaceae family, genus ribes) ripen. Fruit is very tart.

Currants are ready for harvest when the fruit of black currants are black, and red/white/pink currants are brightly colored. Pick them when it is dry otherwise they may mold.

Pick black currants as individual berries. Pick red currants in clusters (strigs). Picking in clusters keeps them fresh longer. Currants will store in the refrigerator for 5-6 days. Currants ripen over a 2-week period in **June**.

Gooseberries are ready for harvest when they are soft to touch, and pink, red or black. For use in cooking pick them under-ripe but firm. Gooseberries are picked as individual fruits. Can put tarp or cloth under bush, and knock bush to get berries to fall on ground. Harvest over 4-6 weeks.

Uses: Fruit can be dried, canned or frozen.

Can **propagate** with stem cuttings or tip layering. (For **planting and propagating**, see April Garden-Plant. For tip layering and details about herbaceous/softwood/semi-hardwood/hardwood cuttings, see "Garden Tips" in back of manual.)

Eat **daylilies** (liliaceae or xanthorrhoeaceae family, same as asparagus; Hemerocallis fulva) also known as common daylily. Be sure you know it is a daylily. There are poisonous irises and other types of lilies that look similar. Start to bloom in **June**. Each bloom lasts only 1 day.

Parts of daylily are edible all year. **Buds, flowers, leaf shoots and roots are edible** either raw or cooked. Though some people get nausea or diarrhea if they eat it raw. Eat only a small amount the first time. Eat in moderation.

Tubers (roots) are edible all year but are best eaten **late fall and or early winter**. Good fried in oil or boiled for 15 minutes. (The tubers are small and potato-like with tiny hair-like roots attached to them. If it is one long, thick rhizome/root without little tubers or is a single bulb, it is not a daylily.)

In **spring** the **young shoots** are edible and delicious as a stir fry. Or use raw in a salad.

In **summer** you can eat the **flower buds and flowers.** They are good raw or dipped in batter and fried.

(See July Garden-Harvest but not much information.)

Late June **elderberries** bloom. Flowers are edible. Eat only flowers from dark blue, dark purple or black berry elders. Do not eat from red ones. Can make into a tea as a blood purifier or pain reliever. Consult your herbalist, doctor, and/or health care specialist.

(For **pruning** see February Garden-Maintenance. See April Garden-Plant. See August/September Garden-Harvest with **harvest and eating details in August.**)

Hay- first cutting is in June. Starting now and through the **summer and fall,** buy enough hay to last until next summer. It is cheaper to buy hay out in the field rather than at a feed store.

Most important is the quality of the pasture, time of day when cut, how windrowed (raked and dried in field), green color, and weather conditions. There is most protein in grass when cut before bud stage (before flowers bloom).

1. First cut spring hay is coarse with a lot of grasses and not much legumes such as alfalfa and clover. It has a lot of stems and **digests slowly** which generates a lot of heat for animals after eaten (good for winter). First cut hay has high sugar and is very rich. It dries quicker. Has less protein than second and third hay cuttings.

2. Second cut hay is very **rich in nutrients**. It is greener, cleaner and smells sweeter than the first cut. It digests faster than first cut hay. It has fewer stems. It is heavier so more likely to rot or catch fire if baled wet. If a bale is really heavy to lift, it may be wet. You can sometimes feel heat building up in the bale.

3. Third cut hay is really thick and green. It has a very sweet smell and is harder to dry. It is tougher. But it has a lot of **legumes** and less grass. It is higher in protein because of the high legume content.

Animal Feed: Goats, cattle and other ruminants are not as picky as horses (not a ruminant) about hay. Pigs are even less picky. Horses like timothy and orchard grass hay. Goats, sheep and cattle like fescue hay. They love alfalfa hay. Rabbits love alfalfa hay but will eat fescue hay. (Alfalfa is a legume.) Pigs will eat hay but do not digest it as well since they are not ruminants.

A cow eats about 35 pounds of hay a day during cold weather. Horses eat about 44 pounds a day. Sheep and goats eat 4 1/2 pounds a day.

Popular Types of Hay:

1. Alfalfa hay is the best type of hay. It is grown mostly in the northern, central and northwestern states. It is a rich legume hay with a very high protein content (15-20%) as well as being high in selenium, calcium and many other nutrients. Because it is so rich, it is usually given with other hays. If fed too much of it, animals can get fat. Since it is rich, less is needed of it than other hays. Do not feed a lot of it all of a sudden. Gradually introduce it so animals do not get colic, laminitis or bloat. If an animal has bloat, feed a coarse hay like fescue. (For **bloat** see April Farm Animals.)

2. Bermudagrass hay is a warm season perennial grass. It has 6-11% protein. Grown in warmer areas.

3. Clover hay is a sweet legume hay that is quite rich with 13-16% protein. It is short stemmed with red or purple bulb-like tops. It usually is mixed with other types of hay. It is darkish brown.

4. Fescue hay is a grass that has a reputation for being low quality. It is 5-9% protein. The quality is better if it is harvested when fescue starts to show a few seed heads. Later harvesting has reduced protein content. Fescue can cause poisoning due to a fungal infection. You may not want to feed it to animals in the late stage of pregnancy.

5. Orchard Grass hay is a bit richer than timothy hay with a blue green color. It has 7-11% protein. The strands of hay are quite long. It is usually mixed with other hays.

6. Timothy hay is a hearty basic hay. It has 7-11% protein. It consists of thick grassy stems with small 1 to 2 inch brush like tops.

(For **how to harvest hay**, see May Garden-Harvest. For **fescue** see March Garden-Plant. For **cover crops and pasture including types of hay,** see "Garden Tips" in back of manual.)

Mushrooms: Shaggy Mane (Agaricaceae family; Coprinus comatus) and other edible mushrooms can be found this time of year in wood chips, grass and bare dirt. Harvest lasts a few weeks then returns again in **September.** Use the same day you pick it.

Get a good book on mushrooms. Ask an expert before you eat any. Some varieties are poisonous. Some people react badly to all varieties of mushrooms. If you eat any, only eat a little to see how you react.

(See September Garden-Harvest.) Study the book "Nature's Garden: A Guide to Identifying, Harvesting, and Preparing Edible Wild Plants" by Samuel Thayer.

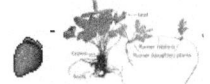 Harvest in June or July early/short season **potatoes** planted in April. (For **sprouting and then planting in April,** see March Greenhouse. For **varieties and lots of planting details,** see April Garden-Plant.)

(**Harvest:** For early/short season potatoes, see July Garden-Harvest. For mid-season potatoes, see August Garden-Harvest. For late/storage potatoes, see September/October Garden-Harvest. **October Garden-Harvest has most of the details about harvesting.**)

 June-bearing **strawberries** harvested **first 3 weeks in June**. Best picked in morning. Do not wash until ready to eat or process. Cook some with rhubarb since they are in season at the same time.

(For **sowing/growing details,** see April Garden-Plant. For **harvest details,** see May Garden-Harvest. See June Garden-Maintenance.)

Harvest **valerian** leaves spring and summer to make sedative tea. Consult your herbalist, doctor or health care specialist.

(For more about **harvesting leaves,** see August Garden-Plant. For **sowing and dividing,** see May Garden-Plant. For **transplanting daughter plants**, see August Garden-Maintenance. For **harvesting roots and how to use**, see October Garden-Harvest. For how to make a **tincture, decoction or infusion,** see April Garden-Harvest.)

Greenhouse, Hoop House or Cold Frame  (June)

 Sow **root crops** such as celeriac, hamburg root parsley, and parsnips for harvest in **October or November.**
(For **celeriac** see May Garden-Plant. For **hamburg parsley** see July Garden-Plant. For **parsnips** see April-Garden Plant.)

Farm Animals  (June)

In June or July start saving chicken, duck and turkey **eggs to eat in the fall and winter.** Peak egg production is **April and May.** Good egg production is **February through July.** Lowest egg production is **September through December.** (The longest day is in June. Molting is in September and the shortest day is in December.)

(For **egg storage methods,** see July Farm Animals.)

 Give flea and tick medicine to dogs and cats.

June p. 13

Honey Beekeeping in June

Unswarmed colonies are full of bees. The queen's rate of egg laying may drop a bit this month. The **main honey flow** happens this month. Inspect hive weekly. Add **honey supers** (boxes) as needed.

On hot, humid nights some bees may stay on the outside of the hive (**bearding**). This is OK.

There may be **mites** in the hive but it is better to treat the bees after the main honey flow so the honey is not contaminated with the chemicals.

(For an **overview of beekeeping,** see January Farm Animals. For **harvesting honey**, see September Farm Animals.)

 # JULY
July 1 daylight is 14 hours 28 minutes. July 31 daylight is 13 hours 54 minutes.

Abundant harvest.
According to the **Farmer's Almanac** the Dog Days are the 40 days from July 3 to August 11 which is when the Dog Star, Sirius, rises (at sunrise) in the eastern horizon. Rainfall is at its lowest for the entire year.

Early July:	Bee balm blooms. Peak day lily blooming.
	Peas stop producing until cooler weather. Weather is drier than last few months.
Mid July:	Some blackberries and blueberries are ripe. Dill and bronze fennel bloom.
	Temperatures are in the high 80's in the shade.
Late July:	Some dill and parsley seeds ready to harvest. A few wooly bear caterpillars appear.

Garden- Maintenance

 Now soil is hot enough to **mulch** around heat loving plants.

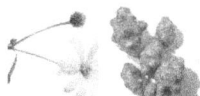 **Asparagus** (liliaceae family; Asparagus officinalis) **production** can be improved with salt (sodium chloride, NaCl). **Apply early spring (before spears appear) or early July.**

Salt also helps it resist diseases such as crown and root rot caused by fusarium fungus. Symptoms are very low yield, ferns falling over, and ferns dying back as early as August. (Recommended by "Rodale's All-New Encyclopedia of Organic Gardening".)

Do not use iodized salt. Use pickling salt. Only apply to plants over a year old. Add 2 pounds **pickling salt** per 100 square feet or per 100 foot row. Salt may build up to dangerous levels in areas with little rain.

(For **planting seeds and crowns,** see April Garden-Plant. See April Garden-Harvest.)

Early July cut **Jerusalem artichoke** (sunchoke) plants to 3 feet tall otherwise falls over later in summer. Feed leaves/stalks to ruminants. (See March/April/October Garden-Harvest with **harvest details in October.** See May Garden-Plant.)

Trim herbs before they flower Unless saving seed, want them to self sow, or want to attract beneficial insects. Beneficial insects eat insects that damage plants. It is a good, natural pest control method.

Trim plants such as **catnip, marjoram, mint, oregano, rosemary, sage, saint johns wort, salad burnet, stinging nettle, wormwood, and yarrow** 3 inches to 1 foot tall depending on type. Can eat, dry or freeze trimmings. Consult your herbalist. (See May Garden-Plant.)

Fruit Trees and Bushes in July:

For most plants it is best to **fertilize in spring**. Do not apply fertilizer after July 4 especially not nitrogen. It is OK to apply rock dusts at any time. (See fertilizers in "Garden Tips" in back of manual.)

 **Fertilize strawberries after fruit harvest** so that is **June, July or August** depending on variety. If fertilize before harvest, it may make fruit soft.

(For **sowing/growing details**, see April Garden-Plant. For **harvest details,** see May Garden-Harvest. See June Garden-Harvest. See June Garden-Maintenance.)

 **Fertilize** apples, apricots, blackberries, blueberries, currants, elderberry, gooseberries, peaches, pears, plums, raspberries **early spring** to promote growth. **Do not apply fertilizer late summer or early fall** because it will encourage new growth that will be susceptible to frost injury. Except rock dusts can be applied any time. (See March Garden-Maintenance.)

🍎 Pruning Apples Trees:

Young apple tree in July: The first summer make sure the top shoot becomes the leader. Pinch back all other shoots. The second summer make sure the top shoot is growing vertically, cut off any competing shoots. (For pruning see February Garden-Maintenance. See April Garden-Plant.)

 **Brambles Pruning:** (For details about **pruning black and purple raspberries, blackberries, Doyle blackberries**, see October Garden-Maintenance. For **propagating red raspberries**, see February Garden-Maintenance. For **propagating black and purple raspberries**, see July Garden-Maintenance. For **propagating blackberries** see April Garden-Plant. For **planting** see April Garden-Plant.)

 Propagate elderberry. Take a 4-6 inch cutting of semi-hardwood stems. Put 2 inches deep in pot in greenhouse or cold frame. In **October** of the next year transplant outside.

(For **pruning** see February Garden Maintenance. See April Garden-Plant and August Garden-Harvest. For **propagation with hardwood**, see March and October Garden-Maintenance.)

(For image and description of cold frame, see January Greenhouse. For how to make **potting mixes** and details about herbaceous/softwood/semi-hardwood/hardwood cuttings, see "Garden Tips" in back of manual.)

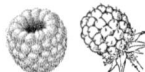 **Propagate fig in July.** In the **summer** take a 6-8 inch cutting from a vigorously growing stem that has some of last year's growth. Remove bottom leaves to expose 1-2 nodes (where leaves or flowers grow). Dip cut end into rooting hormone. Put cuttings in a 4 inch pot filled with a porous potting medium such as peat moss and perlite.

Moisten the soil and cover the pot with a plastic bag to keep cuttings humid. Place the pot in shade. Once vigorous growth appears, remove the bag. Plant **next spring.**

(See April Garden-Plant. See August Garden-Harvest. For covering fig, see November Garden-Maintenance. For how to make **potting mixes, rooting hormone,** and details about herbaceous/softwood/semi-hardwood/hardwood cuttings, see "Garden Tips" in back of manual.)

Propagate hardy kiwi by taking softwood cuttings in **July**. Cut just below a node (where leaves or flower grow) and put in peat/perlite potting medium. Use rooting compound and mist (or cover with plastic). Takes 6-8 weeks to root. Plant 10 feet apart outside.

(For **pruning** see February and June Garden-Maintenance. See April Garden-Plant. See September Garden-Harvest. For how to make **potting mixes, rooting hormone,** and details about herbaceous, softwood, semi-hardwood and hardwood cuttings, see "Garden Tips" in back of manual.)

Propagate black and purple raspberries from tip layering. They have long canes with tips that bend over to the soil. Cover the tips with 2-4 inches of soil to encourage rooting. Can hold tip down with rock. Next **spring,** the rooted tips can be cut from the mother plant with a sharp spade and replanted.

(For **pruning** black and purple raspberries, blackberries, Doyle blackberries, see October Garden-Maintenance. For **planting** see April Garden-Plant. For **tip layering**, see "Garden Tips" in back of manual.)

🌱 Garden- Plant 🌱 (July)

Vegetables: Brussels sprouts, buckwheat, parsleyroot, and peas.

Early to mid July last time of season to **sow warm season plants** such as snap beans, cucumbers, fennel (annual), and summer squash. (For fennel see April Greenhouse. For beans, cucumber and summer squash, see May Garden-Plant.)

Sow cool season root crops and plants for fall/winter harvest. Sow now through September.

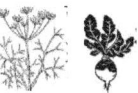**Sow Florence fennel and rutabagas.** Sow fennel 13 weeks before first frost in fall. Sow rutabagas 12 weeks before first frost. (For **fennel** see April Greenhouse.)

Sow kale, kohlrabi, mangel beets, turnips and rutabagas for **special forage crops for animals.**
(For **kale** see March Garden-Plant. For **kohlrabi and turnips**, see August Garden-Plant. For **early beets** and sowing details, see March Garden-Plant. For **rutabagas** see June Garden-Plant. See **Fall and Winter Pasture-Special Forage Crops** in "Garden Tips" in back of manual.)

Sow **brussels sprouts** (brassicaceae family; Brassica oleracea) seeds. They do not like heat. The sprouts become bitter if they develop in temperatures higher than 75 degrees. **The sprouts mature best when temperatures are from the low 60s to low 70s.** They taste best when eaten after light frost. (See "Plant Families" in back of manual.)

Can sow early spring for an early summer harvest (select a variety that matures quickly and is heat resistant). **Sow in greenhouse or outside 90-100 days (13-14 weeks, 3 to 3 1/2 months) before last frost in spring. This is February.** Best to start in greenhouse. Then transplant when established.

Or sow in mid-fall for a winter harvest (this is best). Sow when temperatures are in the 80s or a less. **Sow 3-4 months (12-16 weeks) before the first frost in fall. This is late June or early July.**

Brussels sprouts are **very cold hardy (down to 0 degrees)** and easily survive a hard frost and snow.

Sow seed 1/4 inch deep, 2 inches apart. Poor germination. Germinates in 7-14 days. When 3 inches tall, thin seedlings to 18 inches apart in rows 2 feet apart. Grows to 2-3 feet tall. **Matures in 85-120 days (12-17 weeks, 3 to 4 months).**

Care: Likes full sun. Likes soil pH 6.5-7.5. Make sure it gets enough water through growing season. An **annual**.

As it grows, remove yellow leaves at the bottom. To encourage **early sprouts**, break off all lower branches to about 6-8 inches above ground as soon as you see a sprout begin to form. The leaves can be eaten.

Early September may need to cut off top of plant to encourage development of sprouts rather than leaves. **Harvest** when buds are 1-2 inches across. They should feel firm. (See October Garden-Harvest.)

Sow **buckwheat** cover crop if ground will be bare for over 5 weeks. Or sow short season plants such as amaranth greens, arugula, cress, mache, mizuna, mustard greens, radish, or spinach. (For **buckwheat sowing/growing details** see June Garden-Plant.)

Sow **parsleyroot** (parsley root) (umbelliferae family, Petroselinum crispum, hamburg parsley). Seeds only retain viability for a year or two at most. Poor germination like regular parsley. Seeds need to be stratified. **Or can sow late fall after a few frosts such as late October or November.** Nature stratifies the seeds and they germinate in the spring. (For **how to stratify**, see "Garden Tips".) (The same is true for parsley.)

Germinates better in dark. If not stratifying, soak seed in water 48 hours changing water twice. Discard water.

Sow 1/4 to 1/2 inch deep, every 1 inch. Thin to 5-6 inches apart. Matures in 85-100 days (3 to 3 1/2 months). A **very hardy biennial.**

Care: Care is similar to parsley. Likes full sun but can take some shade. Adding raw manure to soil leads to forked roots. Add a little phosphorus to soil. Likes soil pH 6.3-6.6. Likes moist soil.

(See October Garden-Plant. See October Garden-Harvest. For **parsley**, see April Garden-Plant.)

Late July or early August sow **peas** for fall harvest. **Sow 10 weeks before first frost in fall.** (For **sow/grow**, see March Garden-Plant.)

 **Garden- Harvest**  (July)

Vegetables: Beets, comfrey, daylily, echinacea, flax, garlic, hay, oats, onions, and potatoes.
Fruit: Blackberry, blueberry, cherry bush, ground cherry, raspberry, and strawberry.

Harvest top half of **basil and other leafy herbs** such as arnica, calendula, oregano, rosemary, sage and wormwood.

All herbs should be harvested when oils responsible for **flavor and aroma** are at their peak. Herbs grown for their **leaves** should be harvested before they flower. Herbs grown for their **seeds** should be harvested when seed pods change from green to brown to gray but before they shatter (open). Herbs grown for their **flowers** should be collected just before full flower. Herbs grown for their **roots** should be dug up in the fall after the leaves have died.

For **leaf harvesting** most herbs can have up to 75% of their foliage cut without damaging the plant. It is best to cut in the morning. **Annual** herbs can be harvested until frost. **Perennial** herbs should be harvested until late August so plants have time to store energy for winter. Also late pruning encourages new growth that may be damaged by frost.

Using: For long-term storage herbs can be frozen, dried or made into a tincture. For **freezing** spread the herbs loosely on a cookie sheet. Place in freezer. After frozen put them in a plastic bag. Consult your herbalist.

For **drying** do not wash herbs unless they are dirty. Tie the herbs in small bundles and hang in an airy, warm, dry place. Keep out of sunlight. Can dry in an electric or solar food dehydrator. (For a **solar dehydrator**, see May Garden-Harvest.)

(**Basil:** For **sowing/growing details**, see April Greenhouse. See May Greenhouse. For **transplanting** see June Garden-Plant but details are in April Greenhouse. See September Garden-Harvest.)

(For **how to make a tincture, decoction or infusion**, see April Garden-Harvest.)

Harvest **beets (early)** when bulb is 1 1/2 to 2 1/2 inches in diameter. (For **early beets and sowing/growing details,** see March Garden-Plant. For **late beets** see July Garden-Plant. See September Greenhouse. For **harvesting and how to store**, see October Garden-Harvest.)

 Cut **comfrey** plants and feed to pigs, cattle, goats, horses, sheep, chickens, turkey, rabbits and other animals. (For **planting details** see March Garden-Plant. See June/August/September Garden-Harvest but not much harvest information there except a lot in **June Garden-Harvest**.)

Early July peak **daylily** blooming. Buds, flowers and roots are edible raw or cooked. Started blooming in June. Consult your herbalist. (For details see June Garden-Harvest.)

Harvest **echinacea leaves and flowers** (Echinacea purpurea, coneflower) for medicinal purposes. Consult your herbalist, doctor and/or health care specialist.

Flowers are harvested when they first start to open in the **summer (mid July)**. Pick in the morning. Can make a tincture out of it. Or dry on screen in airy spot out of the sun. Then crumble into a jar to store.

Leaves are harvested in the morning in the summer when flowers are in full bloom (July). Dry in a dark, airy place.

(See April Garden-Plant. For **harvesting roots and how to use**, see October Garden-Harvest. For how to make a **tincture, decoction or infusion**, see April Garden-Harvest.)

 Early July harvest **flax** (linaceae family; Linum usitatissimum) seeds. **Edible seed** is produced in bolls or pods on flowerstalks each containing 4-10 seeds. Harvest seeds when approximately 90% of seeds are brown. Or harvest when seedpods are yellow and starting to split open. There will still be a few green leaves on the plant at that stage.

Cut the pods from the plants. Spread them out so they can dry more but not in direct sun. Once the seeds are dead ripe and can not be dented with a fingernail, separate them from the seed pods.

Threshing: Manually open each pod and take out the 4-6 seeds. Or crush all the seed pods by threshing such as putting them in a bag and beating them against an object or crush them under your foot.

Winnowing: Use a fan or the wind to separate seeds from the chaff. Pour handfuls of the seed mixture from several feet up, and let the wind blow out the lighter pieces. Seeds are heavier and fall to the ground.

If seeds are **dry**, store them for 8 to 10 months at room temperature. Recommended storage moisture is 11%.

Uses: Flax is grown both for its **seeds and fiber.**

Various parts of the plant are used to make fabric, dye, paper, medicine, fishing nets, soap and other products. It has been used for fiber since pre-historic times. It is soft and flexible and used to make linen.

Linseed oil is made from flax seed. It is eaten or used as an ingredient in wood finishing products. The leftover seedcake is fed to animals.

It is an anti-oxidant. It is very high in alpha linolenic acid (**omega-3 fatty acid**) that is essential for humans. It has a high percentage of dietary fiber, both soluble and insoluble. Feeding flaxseed to **laying hens** increases omega-3 fatty acid in eggs by 6 to 8 times, making one egg equal to 4 ounces of cold water fish as a source of omega-3 fatty acids.

The **seeds** can be sprouted and eaten. It can be used in baked bread and other baking. It is good in cereal, biscuits, and soup.

Ground flaxseed is used a bulk food to relieve **constipation**. Mix with water and let soak 15 minutes first. Consult your herbalist, doctor and/or health care specialist.

(See March Garden-Plant. For **general grain harvesting** information, see May Garden-Harvest.)

 Harvest **garlic** when lower leaves are yellow. Or when about half of the leaves are green and the other half turning brown and dying off (or a few brown and 5-6 leaves are green). Dig up, do not pull up by leaves.

If you wait until all the leaves are brown, then the bulbs will be overripe with cloves starting to separate from one another. These do not store as well. But if you harvest too soon the storage time is also reduced.

Pull up a few plants to see if the **bulbs** look large enough for harvest. Garlic is more sensitive to when it is picked than onions are.

Softneck garlic (Allium sativum) is the most common type available. It has a row of somewhat large outer cloves and a row or two of inner small ones. It keeps better than hardneck garlic.

Hardneck garlic (Allium sativum var. ophioscorodon) such as Rocambole has fewer but larger cloves. It is better suited to Northern winters because it has long roots that keep in it the ground during heaving. Hardneck garlic is usually ready to harvest a month before softneck garlic.

How to Harvest: Always handle gently to prevent bruising. Leave the tops on since they cure better this way. Do not leave in the sun. Dig out gently and remove some of the dirt. Cure 1-2 months in airy, dry place.

After being cured, the tops and roots can be cut off. **Store** at cool temperature (around 50-60 degrees) in low to medium humidity (40-60%). An unheated room that does not freeze is good.

Uses: Eat small garlic cloves and plant large cloves. Flowerheads and leaves can be eaten. (See March/April Garden-Plant. For **planting details and more about varieties,** see September Garden-Plant.)

Late July and August is second **hay** cutting. (For **different types of hay cuttings and types,** see June Garden-Harvest. For **hay making,** see May Garden-Harvest. For **cover crops and pasture,** see "Garden Tips".)

Early July harvest **oats (hulled common)** (poaceae/grass family; Avena sativa) or **oats (hulless)** (Avena nuda) for **grain and hay.** Harvest oats when the grain changes from green to cream. Or test it by rubbing seedheads between your hands. If the grain comes out easily, try biting it. If it is hard, it is ready to harvest.

If harvesting for the **hay/straw,** cut when stalks are still green and the grain is in the milk stage. That means it is still soft and runny when you bite on it.

For a small area, cut with a scythe or sickle. Then windrow (rake into short rows) for a few days. Turn once or twice to dry all parts.

Hulling: Hulless or naked oats do not have hulls so are much easier to process. If you have hulled oats, spread the grain on a baking sheet and bake for 90 minutes at 180 degrees. Then run the brittle hulls through a roller mill or meat grinder to crack the hulls. Winnow in front of a fan to get rid of the chaff. You get about 3-11 pounds of oats from 100 square feet.

Uses: Oats are fed to all types of livestock. Most animals except poultry will eat unhulled oats. They are 12-22% protein. Oat straw is more nutritious and easier to digest than wheat straw.

(For **oat sowing details and cover crops**, see March Garden-Plant. For **general grain harvesting** information, see May Garden-Harvest. For **cover crops**, see "Garden Tips" in back of manual.)

Harvest **onions (perennial)** (amaryllidaceae family; allium genus) when tops fall over. Eat onions whose tops did not fall over. **Eat big onions, plant small onions.** Flowerheads and leaves are edible.

Small bulblets at top of stalks can be planted in September or eaten soon. Store bulblets in a mesh bag or other airy container. Hang in a cool, dry place such as an unheated room that does not freeze. Do not store in a root cellar because it is too damp.

Harvest on sunny day. Leave on soil a few sunny days to dry. Then cut off most of the tops. Then cure 1-2 months in airy, dry place. **Store** at cool temperature (around 50 degrees) in low to medium humidity (65-70%). An unheated room that does not freeze is good. Store onions away from apples and tomatoes.

(For **planting seeds of annual onions**, see February Greenhouse. For **fertilizing and cutting seedstalks**, see April Garden-Maintenance. For **planting perennial onions**, see September Garden-Plant.)

In July can dig up new, early **potatoes** as needed to eat right away. Early and mid-season potatoes are ready to be dug up when **plants start to flower.** Potatoes are planted April through June, and harvested June through October. (For when to plant early/mid/late season potatoes, see April Garden-Plant.)

(For **sprouting and then planting in April,** see March Greenhouse.)

(**Harvest:** For early/short season potatoes, see June Garden-Harvest. For mid-season potatoes, see August Garden-Harvest. For late/storage potatoes, see September/October Garden-Harvest. **October Garden-Harvest** has most of the details about harvesting.)

Harvest Fruit in July:

Harvest **blackberries July through August.** Pick when fruit is dull black or deep purple or burgundy but not deep black. They should look full and almost swollen. Pick gently since they are soft. If the fruit comes off easily from the stalk, then it is **ripe**. Taste is the best way to test. Berries do not ripen any more after they are picked.

While picking do not put too many on top of each other or else they will get squashed. You can tie a bucket or basket around your waist so that both hands are free. You pick much more quickly this way. If your variety has thorns, you get less scratched if you wear gloves and long sleeves.

During the height of the season, pick every few days. Best to pick in morning. Poultry love berries.

Processing: They last in the refrigerator for 3-4 days. Can dry or freeze. Freeze in a single layer on a cookie sheet, then put in freezer bags. Or freeze in milk. Or make into jam.

Optional: To pretreat sensitive types of fruit **before drying,** soak in 1 cup water that has 500 mg of vitamin C mixed in. Or soak in 1 cup of water that has 1 tablespoon salt or vinegar. (See April Garden-Plant.)

Blueberries are harvested June through August depending on variety. **Highbush** bears fruit May through July. **Rabbiteye** bears fruit July and August. **Wild blueberries** usually produce fruit in August.

Berries turn blue 3-4 days before they are at maximum ripeness. When ripe, blueberries are completely blue and easily fall off bush. Pick every few days.

Tie a bucket or basket around your waist so that both hands are free. You pick much more quickly this way.

Birds love blueberries so may have to put netting on bushes. Keep well watered when fruit is forming.
(For more information about **types and planting,** see April Garden-Plant.)

 Harvest **cherries from bushes** (sand cherries):
Bush cherries ripen in midsummer. When picking, get stem along with the cherry. There is a tiny fruit spur on the branches. Be sure not to injure this when harvesting. Produces fleshy, purple-black fruit. (See April Garden-Plant.)

Harvest **ground cherries** (solanaceae family; Physalis pubescens or Physalis subglabrata) when light brown outer husk starts to open. Fruit ripens from green to yellow-gold-orange. Fruit drops when ripe. Do not eat when green (immature) because they contain solanine, the substance that makes potatoes toxic. Contact your herbalist. Fruit is size of cherry.

Using: Has very sweet tomato/pineapple taste. Can eat raw or cooked. Will get sweeter if allow fruit to ripen in husk for several weeks after harvest. Bring into airy, dry place to ripen more.

Can leave some fruit on ground, and it will reseed next year. Can store for up to 3 months if left in the husk. Remove husks to dry or freeze. Very prolific producer.

Saving Seed: The seed can be saved the same way as tomato seeds. (For details see "Plant Families" and "How to Save Seed" in back of this manual.) Or this method- Remove husks and put a small amount in a blender with a little liquid. Blend them. The blender will not hurt the seeds. The top layer can also be swished with a wire whisk to release more seed. After blending put the goop in a large bowl. Add some water and swish it around. Good seed goes to the bottom. Carefully dump out most of the water. You may have to add water a few times to get clean seeds on the bottom. Strain in a fine mesh strainer.

(For **sowing** see March Greenhouse. For **transplanting** see May Garden-Plant.)

Harvest **raspberries.** June bearers produce a heavy crop from **June through early July.** Everbearing varieties produce 2 crops, one in **June** and again in fall (**September through first frost**). Pick every day or every other day in early morning.

The raspberry is easily removed from plant without being squashed. It leaves behind the white center portion, the receptacle. Raspberries do not store for long. Can dry or freeze. Birds rarely bother raspberries. (See April Garden-Plant.)

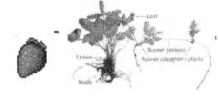 Late July everbearing **strawberries** produce a lot. (See April Garden-Plant.)

 **Greenhouse, Hoop House or Cold Frame**  (July)

 Sow **dandelion** in greenhouse in July, August and September for winter harvest. (See March Garden-Plant.)

 Farm Animals  (July)

In July or June start saving chicken, duck and turkey **eggs to eat in the fall and winter.** Peak egg production is **April and May.** Good egg production is **February through July.** Lowest egg production is **September through December.** (The longest day is in June. Molting is in September and the shortest day is in December.)

Egg Storage Methods in Shell: They work best with fertile, unwashed eggs from your farm, not store bought. To **test eggs** after storage, put in water. Good eggs sink to the bottom, bad eggs float.

Best method: Store in cartons in refrigerator. Turn over once a week. Can coat eggs with lard. Will last many months. For all eggs stored a long time, the white gets more runny but it is OK to eat as long as it smells good.

Good method: When waterglass (sodium silicate) is mixed with water (one part waterglass with 9 parts boiled and cooled water), it forms a gelatinous substance in which eggs are immersed. Store in a crock in a dark, cool location. The solution can be reused. Waterglass can preserve eggs for five months or more without refrigeration.

OK Method with no refrigeration: Cover eggs with lard. The eggs store even better if in an airtight container. Place in cool, dark location. Turn eggs once a week.

Egg Storage by Freezing: Mix raw whites and yolks together. Optional, then add either of the following to each cup of raw eggs: 1/2 teaspoon salt, or 1 tablespoon sugar or corn syrup. They prevent the egg yolks from becoming too gelatinous once frozen. Lasts about 1 year.

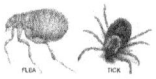 Give **flea and tick medicine** to dogs and cats.

 July 1st stop giving Sweetlix high **magnesium minerals** to goats and other ruminants.

Give regular minerals. The high magnesium minerals are given **April through June.** Do not feed copper to sheep. Consult your veterinarian or County Extension office. (For details see April Farm Animals.)

 **Worm** goats, horses, pigs, sheep, cattle, donkeys, dogs, cats and other animals.

Give garlic water to pigeons, ducks, turkey and chickens. Consult your veterinarian or County Extension office. (For **garlic water,** see January Farm Animals. For **mites and lice**, see March Farm Animals. For **natural wormer recipes**, see April Farm Animals.)

Honey Beekeeping in July

July and June are similar in the care that the bees need. If the weather is good, the **nectar flow continues.**

On hot and humid nights, you may see a huge curtain of bees cooling themselves on the exterior of the hive (**bearding**).

Continue inspection of hives. Add more honey supers (boxes) if needed.

(For an **overview of beekeeping,** see January Farm Animals. For **harvesting honey,** see September Farm Animals.)

 # AUGUST August 1 daylight is 13 hours 53 minutes. August 31 daylight is 12 hours 55 minutes.

Abundant harvest. Dry season.
Early August: Some acorns fall. Some locust tree leaves start to turn brown. Some apples are ready to harvest.
Mid August: Asian pears are ripe. Goldenrod and Joe pye weed bloom.
Late August: Some trees have a few leaves turning yellow. Pick elder berries and rose hips.
Weather starts getting a little cooler at end of month. Some leaves start to fall.

Garden- Maintenance

 **Brambles- Pruning:** Prune old canes (2 year or floricanes) after all fruit is harvested. **Floricanes** are woody, silvery brown with light green leaves. They are growth from the previous year. **Primocanes** are green with darker green leaves. They are this years growth.

Erect Blackberry Pruning: After plants bear fruit in **July or August**, remove all old (2 year or floricane) **canes**. The primocanes that grew this summer are now lying on or near the ground. Cut them to 4-5 feet tall or whatever height to prevent berries from laying on ground, or build simple trellis such as horizontal wire 3 feet above ground. If needed, tie the canes to the trellis/wire with plastic or cloth. (For trellis, see April Garden-Plant.)

Trailing Blackberry Pruning: After fruit has been harvested in July or August, cut out all 2-year (floricane) canes.

Black, Purple and Red Raspberry Pruning: Cut old fruit bearing **canes** (floricanes) to ground after harvest berries in July and August.

(For **details about pruning black and purple raspberries, blackberries, Doyle blackberries,** see October Garden-Maintenance. See April Garden-Plant. See July Garden-Harvest.)

🍒 Do not apply nitrogen to **fruit trees or bushes** in late summer or fall. It is best to fertilize in spring. Rock dusts can be applied at any time. (For **fertilizing information**, see March and July Garden-Maintenance.)

Cut watersprouts on **fruit trees** in August. (For **pruning** see February Garden-Maintenance.)

 Replant **valerian** daughter plants. Plant 1 foot or more apart in rows 2-3 feet apart. Valerian leaves and roots can be made into a sedative tea. Consult your herbalist, doctor and/or health care specialist. (For **sowing and dividing**, see May Garden-Plant. For **harvesting roots**, see October Garden-Harvest.)

Late August **weed eat and mow** along driveway, around trees, berries, grapes, etc.

 **Whitewash trunks of fruit and nut trees** with white latex paint or other formulas.

(For **whitewash formulas and how/why,** see June Garden-Maintenance. See April Garden-Maintenance.)

🌱 Garden- Plant 🌱 (August)

Early August: Creasy greens, kale, lettuce, radish and turnip.
Mid to Late August: Endive, escarole, garden cress, kohlrabi, mustard greens, mullein, radicchio, spinach.

Sow cool season root crops and plants for fall and winter harvest:

Sow beets, collards and Swiss chard. Sow beets 6-8 weeks before first frost in fall. Sow collards and Swiss chard 8-10 weeks before first frost. (For early beets and sowing details, see March Garden-Plant. For collards see March Garden-Plant. For Swiss chard see May Garden-Plant.)

Early August Outside:

Sow **creasy greens** through **early fall** (better than spring). (For **sowing/growing details**, see March Garden-Plant.)

Sow **kale** (brassicaceae or cruciferae family; Brassica oleracea) for harvest through **November to early April. Sow 8-12 weeks before first fall frost. Cold hardy to 20 degrees.** Withstands frost better if not given too much nitrogen. Tastes better after frost.
(For **sowing and growing details**, see March Garden-Plant.)

Sow late **lettuce. Sow 6-8 weeks before first frost in fall.** (For **sow/grow,** see March Garden-Plant. See August/September Greenhouse.)

Succession sow **radish (summer)** every week **until end of September. Last sowing is 4 weeks before first frost in fall.** Harvest in 1 month. Hardy to 25-28 degrees. (For **summer radish and sowing/growing details,** see April Garden-Plant. See September Garden-Plant.)

Sow **radish (winter)** (brassicaceae family; Raphanus sativus) late August (**mid to late summer) 6 weeks before first frost in fall. Hardy to 25-28 degrees.** Matures in 50-65 days. Plant same as summer radish except winter radish is larger and slower growing. Thin to 4-8 inches apart. They are white, black or green.

Types:
1. Black or Spanish radishes (Raphanus sativus var. niger) are 4-5 inches in diameter. Can be 9-12 inches long. They weigh about 2-7 pounds. They have a pungent flavor so use sparingly. Remove greens and roots before storing. Cooked, their flavor becomes mild and dry, like a turnip.
2. Daikon, Japanese or Chinese radishes (Raphanus sativus var. longipinnatus) grow to about 18 inches long and **weigh 2-20 pounds** or more. Some are thin and long, others are short and round. Roots are white.
Some varieties of both types: China Rose (52 days, white); Chinese White (60 days; large, long, white roots); Miyashige; Round Black Spanish (55 days; rough, black skin, white flesh); Tama Hybrid (70 days; daikon type; roots up to 18 inches long with 3 inch diameter; smooth, white).

Harvest when large and mature. They can stay in ground a few weeks after mature and still be good to eat.
Uses: High in vitamin C. Denser, sturdier and more pungent than spring radishes. Eat root raw, cooked or pickled. Good in salad. The leaves are edible.
Storage: They **store well into late winter.** Store the same as carrots or beets. Removes leaves before storing but keep small roots on. Layer in damp sand, peat moss, leaves or sawdust in root cellar, barrel, trench or clamp. (A

clamp is a hole lined with straw, filled with vegetables, then covered with straw.) Or keep in ground and cover with mulch or straw to prevent from freezing.

(For **summer radish and sowing/growing details,** see April Garden-Plant. See September Garden-Plant. For **carrot storage**, see September Garden-Harvest.)

Sow late/storage **turnips** (brassicaceae family; Brassica rapa) **9 weeks before first frost in fall so plant early August**. Succession sow: Sow **late summer** for fall harvest. Sow **early fall** for late fall harvest.

Sow seeds 1/4 to 1/2 inch deep, 1 inch apart in rows 1 foot apart. Germinates in 7 days. Thin early turnips to 3-4 inches apart. Thin fall turnips to 6 inches apart. Can eat thinned plants. Matures in 30-75 days.

Care: Likes full sun but can take some shade. Likes well drained but moist soil. Likes soil pH of 6.5. Likes moderately rich soil.

Keep well **watered**. When turnip roots grow slowly, they become woody and strong flavored.

Care is similar to **rutabaga** but rutabaga needs 4 weeks longer growing time than turnips. (For rutabaga see June Garden-Plant.) It is a hardy, cool weather **biennial in the mustard family.** (See "Plant Families" in back of manual.)

Fertilize: Phosphorous and potassium are needed for good root development. Boron is needed to prevent Brown Heart (water core). Can apply boron separately as a spray 4-6 weeks after planting.

Uses: Can be used as **winter forage** for animals. (See pasture in "Garden Tips" in back of manual.)

Pests: (See "Plant Health" in back of this manual.)
    Aphids can be controlled by removing infested leaves and hosing large populations off the plants.
    Flea Beetles can be controlled by keeping weeds and plant debris under control.
    Root maggots can be stopped early in the spring by covering plants with row covers.

Disease:
    Clubroot fungus can develop where brassicaceae plants have been frequently grown so it is important to rotate crops. Clubroot thrives in acidic soil. Likes soil pH of 5.5-6.8.
    White rust fungus causes small white cottony blisters on upper surface of leaves and a yellowing on the undersides. There are no fungicides that control it.

Varieties Good for Roots: (some have white flesh, others have yellow)
    Just Right (hybrid, 28 days for greens and 60 days for roots; smooth, high quality, mild roots, white)
    Gilfeather (75 days, heirloom; egg-shaped, large; smooth texture, delicate flavor, smooth foliage)
    Golden Ball (60 days, sweet, fine-grained with yellow flesh)
    Market Express (very early, 38 days for baby turnips, pure white roots)
    Purple Top White Globe (55 days, **the most common purple and white turnip**; smooth, globe roots)
    Royal Crown (hybrid, 52 days, purple top, fast growing, uniform roots, resistant to bolting)
    Scarlet Queen (hybrid, 45 days, bright scarlet root, resistant to downy mildew, slow to get pithy)
    White Knight (75 days, smooth, uniform, pure white, flattened globe roots)
    White Lady (hybrid, pure white, sweet, tender, delicious roots, slow to get pithy; smooth tops)

Varieties Good for Greens:
    Alltop (hybrid, 35 days, vigorous, high-yielding, rapid regrowth, resistant to mosaic virus)
    Seven Top (open-pollinated, 40 days; dark green leaves; for tops only)
    Shogoin (42 days; tender, mild; roots good when young)
    Topper (hybrid, 35 days; heavy yield, good bolt resistance; resistant to mosaic; roots edible)

(For **early turnips**, see March Garden-Plant but details are here. For **late/storage turnips and harvest details**, see October Garden-Harvest.)

Mid to Late August Outside:

 Endive and escarole (compositae family; Chrysanthemum endivia, same genus as chicory) is in the **chicory family,** closely related to radicchio. (Chicory, Belgian endive, and radicchio are the plant cichorium intybus. Endive and escarole are the plant cichorium endivia.) Endive has a broad leaf. Escarole has a narrow, curly leaf. **Both are cool season biennials.**

Sow very early spring, 4-6 weeks before the last frost. Better planted late summer or fall, 1-12 weeks before first fall frost.

Cover with 1/2 inch of soil. Takes up to 14 days to germinate. Thin seedlings to 6-12 inches apart with rows 18-24 inches apart. Matures in 85-98 days.

Care: Prefers full sun. Likes soil pH of 5.0-6.0.

Harvest: Tastes better after light frosts. Endive and escarole are ready for harvest when leaves are 5-6 inches tall. Cut off just above soil level. They re-sprout for a continuous harvest.

Uses: Eat leaves raw or slightly cooked. Good in salads. Escarole is less bitter than endive.

(See September Greenhouse but details are here.)

 Succession sow **garden cress** (brassicaceae family; Lepidium sativum, pepperwort cress, broadleaf cress, peppergrass) every 10 days until **4-6 weeks before last frost in spring. Or sow 4 weeks before first frost in fall** and succession sow every 10 days through early fall. It does not like hot weather and becomes bitter then.

Seeds need light to germinate. Germinates in 2-6 days. Thin to 6 inches apart with rows 18 inches apart. Harvest starting in 10 days. Matures in 14-30 days.

It is in the **mustard family**. (See "Plant Families" in back of manual.) Cress is a low growing, reseeding **annual.**

Care: Easy to grow. Likes soil pH of 6.0-6.8. It grows best in partial shade. Do not let roots dry out. It has few diseases or pests.

Can also be **propagated** by stem cuttings.

Harvest in 15-20 days when 4-6 inches tall. Can harvest entire plant or cut off parts and wait for regrowth.

 Sow **kohlrabi** (brassicaceae family; Brassica oleracea, turnip cabbage) in late August for fall and winter eating for people or animals. Sow as special forage crop for animals. (See pasture in "Garden Tips".) **Sow 4-6 weeks before last frost in fall.** A member of the **cabbage family.** (See "Plant Families" in back of manual.) It looks like a cross between a **cabbage and a turnip.** A hardy, cool weather **biennial** that likes temperatures between 40 and 75 degrees.

Sow seeds 1/4 to 1/2 inch deep, 1/2 inch apart in rows 12-18 inches apart. Germinates in 5-15 days. Thin seedlings to 4-6 inches apart. Can transplant seedlings. Good to succession sow. Matures in 55-60 days.

Care: Easy to grow. Prefers soil pH 6.0-7.5. Likes full sun. Keep evenly watered. It becomes woody if it gets too dry. Has same pests and diseases as cabbage.

Varieties include those with green, purple or white skin and stems. Leaves are blue-green. **Some types:** Early Purple Vienna (matures in 60 days), Early White Vienna (55 days), Grand Duke (50 days), and Purple Danube (52 days).

(For **early planted kohlrabi,** see March Greenhouse but there is not much information. For **late kohlrabi and general harvest information,** see October Garden-Harvest.)

 Sow **mullein** (scrophulariaceae or figwort family; Verbascum genus with 250 species; velvet plant) **8 or more weeks before first frost in fall. Or sow 2-4 weeks before last frost in spring.** The seeds need to be stratified (undergo cold temperatures) for 3 weeks before they germinate. Seeds remain viable for decades.

Needs light to germinate. Germinates in 2-25 days. However, it can germinate anytime in spring, summer or fall. Those that germinate in the fall need to be at least 6 inches across to survive winter. Mulch first year plants in winter. Thin to 15-18 inches apart.

Some are biennials and some are short-lived perennials. They form a rosette of leaves near the ground.

Biennials flower the second year. Most types flower June through November. The small flowers are purple, red, rose, violet or yellow. With the flower stalk they grow 3-4 feet tall.

They have a long tap root that **brings up minerals** from deep in the soil. They break up compacted soil.

The type usually found in **North Carolina** (Verbascum thapsus; **Great mullein or Common mullein**) is a biennial with yellow flowers the second year. Flowers June to August.

Care: Likes full sun. Fairly drought tolerant. Likes well drained soil that is sandy, gravely or rocky soil. Likes soil slightly alkaline.

Propagation: Can propagate from **root cuttings** of mature plants in the spring or fall. Cut the root into pieces 1-2 inches long. Put in growing medium in a horizontal position and put 1/2 inch of the medium over them.

Can propagate from **softwood stem cuttings.**

Has a long taproot so does not like being transplanted.

Harvest: Harvest **flowers** late summer after they have opened. You collect a few flowers every few days to a week since they do not mature at the same time. Collect **leaves** in late summer or early fall of the first year. Or collect early spring of the second year. The leaves have short, thin bristles (hairs).

Uses: The seeds and leaves contains high levels of **rotenone, an insecticide**. So use in moderation. Consult your veterinarian or County Extension office.

Leaves are used as a remedy for sore throat and coughs by making a tea with them. **Strain through coffee filter to get out little hairs.** Consult your herbalist, doctor and/or health care specialist.

Leaves can be made into a tea and used on **skin problems** such as wounds, burns, hemorrhoids and bruises.

Also used for coughs or lung infections by **smoking the leaves**. It is better to use leaves from plants 1 year old or less. Helps to break up congestion and promote effective coughing. The smoke is very light. Smokers find this easier to do. If you are not a smoker, use a tincture or tea. By the way, it can be mixed with tobacco. It has almost no flavor.

The **flowers and unopened buds** are infused in olive oil for **earache**. Pick the flowers and let them wilt for several hours so they lose some moisture. Fill a small glass jar with the flowers. Then add an oil such as olive oil to fill the jar. Put on lid. Let it sit in the sun for 1-2 months. Then strain into clean glass bottles. Use a Q-tip to apply to the ear.

Flowers can be used to make a yellow dye.

It is excellent for use as a hand drill in friction **fire lighting**. The dried stalk can be used as a torch. The leaves can be used as candle wicks or toilet paper.

(For **how to make a tincture, decoction or infusion**, see April Garden-Harvest. For **stratification**, see "Garden Tips" in back of this manual. For **propagation methods**, see "Garden Tips" in back of this manual.)

Succession sow late **mustard greens. Sow 8-10 weeks before first frost in fall.** Better planted in late summer rather than in spring. Ready to harvest in 1 month. (For **sowing/growing details** see February Greenhouse. For **early mustard** see March Garden-Plant.)

Radicchio is frost hardy. Succession plant every 2 weeks for 1 month from **late August through September. Sow 8 weeks before the first frost in fall.** (For **sowing/growing details**, see February Greenhouse. See September Garden-Plant.)

Succession sow **spinach** (winter / late). **Best planted late summer (August or September), 6-9 weeks before first fall frost. Very cold hardy down to 15 to 20 degrees.** Matures in 1 1/2 months (October, November and December).

(For **sowing/growing details and information about early/late spinach**, see March Garden-Plant. For early spinach harvest, see May/June Garden-Harvest. **For all harvest details, see October Garden-Harvest.** See August/September/October Greenhouse.)

 Garden- Harvest (August)

Fruit: Apples, apricots, Asian pear, elderberry, fig, melon/watermelon, peach/nectarine, pear, plum.
Vegetables: Beans, celery, comfrey, corn (sweet), globe artichokes, hay, hops, onion/garlic, potatoes, tobacco, tomatillo, tomato, and valerian.

Recommended Food Storage and Seed Saving Books

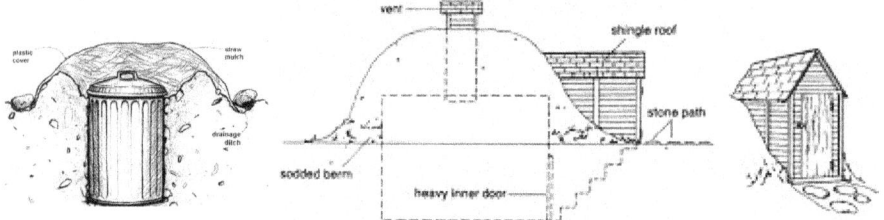

For **storing fruits and vegetables**, read the excellent book "Root Cellaring" by Mike and Nancy Bubel. Also the classic "Putting Food By" by Greene/Hertzberg/Vaughan.

For **canning** read the classic "Ball Blue Book of Preserving". Also "Complete Guide to Home Canning and Preserving" by the United States Department of Agriculture.

For **drying food** read "Making and Using Dried Foods" by Phyllis Hobson.

For **preserving foods by fermentation** such as sauerkraut and yogurt, read "Wild Fermentation: The Flavor, Nutrition and Craft of Live-Culture Foods" by Sandor Katz and "The Permaculture Book of Ferment and Human Nutrition" by Bill Mollison.

For **cooking** the old-fashioned, healthy way, see "Nourishing Traditions" by Sally Fallon.

For **saving your own seed**, read "Saving Seeds: The Gardener's Guide to Growing and Storing Vegetable and Flower Seeds" by Marc Rogers. For a detailed, professional guide, read "Seed to Seed: Seed Saving and Growing Techniques for Vegetable Gardeners" by Suzanne Ashworth and Kent Whealy.

(See "How to Save Seed" at the end of this manual.)

Fruit Harvest in August:

Pick **apples August through October** depending on variety. Remove dropped fruit from ground weekly so does not harbor insects.

Varieties: Early apples only keep for a few weeks. Mid-season apples keep for 1-2 months. Late varieties that ripen late September or after keep 3-8 months. Do not store mid-season with late-season apples.

When to Pick: Apples are ready to pick when seeds are dark brown. For green varieties the ground or base skin color changes from green to yellow. The flesh loses its greenish tint and turns yellow or white. **(Except very late apples for storage are picked unripe.)**

How to Pick: The apple will come off the tree easily with the stem still on. Pick apples by lifting and twisting. Be careful not to damage or pull off fruit spur. Pick apples carefully so it does not bruise. Place gently into lined basket. Keep out of sun. Cool as soon as possible.

Storing: Can let apples sit for 1-2 weeks before storing to let them sweat. Then wrap each apple in 8 inch square of newspaper or oiled paper. You do not have to wrap apples if putting in trays that completely separate the fruit. It is best if apples do not rest on top of each other. Put in container with air circulation. Store apples on bottom shelf in coldest part of root cellar at 37-40 degrees. Do not let freeze. Check fruit regularly and throw out bad apples.

Uses: Can make apple cider or cider vinegar. Can make into applesauce, then can or freeze. Can dry apple slices.

(For **pruning** see February Garden-Maintenance. See April Garden-Plant. For **thinning** see June Garden-Maintenance. See September Garden-Harvest but most information is in this section.)

 Harvest **apricots** when fully ripe on tree. Some varieties ripen over a 3-week period. Pick fruit when yellow and starting to soften, yet still firm. Fruit keeps 1-3 weeks if stored in a cool location.

(For **pruning** see February Garden-Maintenance. See April Garden-Plant. For **thinning** see June Garden-Maintenance.)

 **Asian pears** are picked when ripe. Unlike other pears that yield to gentle pressure when ripe, Asian pears are ripe even when they are extremely firm. Look for a fairly strong and sweet aroma.

Asian pears from **Japan** are ripe when a yellow-brown or yellow. Those from **China** are ripe when greenish-yellow. The color should be mostly even all over. Ripe Asian pears have a thinner skin than unripe ones. The skin should be translucent.

To take off the tree, lift the fruit and twist. It should come away from the fruit spur easily. Be careful not to damage fruit spur.

They ripen **August through September.** Store like regular pears. Wrap each pear in paper individually. Keep between 33-40 degrees in 80-90% humidity. Do not let freeze. A root cellar is good. Pears store a few months. They are crisp and crunchy. (For **pruning** see February Garden-Maintenance. See April Garden-Plant.)

Elderberry (American Elder) berries are ripe when they are dark purple, dark blue or black. **Do not eat red elderberry varieties, as they are poisonous**. Only dark varieties are edible. All berries must be cooked because it is poisonous to some people. Consult your herbalist.

Many **European settlers** to the United States ate elderberries before they came to America. They continued to do so during colonial times and up to the 1900s. They were made into jams, jellies and wine.

Raw berries contain enough **cyanide** to cause stomach pain and other symptoms. However, when **dark blue, dark purple or black elder berries are boiled** for a few minutes, the cyanide is released in the steam and the fruit is safe to eat. However, **red elder berries are still toxic** to humans even when cooked.

How to Pick: Berries grow in a cluster of small berries. Pick or cut off entire cluster at base of stem and put in bucket or basket for processing later. You can tie bucket/basket to waist so hands are free. Later can pull berries off twig (cluster) by hand or with comb that has big teeth.

Uses: The fruit is tart. Freeze or dry berries. Helps fight flu and infection. Consult your herbalist, doctor and/or health care specialist.

Stems are hollow and can be used as straws. Can be made into flutes by cutting stems in spring and removing inside pith with a hot stick.

Bark and root are used to tan leather. Flowers are used to make dye and wine.

(For **pruning** see February Garden-Maintenance. See April Garden-Plant. For **propagating** see March/July Garden-Maintenance. See September Garden-Harvest but details are in this section.)

 In greenhouse or enclosed sunporch, **figs** start getting ripe early August. They ripen through October. Figs are ripe when soft and hanging. There may be slight splits in skin and a drop of nectar. To store, dry them.

(See April Garden-Plant. For **covering fig,** see November Garden-Maintenance.)

 Harvest **Melons and Watermelons** in August.

Melons taste best when ripened on vine. They do not get any sweeter after they are harvested. Melons start to ripen about 30 days after flowering. A Brix refractometer can test sweetness. After being picked it keeps for about 2 weeks.

How to Tell If Ripe: Most melons except watermelon are ripe when rind changes from gray-green to yellow-buff. Fruit is still firm and gentle pressure easily separates stem from vine (for most varieties but not all). (For **which varieties separate easily**, see May Garden-Plant.)

If they are very soft, then it is overripe. They are **fragrant** when ripe for most varieties. The stem near the melon cracks all the way around when it is ripe. Melons in the reticulatus group such as muskmelons and Persians that have netting are better when the netting is raised and hard.

Touch the **blossom end** (opposite from the stem side) of the melon. It should be fairly easy to push down a little but it should not go through completely. (This test does not work with watermelon or winter melon.) **Cantaloupe** is mature when the rind changes from green to tan-yellow between the veins. **Melons in the Inodorus group** such as canary or casaba are hard to tell whether or not they are ripe. (See May Garden-Plant.)

Watermelon is hard to determine when ripe. It is ripe when tendrils nearest melon turn from green to brown. Or underside of melon is yellow or cream but not white. Or you hear low pitched thump instead of high pitch when you hit it. The rind should be dull, not shiny. Can pick any time but if harvest in the early morning it may extend storage time.

(For **sowing and growing details,** see April Greenhouse. See May Garden-Plant.)

 Harvest **peaches and nectarines**.

Pick peaches after green tinge disappears. The reddish tint some varieties have is not a good indicator of ripeness. Yellow fleshed peaches change from green to orange. Other varieties become redder, or combination of orange/red.

Should be slightly firm. Comes off branch with a slight twist. Be careful not to remove the fruit spur from the tree. Handle gently. It should have a pleasant, peachy smell. Green peaches do not have much of an odor. Unlike pears, they do not ripen further once picked.

Storage: Ripe peaches will keep for 4-6 days in the refrigerator. Refrigerate them unwashed, and put in a paper bag. Do not pack them too closely.

Can dry, freeze or can peaches and nectarines.

(For **pruning** see February Garden-Maintenance. See April Garden-Plant.)

Pears are picked when easy to separate from tree but not yet ripe. If allowed to ripen on tree, they are gritty and core turns brown. Pick when firm and skin starts changing from green to yellow-green. Keep stems on.

A pear ready to be picked should feel springy. Close your hand around it and squeeze. If it feels rock hard, it is not ready. There should be a slight feeling of give, but not too much.

To take off the tree, lift the fruit and twist. It should come away from the fruit spur easily. Be careful not to damage the fruit spur.

Storage: Wrap each pear in paper individually. Keep between 33-40 degrees in 80-90% humidity. Do not let freeze. A root cellar is good. Pears store a few months.

When you want to eat them, bring into room 60-65 degrees to ripen. They continue to **ripen after being picked. Good storage varieties** are anjou, bartlett, bosc, devoe, kieffer (for cooking), and winter nelis.

(For **pruning** see February Garden-Maintenance. See April Garden-Plant.)

Plums are picked when ripe. They are covered with powdery bloom. With both Japanese and European plums, the taste test is the best way to determine ripeness. They should just be beginning to soften.

Plums only store a short time. Late ripening plums store the best. Can be dried into prunes.

(For **pruning** see February Garden-Maintenance. See April Garden-Plant.)

Harvest Herbs and Vegetables in August:

 Process vegetables for storage over winter.

Pickle and can vegetables. For pickling use grape leaves with immature cucumbers.
Dehydrate vegetables in solar/electric dehydrator, enclosed sunporch, car or greenhouse.
Freeze fruits and vegetables. They store longer if vacuum packed.

 Pick **beans (snap or green)** regularly to promote formation of new beans. (**Shell beans** are picked when the beans in the pod are mature. They can be dried on the vine or removed when full size and dried in a warm, dry spot with good ventilation.)

Picking: Pick immature, green beans. They are green, not yellowish (unless a yellow snap bean). They should not have well developed beans inside so therefore should not be lumpy.

You have less work later if you snap the bean from the vine so stem is not attached to the pod. Otherwise you have to remove it later. Keep picked beans out of sun. Cool as soon as possible. They stay fresh in the refrigerator for 3-4 days.

Drying: You can dry beans with the pods on or dry just the shelled beans. (If drying shelled beans, they need to be mature.) If drying with pod on, break pod into 3 or so pieces and let dry in a shallow cardboard box or on a screen in a warm, airy place. Or remove pods and then dry beans. Once or twice a day stir the box/screen to dry all parts.

Using: Some beans have strings and some beans are stringless. The string is a tough filament along the outside of the pod from one end to the other. It should be removed before cooking beans that are in their pods. They can be frozen, dried or canned.

(For **planting and harvest details,** see May Garden-Plant. For **shell beans,** see September Garden-Harvest.)

To harvest **celery** cut whole plant at ground level before flower/seed stalks appear. Stalks, leaves and seeds are edible. Can leave in garden longer until need to eat it. (For **sowing/growing information,** see March Greenhouse. For **transplanting,** see May Garden-Plant.)

 Cut **comfrey** plants and **feed** to pigs, sheep, cattle, horses, goats, turkeys, chickens, rabbits and other animals. Consult your herbalist. (See March Garden-Plant. See June/July/September Garden-Harvest with most **harvest information in June.**)

Harvest **corn (sweet)** at milk stage when the **ears are fully formed but not yet ripe.** This usually lasts about a week. The silks should be brown and starting to dry, which occurs about 3 weeks after the silks first appear. The husks should hold tightly to the ear, and the kernels should produce a little **milky fluid** when pierced. If your fingernail punches into the kernel very easily, the corn is a little green. If you have to press hard, it is too old.

Older ears can be left on the corn plant to dry for cornmeal or other similar uses.

Remove ears of corn by twisting them at base and pulling them off. (See **harvesting tools** in field corn in September Garden-Harvest.) To store, freeze or can.

Soaking Corn in Alkali Water to Make More Nutritious

Traditional American Indian recipes call for soaking corn or corn flour in lime water (slaked calcium hydroxide, called cal or tequesquite, builder's lime, hydrated lime, slaked lime or Mrs Wages pickling lime). It is a strong alkali. Mrs. Wages pickling lime is food grade calcium hydroxide with no additives or preservatives.

Traditional southern recipes soak corn with hardwood ash water which creates lye, a really strong alkali (sodium hydroxide).

Do not use dolomite powder (calcium carbonate, $CaCO_3$) which is the same as limestone and chalk.

This soaking process is called nixtamalization. It releases nicotinamide (niacinamide or Vitamin B-3) which otherwise remains bound up in the grain. This prevents the vitamin B-3 deficiency disease Pellegra with it's symptoms of sore skin, fatigue and mental disorders. It also improves the amino acid quality of proteins and increases calcium absorption. It helps preserve hominy which would quickly become sour without it. It makes corn easier to digest.

From the book "Nourishing Traditions" by Sally Fallon: "**To make lime water** place about 1 inch of pickling lime in a 2 quart jar. Fill with water, shake well, cover tightly, and let stand over night. The powder will settle and the resultant clear water is lime water. Store in a cool place (it is not necessary to refrigerate) and use for soaking cornmeal by pouring out carefully."

One nixtamalization recipe: Use 2 pounds (1 quart) corn kernels or dried corn flour. Plus 1/4 cup pickling lime (food-grade calcium hydroxide) and 3 quarts water.

In a stainless steel pot, dissolve the lime in the water. Add the corn and remove any floating kernels. Bring to a boil. Reduce heat to low, and cook uncovered for 15 minutes. Turn off heat and let sit uncovered for 4 hours at room temperature or overnight in the refrigerator.

Pour the corn into a colander and run cold water over it, rubbing the kernels to remove the hulls. Rinse thoroughly between 4-11 times. Drain. Use the whole kernels in soup or stew. Or grind them to make masa that can be made into tamales or tortillas.

Hominy or nixtamal was traditionally made by boiling dry corn kernels in a dilute lye solution made from wood-ash leachings until the hulls were easily removed by hand. The lye causes the corn kernels to expand and lose their outer skin. Hominy was washed at least 7 times to remove the lye before being cooked.

(For **sowing/growing corn,** see May Garden-Plant. For **field corn and more general corn harvest,** see September Garden-Harvest. For **general grain harvesting,** see May Garden-Harvest.)

Harvest **globe artichokes** in August and September. Each plant has a primary flowering stalk (with bud) plus more buds below it. They ripen at different times.

When to Harvest: There is a subtle change in the color to less bright green. Cut before leaves on the bud start to open. Leaves start to open at the bottom first.

To harvest cut stem below the bud with a sharp knife. Can store in refrigerator for 2 weeks.

(For **growing system and sowing,** see February Greenhouse. For **second possible sowing date,** see March Greenhouse. To **move from house to greenhouse,** see April Greenhouse. To **move from greenhouse to outside,** see June Garden-Plant. See September Garden-Harvest but details are in this section.)

August and late July is second **hay** cutting. (For **different types of hay cuttings and types,** see June Garden-Harvest. For **hay making,** see May Garden-Harvest. For **cover crops and pasture,** see "Garden Tips".)

Harvest **hops** (cannabaceae family; Humulus lupulus). **Hop flowers (cones)** are ready for harvest in **late August and September.** Squeeze a cone in your finger, if it is damp, very green, and stays compressed after you squeeze it, then they are not yet ready. If it is dry, springs back to its original shape when you press it, and you see/feel more sticky lupulin (yellow powder), then it is ready.

Dry flowers or leaves in a warm, airy place out of the sun.

General Uses: Hops flowers are used in beer brewing. Young leaves, roots and shoots are edible cooked.

Medicinal: Flowers are used as a bitter, sedative tea. It is antibiotic. Good for digestion. It is antispasmodic (relieve cramps or spasms of the stomach, intestines, and bladder). As an infusion, drink one cup at night to aid sleep. As a tincture for anxiety, take 20 drops in a glass of water 3 times daily. To help digestion take 10 drops with water up to 5 times daily. Consult your herbalist, doctor and/or health care specialist.

(See April Garden-Plant. For how to make a **tincture, decoction or infusion,** see April Garden-Harvest.)

Bring cured **onions and garlic** into house in August. Store in cool, dark, dry place such as an unheated room that does not freeze. Can store in a cardboard box. Do not store in a root cellar, trench or clamp that has high humidity. Best humidity for onions is 65-70%. Best humidity for garlic is 40-60%. (For **garlic and perennial onions** see September Garden-Plant. For **harvesting, curing and storing,** see July Garden-Harvest.)

In August can dig up new, early **potatoes** as needed to eat right away. Early and mid-season potatoes are ready to be dug up when plants start to flower.

Planting: Potatoes are planted April through June, and harvested June through October. (For **when to plant early/mid/late season potatoes,** see April Garden-Plant. For **sprouting and then planting in April,** see March Greenhouse. For **late/storage potatoes** see September/October Garden-Harvest with most **harvest information in October.**)

 Harvest **tobacco** leaves when buds start to form. Cut off buds unless saving seed. Bottom leaves are harvested earlier. Harvest continues for a few weeks as leaves develop.

To Dry: Cut a slit near the end of the center rib of each leaf, put a stick or wire through it and hang the leaves. Leaves should not touch when hanging. Throw away any leaves that have mold or fungus on them or that smells moldy.

Air Curing: Hang somewhere dry, out of the sun, and airy. Let hang for 2-3 months to 2 years. It is low in sugar (for light, sweet flavor) and high in nicotine. Cigar and burley tobaccos are air cured.

Fire Curing: Smoke from a slow burning fire permeates the leaves. It takes 3-10 weeks. It is low in sugar and high in nicotine. Pipe, chewing, and snuff (inhaled through nose) are usually fire cured.

Flue Curing: Keep it in an enclosed heated area. It takes about 1 week. It is high in sugar. It is medium to high in nicotine. Virginia tobacco is usually flue cured.

Sun-Curing: It dries uncovered in the sun. This method is used in Mediterranean countries to produce oriental tobacco. Sun-cured tobacco is low in sugar and nicotine. It is used in cigarettes.

To prepare, cut out the center rib and any large ribs. Then cut into pieces. To shred can put in a food processor. Or use other similar method.

Storage: Store cured tobacco in jars or vacuum sealed bags. It improves in richness, color and smoothness with age. It can be stored 4-5 years or longer.

Uses: Consult your herbalist, doctor and/or health care specialist. The active ingredient in tobacco is **nicotine** which is very alkaline. **In large doses it is very poisonous.** In large doses it produces nausea, vomiting, sweats and muscular weakness. Smoking or chewing it can cause disease.

In small doses it increases blood pressure and activity of gastrointestinal muscles. In large doses it lowers blood pressure and activity of gastrointestinal muscles. Use only with advice of a doctor.

Tobacco is a **sedative, diuretic** (increases urination), **expectorant** (treats cough), **laxative, and sialagogue** (increase saliva). When injected into the rectum as smoke or rolled leaves, it acts as a cathartic (laxative). A wet leaf applied to piles (hemorrhoids) is supposed to cure it. Apache Indians used it to treat toothache by chewing or smoking it.

It reduces the pain and swelling of **bee stings, mosquito bites** and other bites/stings. It neutralizes the acidic nature of a bee or other insect sting. Plus it reduces swelling. **To use,** wet a small amount of tobacco and rub it on the sting. Hold on skin or wrap with bandage for 5-10 minutes. Repeat if needed.

It is used to treat **mites and lice in poultry**. Consult your veterinarian or County Extension office. (See March Farm Animals.)

(For **sowing and growing details,** see April Greenhouse. See June Garden-Plant.)

 Read the book "How to Grow Your Own Tobacco: From Seed to Smoke" by Ray French.)

 **Tomatillos** are ripe when husk splits or when fruit begins to soften and fills out husk.

Bears small, round and green or green-purple fruit. Fruit is 1-2 inches in diameter and sometimes as big as a golf ball. As the fruit matures, it fills the paper-like husk and can split it open by harvest.

Fruit **ripens** from green to yellow-green to pale yellow (for yellow variety). The husk turns brown. The final fruit can be yellow, red, green, or purple. Fruit falls to ground when ripe. It is tart and smells lemony.

Uses: Remove husks before eating. Can eat raw or cooked. Used in salsa.

Storing: Store in a cool, dry, airy place. Store with husks on. Will last about 2 months.

(See March/April/May Greenhouse with details in **March**. See June Garden-Plant.)

 Harvest an abundance of **tomatoes** in August. Harvest every day or every few days.

Daily summer temperatures of an average 75 degrees is best. For indeterminate plants you can **harvest until first frost** when the tomatoes will be damaged. However, when daytime temperatures are consistently **below 60 degrees**, fruit no longer ripens on the vine so it is best to pick all green and red tomatoes then.

Ripe tomatoes have an even color except some heirloom varieties have colors that are not even. It is a little bit soft when squeezed. Cherry tomatoes may crack if left on the vine too long.

To harvest grasp the tomato gently and twist. Or cut off with clippers.

Storage: They continue to ripen after picked. They need warmth to ripen but do not need sunlight. Sunshine may cause them to ripen too quickly and rot. Fresh tomatoes are best **stored around 56 degrees** such as in an unheated room that does not freeze. They can turn mushy if stored in the refrigerator. They can be stored on countertop for a week.

Green Tomatoes: Harvest before first frost. A mature green tomato is shiny, well developed and medium/deep green. May show a light yellowing at shoulders. **To ripen** put in room at 60-70 degrees out of direct sunlight. It takes about 2 weeks to ripen.

If you have a lot of green tomatoes, save some for ripening later. Store at 55-60 degrees but never below 50 degrees or they may never ripen. They ripen in 25-28 days. Check regularly to remove bad tomatoes.

You can pull up the **whole tomato plant** and hang in a dark, dry place to let the fruit ripen. A temperature of 60-65 degrees is best.

Store without stems. Light is not needed to ripen green tomatoes. Put them in a **paper bag**. The tomatoes emit ethylene gas that speeds up ripening. You can put other fruit in the bag such as bananas that emit a lot of ethylene.

Processing: Since tomatoes are acidic, they can be canned in a water bath. Or you can dry them in a dehydrator either electric or solar. Or cook into sauce, and freeze or can.

(See April Greenhouse, June Garden-Plant. See October Garden-Harvest but most details are here.)

 Harvest **valerian** leaves spring and summer to make sedative tea. Can use now or dry to use later. Dry leaves in warm, airy place out of the sun. You can make a tincture from the **flower**.

Using: For tea add 1 cup dried valerian leaves to 1/2 gallon water. Do not use all the time. Use in moderation. Simmer briefly and let sit for 30 minutes. Then drink. Cats like valerian in the same way they like catnip.

Medicinal: The leaves and flowers are not as strong medicinally as the roots. Reduces anxiety. Good for headaches. Can be a stimulant to some people. Consult your herbalist, doctor and/or health care specialist.

(For **sowing and dividing,** see May Garden-Plant. For **transplanting daughter plants**, see August Garden-Maintenance. For **harvesting leaves**, see June Garden-Harvest but details are here. For **harvesting roots**, see October Garden-Harvest. For **how to make a tincture, decoction or infusion**, see April Garden-Harvest.)

Greenhouse, Hoop House or Cold Frame  (August)

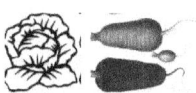 Early August plant **carrots** for baby carrots to harvest in October or November. (For **sowing/growing details,** see June Garden-Plant. For **harvest and storage details,** see September Garden-Harvest.)

Mid August through September sow **lettuce and radish** (summer or winter) every 7-10 days. (For radish and lettuce, see April Garden-Plant. For **radish,** see August Garden-Plant.)

August and September sow in greenhouse for **November and December harvest:**

 Sow **dandelion** in July, August and September for winter harvest. (See March Garden-Plant.)

 Sow **collard greens** to harvest fall/winter. (For sow/grow details, see March Garden-Plant.)

 Sow **kale.** (See March Garden-Plant.)

 Sow **parsley**. (See March Greenhouse. For sow/grow details, see April Garden-Plant.)

 Sow **sorrel and Swiss chard.** (See May Garden-Plant.)

 Sow **spinach.** (See March/August Garden-Plant with sow/grow details in March.)

Farm Animals  (August)

 Give flea and tick medicine to dogs and cats.

  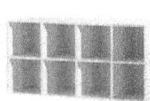  No more new **homing pigeon babies** by the end of August. Remove nest bowls. (Second image above is pigeon in nest bowl.)

A **homing pigeon, racing pigeon, or messenger dove** is a variety of domestic pigeon derived from the Rock Pigeon that was bred to find its way home over very long distances. Flights as long as 1,120 miles have been recorded. Their average flying speed over medium distances is around 30 miles per hour, but they can go as fast as 60 miles per hour.

Their **home base** or where they live most of the time is where they return when they are released. Their owner transports them in a cage to another location either a few or many hundreds of miles away. When they are released, they return to home base.

They are called **carrier pigeons** when they carry a message. The message is written on thin paper, put in a small tube, and attached to their leg. They have been used to carry messages for 3,000 years and were used in World War II.

Pigeon housing is called a **loft** (rather than a coop). (The third image above is where the pigeons roost and nest. The last image is a simple loft with an outside pen enclosed in chicken wire.)

Basic care is similar to that of poultry. They mostly eat grain but do eat greens. They do not eat insects. Feed needs to be 12-15% protein. The most common feeds include wheat, peas (yellow or green), barley, millet, corn, flax seeds, oats, sesame seeds, and brown rice. They like black sunflower seeds.

Give garlic water sometimes. They like wheat germ oil when they are molting. They need pigeon grit to help digest food. It consists of oyster shells, charcoal, salts and other minerals.

They can **breed** at 5-6 months old. They do not roost in trees like chickens. They prefer a solid surface. A hen lays 2 eggs. It takes 18 days to hatch. Both parents (hen and cock) take care of the babies. Babies fledge (leave the nest) at 35 days old.

If you have **found a lost pigeon**, please contact the American Racing Pigeon Union at http://www.pigeon.org/lostbirdinfo.htm or 405-848-5801 in Oklahoma.

(See April Farm Animals. For **chicken care**, see February Farm Animals. For **garlic water,** see January Farm Animals. For **poultry molting**, see September Farm Animals.)

 In August or July start saving chicken, duck and turkey **eggs to eat in the winter.**

Peak egg production is **April and May**. Good egg production is **February through July.** Lowest egg production is **September through December.** (The longest day is in June. Molting is in September and the shortest day is in December.)

(For **egg storage methods,** see July Farm Animals.)

Honey Beekeeping in August

The **nectar and pollen collection slows** down this month. Bees have to fly longer distances to get food. The colony's **growth is also slowing.** Drones (male bees) are still around, but outside activity slows as the nectar flow slows.

There is no more chance of swarming. Watch for **honey robbing** by wasps or other bees. Do not leave hives open for any length of time since this encourages other bees to rob it.

Honey harvest may begin this month and continues through October.

(For an **overview of beekeeping,** see January Farm Animals. For **harvesting honey**, see September Farm Animals.)

 # SEPTEMBER
September 1 daylight is 12 hours 53 min. Sept 30 daylight is 11 hours 50 min.

September 22 or 23 is **Autumnal Equinox, night and day are about the same length.**
Abundant harvest. Dry season.
Mid September: Lots of acorns fall.
Lots of Wooly Bear caterpillars. The brown stripe in the middle gets longer, the older it is.
Late September: Most leaves on trees are still green but trees such as sourwood, poplar and locust have almost all turned colors and started falling.

 Garden- Maintenance

 Mid September **cut off tips, tops, flowers, and small fruit** that will not ripen before frost from melon, broccoli, brussels sprouts, eggplant, tomatoes, winter squash, pumpkin, etc. Remove suckers from tomatoes and tomatillos. This way all energy goes to the remaining parts of the plants.

Look for round-headed **apple tree borers** in apple and other fruit tree trunks. (For details, see May Garden-Maintenance.)

Asparagus (liliaceae family; Asparagus officinalis) **seed can be saved from the female plant.** Cut the ferny plant top in **late fall** when berries are red. Hang fern to dry. Or let seed dry outside on plant.
Optional: Can rub berries on screen to release seeds, usually 6 in each berry. Or soak berries in water to soften skins. Squeeze out seeds and rinse off pulp. Dry seeds between paper towels or paper for several days. Store in plastic bag in refrigerator until **spring.**
Or just dry whole berry, and **sow whole berry in spring.**
(For **planting crowns and sowing seed,** see April Garden-Plant. For **salt and production,** see July Garden-Maintenance. See April Garden-Harvest.)

 Late September **divide perennial plants** every 2-4 years.

 Put **tree guards on fruit trees** to protect from rodents chewing on bark when snow surrounds trunk. And to protect from sun scald. Remove in **spring** unless it is well ventilated. (**Whitewashing trunks** also protects bark from sun scald. For whitewashing, see June Garden-Maintenance.)

 Garden- Plant (September)

Early September: Lovage, radicchio, radish, winter cover crops (oats, ryegrass, vetch, wheat), and garlic/shallots/perennial onions.

Fall Sowing: Some crops (perhaps many crops) such as parsnips and salsify will come up early spring if you **sow seed in the fall.** This can happen if you are saving seed and let some plants go to seed. When you harvest the seeds, some accidentally fall to the ground. If you leave them there, they grow on their own in the spring.

You can do this on purpose by sowing frost hardy plants in the fall. It can work with tender plants such as tomatoes but is more reliable with hardy plants. Experiment with it if you have extra seed. (For a **list of frost hardy plants**, see February Greenhouse.)

You do not have to figure out what time to sow in **spring**. You let nature decide when it is time for seeds to grow. Thin or transplant in spring. (For **fall sowing of parsley and parsleyroot,** see October Garden-Plant.)

Early September:

Sow **lovage** (apiaceae family, Levisticum officinale). It is an aromatic **perennial** in the **carrot family related to celery** (See "Plant Families" in back of this manual.) Can propagate by root cuttings. Easy to grow.

Sow seeds **late summer or early fall**. The alternating freezing/thawing helps seeds to germinate in **spring**. If sown in spring, it can take up to 4 weeks to germinate. Seeds do not retain viability for more than 1-2 years.

Lovage plants should be thinned to 2-3 feet apart. The first year it grows to 2 feet tall. Grows 5-7 feet tall in 5 years. Growth starts **early spring** for plants grown last summer.

Care: Likes soil pH of 5.5-7.5. Prefers full sun but tolerates partial shade.

Uses: The leaves, stems, roots and seeds are edible. Has a strong celery flavor with a hint of anise or licorice.

Leaves: The leaves look like celery leaves. Can eat leaves raw or cooked but is better cooked. Good in soup. **Leaves are best eaten in spring** because it becomes bitter later.

Roots / Rhizomes: It was originally cultivated for its aromatic roots. They are diuretic (increasing urination) and carminative (relieving flatulence or colic by expelling gas). Since the 14th century it was used for treating boils and sore throats. In Europe it is used for stomach upsets, menstrual disorders, rheumatism and migraine headaches. Consult your herbalist, doctor and/or health care specialist.

Seeds: Can be used in an infusion. Whole or ground seeds can be added to candy, meat, bread, crackers or biscuits.

(See October Garden-Maintenance. For **how to make a tincture, decoction, or infusion,** see April Garden-Harvest.)

Radicchio is frost hardy. Succession plant every 2 weeks for 1 month from **late August through September. Sow 8 weeks before the first frost in fall.** (For **sowing/growing details**, see February Greenhouse. See August Garden-Plant.)

Succession sow **radish (summer)** every week **until end of September. Last sowing is 4 weeks before first frost in fall.** Harvest in 1 month. Hardy to 25-28 degrees. (For **summer radish and sowing/growing details**, see April Garden-Plant. See August Garden-Plant.)

Sow **radish (winter)** (brassicaceae family; Raphanus sativus) late August or early September **(mid to late summer) 6 weeks before first frost in fall. Hardy to 25-28 degrees.** Matures in 50-65 days. Plant same as summer radish except winter radish is larger and slower growing. Thin to 4-8 inches apart.

(For **summer radish and sowing/growing details**, see April Garden-Plant. See August Garden-Plant. See August/September Greenhouse.)

Sow **cool-season winter cover/compost crops** such as barley, crimson clover, fall oats, fava beans, fescue, field peas, grasses such as ryegrass (annual), rye, vetch, and wheat. **Fescue** is best planted September through November. (See **Garden Seeders and Broadcast Spreaders** in "Garden Tips" in back of manual.)

(For **clover, fava beans, fescue, oats and peas,** see March Garden-Plant. For details about **barley, rye and wheat,** see October Garden-Plant. For **vetch** see October Garden-Plant but details in this section. For information about **cool weather cover crops,** see March Garden-Plant. For **warm weather summer cover crops,** see June Garden-Plant. For more about **cover crops and green manures,** see "Garden Tips" in the back of this manual.)

 Sow **oats (fall)** (poaceae/grass family; Avena sativa or Avena nuda). **Sow 6-8 weeks before hard frost in fall which means sowing mid August to early September. Can sow 10 weeks before fall frost.** They die in the winter. (For details see March Garden-Plant and July Garden-Harvest.)

 Ryegrass (annual) (poaceae/grass family, Lolium genus with 9 species of tufted grass).

Sow 6-8 weeks before hard frost in fall. It is one of the most popular seeds sold for **lawn and pastures.** It is a **cool season grass** with fast germination (in 1 week). Sow 1/4 to 3/8 inch deep. Sow 20-25 pounds per acre if broadcasting seed. Grows well with clover. (For **clover,** see March Garden-Plant.)

Care: Likes soil pH of 6-7. Likes full sun but does OK in light shade. Does OK in moist conditions. Roots grow down 60 inches. It has prolific growing rates.

Used as temporary planting for overseeding warm or cool season grass. Good forage crop and for erosion control.

 **Vetch (hairy)** (fabaceae family; Vicia villosa). **Sow 6-8 weeks before hard frost in fall.** It is a hardy vine, **annual or biennial legume** that fixes a large amount of **nitrogen,** around 150 pounds per acre. It grows 2-3 feet tall when planted alone and higher when planted where it can climb. Vines can grow 8 feet long.

Sow 1-2 pounds per 1000 square feet or 40-60 pounds per acre. Does better if use alfalfa-clover inoculant to improve nitrogen fixing. (For **inoculants** see "Garden Tips" in back of manual.) Some seed may germinate in a few years.

It is the best **cold climate vetch.** If well-established in fall, it tolerates frozen soil down to 0 degrees, remaining dormant until spring. Fall-planted hairy vetch flowers in **April** and seeds ripen in **May-June.**

Uses: Used as cover crop, green manure, pasture, silage and hay. (See October Garden-Plant but details here.)

 Sow **wheat (winter)** (poaceae/gramineae/grass family; pooideae subfamily, triticeae tribe, Triticum aestivum, cereal wheat) in **September or October. Sow at least 6 weeks before your first fall frost. Can sow 8-10 weeks before first fall frost.** Can also plant **triticale** (a wheat/rye hybrid) at this time.

(For **sow/grow details,** see October Garden-Plant. See May Garden-Harvest. For **cover crops,** see March Garden-Plant. For **general grain harvesting,** see May Garden-Harvest.)

 **Plant garlic, shallots and perennial onions** (amaryllidaceae family; Allium genus). They can be **planted in fall or spring** but fall produces larger harvest. However, there is risk of death due to severe freezing in the winter.

Plant **late September through October** before ground freezes. Plant 2-4 inches deep, 4-8 inches apart depending on bulb size. Be sure to plant them with the root down.

It grows some in the fall and then remains dormant all winter. Green tops are killed by hard freezes but bulb and roots stay alive. Mulch late fall with leaves, hay or straw.

Care: Likes full sun. Likes soil pH 6.5-7.0. Soil should have good drainage. Keep well weeded.

Add rock phosphate, bonemeal, gypsum (calcium sulfate for sulphur), and/or greensand.

(See below for details about garlic and perennial onions. For **fertilizers** see "Garden Tips" in back of manual. For **fungal diseases and onion pests,** see Amaryllidaceae Family in "Plant Families" in back of manual.)

( Read "Growing Great Garlic" by Ron Engeland, and "Garlic, Onion and Other Alliums" by Ellen Platt.)

 Plant **garlic** (amaryllidaceae family; Allium genus) **early fall (September or October), or early spring (March or early April) about 4 to 6 weeks before the last frost date.** It is better to plant in

the fall because you get bigger bulbs. However, there is the risk of death by freezing over the winter. Mulch plants in the winter.

In the fall plant a little earlier than onions. Eat small garlic cloves, and **plant the biggest and best cloves.** It is easy to grow and **very hardy.** It takes up only a small amount of space to get a good yield. It has few pests or diseases.

Plant in easily dug soil that has few rocks. Pull apart a bulb and plant each clove separately. (See second image above of clove with root side facing down.) **Plant with the root facing down**, covering with about 2 inches soil. Garlic planted in the spring can be planted less deep since it does not have to deal with cold winter temperatures. Plant about 8 inches apart, varying the distance based on the size of the clove.

Green shoots grow that will later be killed by hard freezing. But the clove and roots remains alive. Cover with mulch such as leaves or straw to protect them during winter. In the spring remove the mulch and give them some fertilizer.

Garlic is **day-length sensitive** meaning it needs days to be at least a certain number of hours long before they develop properly. So latitude affects the type of garlic that is grown. **Hardneck garlic** is usually grown in cooler climates including the mountains of North Carolina. **Softneck garlic** is usually grown closer to the equator.

There are 3 types of garlic:
1. Hardneck garlic (Allium sativum ophioscorodon) such as porcelain, purple stripe, and rocambole. Their cloves are large and easy to peel. They grow well in cold climates. Porcelain will store for 8-10 months. Purple stripe and rocambole will store for 6 months. Rocambole dries up too much if humidity is less than 50%.

Hardneck garlic develops a long **flower stalk** about 1-2 months before maturity. Cut it off unless you are saving seed but most garlic is propagated by planting cloves. (The flower can be cooked and eaten.) You want the plants energy going into the bulb, not the flower.

2. Softneck garlic (Allium sativum var sativum or common garlic) such as artichoke, Asiatic, creole, turban, and silverskin. Softneck garlic is commonly found in the grocery store. Multilayer white or rose parchment covers the entire bulb. It has a soft, pliable stalk that is easy to braid. It usually has several layers of cloves around the center of the bulb. The outer bulbs are the largest.

They grow well in many types of climates. Artichoke will store for 8-10 months. Asiatic and turban will store for a few months. Creole and silverskin will store for up to a year.

3. Elephant garlic (Allium ampeloprasum) is not really garlic but is a bulb-forming leek. It has large cloves but not much garlic taste. Stores up to a year. (For **leeks**, see February Greenhouse.)

(See March/April Garden-Plant. For **harvest and storage details**, see July Garden-Harvest. For **fertilizing** see April Garden-Maintenance.)

 Plant **perennial/walking/multiplier/potato onions** (amaryllidaceae family; Allium genus) Common seed onion, shallot/small, multiplier onion, and topsetting onion are Allium cepa). Plant in **September or October.**

Eat big onions, plant small onions. Small bulblets at top of stalks can be planted in fall or eaten. It is better to cut off the buds/flowers unless you are saving seed. The flower is edible. Onion seed only remains viable for a year or two.

Care: Likes full sun. Likes well drained, loose soil. Likes pH 6.5-7.0. Keep onion plants well weeded.

Fertilizer: Likes rich soil with lot of phosphorus and potassium. Rock phosphate is good. Soils with nitrogen deficiency have yellow-green plants with necks that do not collapse. Phosphorous deficiency has light green plants which mature slowly. Potassium deficiency has light green plants with brown tips and poor bulb formation.

(For **sowing seeds of annual onions and general onion planting**, see February Greenhouse. For **planting perennial onions in spring**, see April Garden-Plant. For **fertilizing** see April Garden-Maintenance. For **perennial onions**, see July Garden-Harvest. For **fungal diseases and onion pests**, see Amaryllidaceae Family in "Plant Families".)

Read the book "Growing Great Garlic" by Ron Engeland.)

🧺 Garden- Harvest 🧺 (September)

All September: Basil, beans, cauliflower, comfrey, corn, cotton, ginseng, globe artichoke, mushrooms and potatoes.
Late September or Early October: Amaranth, carrots, perennial herbs, and sunflower.
Early Fall through Early Spring: Gourds, hay, leeks, mushrooms, parsnips and salsify.
Fruit and Nut Trees/Bushes: Apples, elderberry, grapes, hazelnuts, hickory nuts, kiwi, oak (acorns), rosa rugosa, and walnut.

 **Appalachian Folklore:**

Harvest most crops, and can/preserve vegetables during a waning moon (illuminated area is decreasing, going from full moon to new moon). They keep longer and in better condition.
Slaughter livestock in 4th quarter before the new moon.
Full moon is a good time to harvest above ground plants. Herb leaves should be gathered during the full moon prior to the plant going to flower.
New moon is the best time for planting seeds. Herb roots should be dug during the new moon.
Pick apples and pears
(For more **Appalachian folklore, and images of phases of the moon**, see "Garden Tips" in back of manual.)

 **For how to store food** read these books: "Root Cellaring: Natural Cold Storage of Fruits and Vegetables" by Mike and Nancy Bubel, "Mary Bell's Complete Dehydrator Cookbook", "Making & Using Dried Foods" by Phyllis Hobson, "Ball Blue Book of Preserving" (how to can), "The Joy of Pickling" by Linda Ziedrich, and "Preserving Food without Freezing or Canning: Traditional Techniques Using Salt, Oil, Sugar, Alcohol, Vinegar, Drying, Cold Storage, and Lactic Fermentation" by Deborah Madison and Eliot Coleman.)

For how to make a tincture, decoction or infusion, see April Garden-Harvest.

Harvest **basil and other herbs.** Plants are better harvested in **July**. (**Basil:** For **sowing details**, see April Greenhouse. See May Greenhouse. For **transplanting** see June Garden-Plant. See July Garden-Harvest.)

Harvest storage **beans (shell)**. **Bush beans** have a large harvest for a short time. **Pole beans** have a gradual harvest that lasts until frost. Can sometimes eat pods when young though some varieties are too tough.
Shell beans are ready for harvest when they **rattle in the pod**. The pod is dry and brown. You can also bite a bean to see if it is dry. It should be hard.
You can pick each **bean pod** off the plant in the field, or you can pull up the entire plant and bring it to your threshing area. Harvest before very mature pods split open and beans fall to the ground.
In the fall you can let the beans **dry on vine.** Or you can pick almost mature beans and let them dry someplace else. If pods are withered but still moist, pick them and spread them on a screen or put in shallow cardboard box in a warm, airy place to dry.
Threshing Beans: You can get beans out of pods by squeezing pods open. Shell beans are **easiest to shell** when they are dry but not over dry (where the pod is rock hard). So it is better to shell soon after the pod dries rather than waiting to do it months later. Or thresh by holding the plant by the roots and bang it against the inside of a barrel.
Winnowing Beans: After threshing, clean the beans by **winnowing**. There are many methods such as using a screen and a hair dryer, fan or air compressor to blow off debris (chaff).
Drying: After shelling/threshing if they are not completely dry, you can dry beans in a shallow cardboard box or on a screen in a warm place with good ventilation. Can dry in the sun. Once or twice a day stir the box/screen to dry all parts.

Storage and Uses: Split beans can be eaten right away or fed to farm animals. **Store** in a cool, dark, dry place in a container that rodents can not get into. Beans will last years however beans are best when eaten within a year. After that they gradually get harder and harder so that even many hours of cooking does not make them completely soft.

Save beans for **planting next year** and eating over the winter. Beans are viable (will sprout) about 4 years. (See May Garden-Plant. For **harvesting green / snap beans**, see August Garden-Harvest.)

 Harvest **cauliflower.** With good growing conditions heads develop rapidly to 6-8 inches in diameter. When mature the head is white, compact and firm. Cut the whole head from the main stem. It will keep 1 week in the refrigerator. Leaves can be cooked like cabbage.

(For **sowing and sowing details** see April Greenhouse. For **early cauliflower,** see May Garden-Plant. For **late cauliflower,** see June Garden-Plant.)

  This is the last month to harvest **comfrey**. Do not harvest comfrey in October or later. Let grow so plant can build up roots to prepare for winter. (See March Garden-Plant, June/July/August Garden-Harvest with **most harvest information in June**.)

  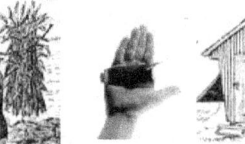 Harvest **corn (field dent or flint or popcorn).** Allow husks to dry completely in field. Can be harvested after light frosts. Pick after husks turn brown. **Field corn** is harvested when kernels are between 20-30% moisture content. Then they are dried to a storage moisture content around 15%.

Bundling and Shocking corn: One method of harvesting corn by hand is to cut the whole stalk with the ears attached and arranging them in bundles and then shocks. To do this, use a **corn knife** (see second and third images above) and cut the stalks off with short downward strokes. Leave 4 inches of stalk in the ground. Gather the stalks in one arm. When your arm is full, drop them in a bundle on the ground. Tie the bundles with baler twine, and then **shock** (see fourth image above) them by leaning 4 together looking like a tepee. Arrange the rest around this group. You can tie baler twine around the shock to keep it standing up. Later in the fall, husk out the corn.

Husking corn: Husking pegs (see fifth image above) attach to your hand to help strip the husk from the ear.
Corn Crib: (See last image above.) Can store with corn still on cob with or without husks in corn crib (corn house). The typical corn crib has slatted walls that allow air to circulate through the corn. It is about 4 feet wide. Corn cribs are elevated above the ground beyond reach of rodents. It can be rodent-proofed with hardware cloth or wire mesh.

Drying Corn with Husk: Remove the ears from the stalk but leave the husks attached. Pull back the husks, and tie the husks of several ears together. Hang on a wire or rafter to dry. To keep rodents from getting to the corn, poke a hole in a large tin can lid and slide it over the wire.

Fodder and Feed: You can feed stalks, leaves, husks, cobs and whole kernels to goats, cattle, horses, sheep and pigs though ruminants do better with the stalks. Poultry need shelled, coarsely ground kernels of corn (scratch size). **Hand-cranked shellers** are available to remove dry kernels from the cob.

Hand-cranked corn crackers crush the dry shelled corn into small bits the size of chicken scratch. Electric or manual grain mills turn whole kernels or small bits into flour or meal for human use.

Using Husks: One of the most common uses for corn husks is **cooking**. They are used for wrapping tamales (a Mexican dish made of cornmeal batter and shredded meat), and for wrapping fish for grilling.

They are used in **craft projects** such as corn husk dolls (save silks for hair), puppets and flowers. Pioneers and Native-Americans stuffed mattresses and cushions with corn husks, wrapped up food, and used them as kindling for starting fires.

It is easier to mold green, fresh husks into shapes than dry husks. But if you only have dry husks, soak for 10 minutes before creating craft project.

Using Cobs: Cobs can be made into pipes. It can be used to make charcoal or used in a fire. Toys are made with corn cobs. You can make corn cob jelly. Add to soup while cooking to add flavor, then remove before serving.

(For **sowing and types of corn**, see May Garden-Plant. For **harvesting sweet corn and soaking corn for greater nutrition,** see August Garden-Harvest. For **general grain harvesting**, see May Garden-Harvest.)

(Read "Homegrown Whole Grains" by Sara Pitzer and "Small Scale Grain Raising" by Gene Logsdon.)

 Harvest **cotton** in September, October, or November. (See May Garden-Plant for details.)

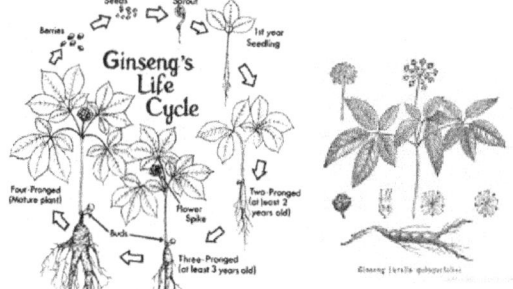  Forage **Ginseng (American) root** (araliaceae family; Panax quinquefolium) in **late August through October.** Some say it is best dug after first frost. It is a **perennial** that grows throughout Appalachia and much of the eastern United States. It likes rich, moist, shady areas. It likes to grow near walnut, lind and poplar trees.

It takes 18-24 months for seeds to germinate. It has a small umbel (flat-topped or rounded flower cluster with individual flower stalks arising from the same point, i.e, looks like an umbrella) flower with very small **yellow-green flowers** that turn into **red berries**. It first grows as 3 leaves. Then in 1 or 2 years, the leaves are 2-prong (2 branches). In 2-3 years, it is 3-prong. It grows to 4 or 5 prongs. It grows to 1-2 feet tall.

In the **fall** the leaves turn a bright golden yellow. The berries turn bright red and drop. Then the plant dies back to the root. The root ranges from 1/2 to 8 inches long. It takes 7 years for the roots to be big enough for harvesting. **Wild ginseng** is more potent than cultivated.

In **Appalachia** it is called sang hunting. It is usually dug with a stick or a sang-hoe made of steel with a narrow blade. When you dig, replant the berries so there will be plants in the future. Try to keep as many small, feeder roots on the main root as possible. Only dig older plants. You can tell the age by the number of prongs the root has. It takes 5 pounds of fresh root to make 1 pound of dry root.

Uses: Roots are used medicinally. It is a mild tonic, stimulant, nervine (calms nerves), and stomachic. Leaves are made into a tea. Consult your herbalist, doctor and/or health care specialist.

Harvest **globe artichokes** in September and August. To harvest cut stem below the bud with a sharp knife. Cut before the leaves on the bud start to open. Can store in refrigerator for 2 weeks.

(For **growing system and sowing,** see February Greenhouse. For **second possible sowing date,** see March Greenhouse. To **move from house to greenhouse**, see April Greenhouse. To **move from greenhouse to outside**, see June Garden-Plant. See August Garden-Harvest.)

 Harvest **Shiitake Mushrooms**

Logs inoculated with Shiitake spawn/spores the previous winter (usually February) may have some mushrooms to eat **September through November**. Consult an expert before eating any mushrooms.

(For many details, see February Garden-Plant.)

 In September harvest **potatoes** (white/red/blue):

Potatoes are planted April through June, and harvested June through October.

(Planting: For when to plant early/mid/late season potatoes, see April Garden-Plant. For sprouting and then planting in April, see March Greenhouse.)

     (Harvest: For early and mid season potatoes see August Garden-Harvest. For late/storage potatoes see **October Garden-Harvest for most harvest information**.)

Late September or early October:

 Harvest **amaranth** (amaranthaceae family) grain. Amaranth keeps on flowering until first hard frost. It is best to **have a killing frost before harvest.** Then a week of good drying weather is needed. (See Amaranthaceae Family in "Plant Families" in back of manual.)

     To tell if seeds are **ripe,** gently but briskly shake or rub flower heads between your hands. If seeds fall readily, they are ripe. Gather ripe grain in dry weather by bending plants over a bucket and rub seedheads between your hands. Can rub the seedheads through screening into a wheelbarrow and then blow away finer chaff using wind, air compressor or fan.

     Uses: Leaves and grain are edible. **Black-seeded varieties** of amaranth are gritty even when cooked, so use these varieties just for their leaves. All **golden and light-colored** amaranths are good cooked as whole grains. Unlike true grains, amaranth has no hulls to remove. Ruminant animals like to eat the stalks and leaves.

     (See June Garden-Plant. For **general grain harvesting information,** see May Garden-Harvest.)

Harvest late / storage **carrots** (umbelliferae family). Carrots can be harvested as baby carrots before they are mature but **storage carrots need to be mature**. Mature carrots have more sugar.

     To harvest grasp top and pull gently. Or dig around it in hard soil. Cut off greens leaving 1/2 to 1 inch stem.

     Black Plastic: You can continue to harvest carrots until the ground freezes. To protect them wait until the tops die back, then cover with black plastic. Then cover that with leaves or mulch that is 12 inches or more deep. Then cover with another sheet of black plastic. Place rocks on top of plastic so it does not blow away. Temporarily remove plastic to harvest what you need.

     Storage: Carrots grown in heavy clay soil store better than those grown in sandy soil. They will store up to 8 weeks in a perforated bag in the refrigerator. They will store 4-5 months at 32-40 degrees and 90-95% humidity.

     Layer carrots in damp sand, peat moss, leaves or sawdust in root cellar, clamp, barrel or trench. (A clamp is a hole lined with straw, filled with vegetables, then covered with straw.) Have about 1 inch sand / sawdust / leaves between each row. Or **overwinter** in garden with 1-2 feet of mulch.

     Uses: They can be dried, canned or frozen.

     (For **early carrots** see March Garden-Plant. See June Garden-Harvest. For **sowing/growing details and late carrots,** see June Garden-Plant. See October Garden-Harvest but not much harvest information. See August/December Greenhouse.)

 Cut **perennial herbs** such as good king henry, oregano, sage, salad burnet, thyme, and winter savory 6 inches to 1 foot above ground to prepare for winter. Can dry leaves or make tincture. Can cover with mulch to help plants survive cold weather.

     (For **planting perennial herbs** see May Garden-Plant. For **how to make a tincture, decoction or infusion**, see April Garden-Harvest. For **general herb harvesting,** see July Garden-Harvest.)

 Harvest **sunflower** (asteraceae family; Helianthus annuus). Dry on stalk or dry in covered area.

     To dry on stalk: As the flower petals fall off, the kernels get bigger. The back of the seedhead turns from green to yellow. This is a good time to **cover the flower head** with a mesh, burlap or paper bag to stop birds from eating the seeds.

     Harvest when back of seedhead turns from yellow to brown. The flowers have begun to die back and most or all petals have fallen off. Cut head from stalk leaving a few inches of stalk.

     To dry in covered area: When you see the backside of the sunflower head getting yellow, cut the head leaving one foot of stem. **Hang or store** in a warm and dry place with good air flow to prevent molding. You may still want to cover the seed heads with bags if drying in an outdoor shed to protect from rodents and birds.

If seedheads are left outside in light frost, the meal from the seed is more oily than those not frosted.

Cure in warm, dry room for 1 week or until hulls are dry and hard. Or cure in sun with the heads facing down. The sun dries the back of the heads which shrinks it and loosens the seeds.

Processing: When seeds are dry but still in seedhead, rub facedown on half inch hardware cloth or wire mesh. Or break up with rolling pin, hammer, stick or brief time in blender.

Shell sunflower seeds by **grinding** them in a seed mill (coffee, flax or poppy seed grinder). Or crush with a rolling pin or hammer. Place in cold water. The shells **float** to the top and can be skimmed off with a slotted spoon. (Stirring the water helps bring more hulls to the surface.) Strain the seeds and dry on paper towel or other surface. You can pick out any remaining shells.

Another method that is not as efficient is to put some seeds in the bowl of an electric mixer or blender, pulse the mixer/blender on and off a few times for a few seconds each time. The shells separate and most seeds are not crushed. Then put seeds into cold water as described above.

Eating: Can eat raw, cooked or roasted. Seeds can be sprouted and eaten. Soak raw, unhulled seeds overnight in water. Then treat like making other sprouts. (For **how to make sprouts**, see December Greenhouse.)

Storage: Unhulled seeds last 12 months in a cool, dark location. Hulled, toasted seeds last a few months. (See June Garden-Plant.)

Early fall through early spring:

Harvest **gourds.** Harvest date depends upon type. Always handle carefully so skin is not injured.

Harden off plants prior to harvesting by stopping water and fertilizer, and letting fruit ripen and harden. A good indication it is ready to pick are brown, dried stems. Leave a few inches of stem on when cut off vine.

Hang in warm, dry place 3-4 weeks (or sometimes months) until seeds rattle. They should not touch while drying. Gourds take a long time to dry. After dry, they can be painted, waxed, cut and decorated.

When to Harvest:

Cucurbita gourds are fleshy and usually injured when left outside in frosty or freezing weather. Harvest when mature and before cold weather. They are more difficult to cure and must be kept in a cool, dry place for several weeks to several months.

Lagenaria gourds are not damaged by light frost so can harvest later in fall. They should be fully mature.

Luffa gourds may be harvested when the fruit turns brown.

How to Cure:

Cucurbita gourds should be harvested only when fully mature. Wash in a solution of non-bleaching detergent (such as sulphonaphthol) with a soft brush. (Liquid general-purpose detergents contain no bleaching agents. Most powdered detergents have bleaching agents.) Then put in a dry place out of direct sunlight with good air circulation. Leave until thoroughly dry and hard.

Lagenaria gourds are washed in a similar manner as cucurbita gourds. They need a very long drying period to fully harden. A barn attic with a metal roof works. It must have lots of ventilation.

Luffa gourds are of 2 types: L. acutangula (sharply ridged) and L. aegyptica (smooth surface). Aegyptica have a greater spongy interior. They are washed similarly to lagenaria and cucurbita. They must be dried thoroughly. The exterior skin is then removed by soaking in water for several days. Then the sponge is shaped and dried.

Saving Seeds: Gourds freely **interpollinate** within a given group. (For more about **squash/gourd/pumpkin cross pollination**, see winter squash in May Garden-Plant. For **saving pure seed**, see "How to Save Seed" in back of this manual.) Seeds that are dry keep in a cool, dry place for 3-4 years with little loss in vitality.

(See April Greenhouse. See May Garden-Plant which has most **sowing and growing information**.)

September is third **hay** cutting. (For **types of hay cuttings,** see June Garden-Harvest. For **cover crops,** see "Garden Tips" in back of manual.)

Harvest long-season (winter) **leeks** in September and October. They can be harvested once they reach about the size of your thumb or wait until they are 2 inches in diameter. It has a 6-10 inch stem. You can pull up out of the ground or for hard soil dig them up. Small side growths can be discarded.

Do not trim off roots except on those that are going to be put in the refrigerator and eaten soon. (They last about a month.) For storing trim off the tops to where the leaves start to get tough.

Harvest until **first hard freeze or even harvest through winter into early spring**. About half will make it through winter if mulched.

Storage: Dig up before hard freeze. Put 2-3 inches of peat moss, sand or sawdust in a bucket and dampen it. But it should not be too wet since it will cause the leeks to rot. Push the leek roots into the peat a little. Keep the leek bulb and stalk above the peat. The tops should have been trimmed but do not trim off outer leaves. Store in a cool, dark place with high humidity (90-95%). Stores best at 32-40 degrees. May store up to 2 months.

Uses: All parts of a leek are edible. Can eat raw or cooked. However, remove tough leaves before eating. Cook in soups, stews and other ways like onions.

(For **sowing/growing details,** see February Greenhouse. See May Garden-Plant. See October Garden-Harvest but most information is here.)

Harvest **mushrooms** such as **Shaggy Mane** and other edible mushrooms that can be found this time of year in wood chips, grass and bare dirt. Harvest lasts a few weeks then returns again in **June**.

Eat it the same day you pick it. Get a good book on mushrooms and ask an expert before you eat any. Some mushrooms are poisonous so be sure about what you pick. Some people are allergic to all mushrooms.

(See June Garden-Harvest.) Study the book "Nature's Garden: A Guide to Identifying, Harvesting, and Preparing Edible Wild Plants" by Samuel Thayer.)

Harvest **parsnips**. Very hardy. Frost improves flavor. Cold temperatures change starches in the roots into sugar especially **after 2-4 weeks of near-freezing temperatures**.

Roots are 1 1/2 to 2 inches in diameter, and 8 to 12 inches long. Before ground freezes, dig up ones to eat right away. Dig up carefully with spading fork.

As cold weather continues, mulch parsnips with 1 foot of leaves/straw/hay to keep ground from freezing for a few more weeks and to protect all winter.

Storage: Then before the ground freezes (usually in **early December**), dig up some to store in root cellar, barrel, clamp or trench. (A clamp is a hole lined with straw, filled with vegetables, then covered with straw.) Then **early next spring,** dig up the rest as needed when the ground thaws. Roots turn bitter once leaves start to grow.

To store in **root cellar,** trim off leaves leaving 2 inch stem. Will store for 2-3 weeks uncovered. For longer storage up to 6 months, put in damp sand, leaves or sawdust, same as **carrots**. Keep at 32-35 degrees and 90-95% humidity.

Using: Eat right away any that got damaged while being dug up.

(See April Garden-Plant. For **carrot storage** see September Garden-Harvest.)

Harvest **salsify.** Harvested and stored like carrots and parsnips. (See **carrot storage** in September Garden-Harvest.) The thin root can be 1 foot long or longer. Eat, harvest and store like parsnips above.

Roots taste best after frost. Roots are 9-12 inches long and 2 inches wide at top. Can eat raw or cooked. Cooked like carrots. Like Jerusalem artichokes, they contain inulin which can be hard to digest. (See May Garden-Plant. For cooking vegetables with inulin, see Jerusalem artichokes in October Garden-Harvest.)

Fruit and Nut Trees/Bushes Harvest in September:

🍎 Late September through October harvest late season **apples**. Very late apples are picked unripe. (For **pruning** see February Garden-Maintenance. See April Garden-Plant. For **thinning** see June Garden-Maintenance. **For picking and storage,** see August Garden-Harvest.)

Continue harvest of **elderberry** berries through early September. Eat only dark berries. Do not eat red berries. Consult your herbalist, doctor and/or health care specialist. (See April Garden-Plant. **For eating and harvest,** see August Garden-Harvest. See October Garden-Maintenance.)

Grapes ripen late August through late October. They need sun to ripen. Grapes **change color** several weeks before they are at maximum sweetness. The grapes of black varieties lose their red color when ripe. White varieties change from green to golden yellow. When ripe, seeds change color from green to brown.

Grapes are **sweet when ripe**. When pick, hold by stems to avoid bruising.

Storage: Store grapes that are dead ripe but not overripe. This is when stems turn from green to brownish or at least dark green. Pack in boxes. Only pack totally dry grapes. Do not squash grapes. Store boxes in 80% humidity around 40 degrees. Should last in root cellar 1-2 months. Catawba, sheridan and steuben store longer than most.

Can dry grapes into **raisins**. It is better to dry seedless grapes or else you have to remove the seeds. Dip the grapes in boiling water for 30 seconds to break the skins. Dry in a dehydrator for about 12 to 24 hours. Or dry in the sun for 3-4 days. (Boiling grapes is optional for sun drying.) Needs to be in airy container or on flat tray. May need to cover with cheesecloth or other thin cloth to keep bugs off them.

(For **pruning** see February Garden-Maintenance. See April Garden-Plant.)

Harvest **hardy kiwi** from mid-September to early November. As it ripens the seed changes from white to brown to black. The fruit is soft when ripe. Keep stem attached. Fruit is fragile. Can store up to 2 months refrigerated. A mature vine can produce 200 pounds of fruit. (For **pruning** see February/June Garden-Maintenance. See April Garden-Plant. For **propagation** see July Garden-Maintenance.)

Harvest **hazelnuts** (betulaceae or birch family; Corylus genus with 14-18 species, filberts). Nuts form in clusters of 1-12 nuts called burrs.

Parts of a Hazelnut Nut: The nut has 3 layers.
   1. First is the **outer husk (hull)** or burr. In the summer it is green and leafy. In the fall it turns dry and brown. (See third image above.) Shuck the nut by pulling the 2 parts of the burr apart. Or cut off with small garden shears.
   2. Next is the **inner shell** that is hard and brittle. It can be hard to remove.
      a. Crack the hazelnut with a nutcracker. They have a harder shell than pecans or walnuts.
      b. Or boil them. Add 4 tablespoons baking soda to a quart of boiling water. Boil nuts for 3-4 minutes. Then drain and cool. The shells may pop off, or you may need to rub them off with a towel.
   3. Then on the inside is the **nut meat (kernel) that is covered with a thin skin or membrane**.

Harvest: First harvest at about 4 years of age in **mid to late fall**. All species of hazelnut are edible. Harvest after the husks have dried, split open and dropped to the ground. Or pick off the bush when leaves and nuts start turning brown. Remove the entire burr (prickly seed covering).

Check daily so squirrels, mice and birds do not get them. You can lay a tarp on the ground to collect them and then shake the bush a little or hit it lightly with a broom.

A 30 foot hedge of bushes produces about 2 bushels of burrs which produce about 3 gallons of hulled nuts.

Dry in a warm, airy place for a few weeks. The white nutmeat will darken. They dry faster if shelled. They last a few months or more in a cool spot.

Uses: Can be eaten as is or ground up and put in candy. The oil is used in syrup. Can be eaten raw or roasted. (See April Garden-Plant.)

 Forage for **hickory nuts September through November**

(juglandaceae family; juglandoideae subfamily, juglandeae tribe, Carya genus). Gather every 3 days, or else squirrels, raccoons, chipmunks, mice and other animals will get them all.

Hickory species belong to the **walnut family**, which includes pecans and other nut-bearing trees. Hickory trees naturally cross breed so identifying species can be a challenge. Flowers April or May.

Some Varieties:

1. Bitternut hickory (Carya cordiformis) is widely distributed in the **eastern United Sates** but is not eaten because it is too bitter. It has a thin shell with lots of nut meat. It grows to about 115 feet tall. It lives about 200 years. It has bright sulfur-yellow winter buds.

2. Pignut hickory (Carya glabra) is very common in **Appalachia and the eastern United States**. It has thin shelled nuts that usually have a lot of meat. The taste when raw varies from bitter and nearly inedible to sweet and tasty. It grows to about 80-90 feet tall but can grow to 120 feet.

3. Shagbark hickory (Carya ovata) is **common in the eastern United States**. Its nuts are generally small, thick shelled and excellent flavor. Grows to about 90 feet tall.

4. Shellbark hickory (Carya laciniosa) Not very common in North Carolina. Its nuts are usually much larger, and also thick shelled and excellent flavor. The nuts of shellbark and shagbark have relatively less meat and more shell than the others. The taste is excellent. It grows up to 120 feet tall.

Parts of a Hickory Nut: The nut has 3 layers. First is the segmented **outer husk (hull)** that is easy to remove. Next is the **inner shell** that can be hard to remove. Then on the inside is the **nut meat**.

Pick the nuts off the ground while they are still in their **green husks or hulls**. To test whether nuts are rotten, put in water. Nuts that float are bad. Dry until the hulls (husks) are brown. They have a four-sectioned husk that loosens as it dries so it comes off easily. Dry away from rodents.

Save the hulls (husks) and let them dry until hard. Use as you would hickory wood for smoking or grilling meats.

Dry the nuts with the **shells** still on.

Method 1 to Prepare: Pound the nuts, shells and all (but not the segmented outer husk) to a coarse meal. For large quantities, make a large **wooden mortar and pestle**. (See images below.) It is very efficient and worth the effort required to make if mass production is the goal. Put a buckskin or cloth cover with a hole cut in the middle for the pestle to keep the nuts and fragments from flying out while pounding.

With the nuts crushed, put them in a **cooking pot**. Add water about twice the volume of the nut meal and **bring to a boil**. After 2-3 hours of simmering, the oils begin to separate and float to the surface and the broth gets a hint of sweetness and a nutty flavor. It can be cooked for up to 6 hours. The longer it is cooked the more oil separates.

The nut meats and shells sink. This lets you separate the solids from the broth by ladling or slowly pouring off the broth. This broth is a cream-colored oily, pasty substance that is called **hickory milk or nut broth.** Use like butter. It is delicious and nutritious. (In colonial trade, 1 quart of hickory milk was exchanged for 19 pounds of pork.)

The **oil** that is skimmed off can be processed further for storage by re-boiling the oil alone to drive off any water that is mixed with it. It should be refrigerated (preferred) or kept in a cool place.

Dump the shells and meats into a cloth and wring the rest of the broth out. The remaining solids can be dried and passed through a small mesh sifter. Most of what remains is **nut meat** with a small amount of shells that are small enough in size that it is OK to eat. But it is not very tasty since most oils have been removed.

   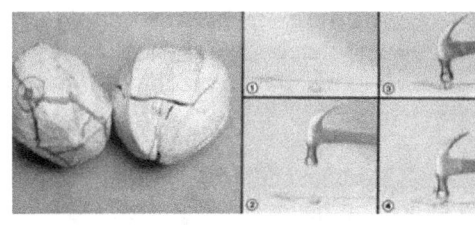

Method 2 to Prepare: After the nuts are dry, **crack them open**. Do not use a lever type nutcracker since this will shatter the nuts into little pieces. Place a brick on a hard, level surface. Then grasp a nut between your left thumb and forefinger, with its stem end pointing toward the right. Balance the hickory on top of the brick (narrow edge downward) and aim your **hammer** at a spot about 1/3 of the way down from the stem. (See above images on right.) Whack that spot with a short, sharp blow and the nut will pop right open. Use a nut or dental pick to remove the nut meat. They store a few months or sometimes until next year's harvest.

Storage: Rather than drying the nuts, they can be placed in **damp storage**. They taste sweeter in the spring after a winter of cool storage in a storage pit in the ground or in a bucket sealed to keep the moisture in. For **bucket storage,** soak the nuts for about two days with a few changes of water and then drain the water and let them dry until the surface is not wet. Then seal them in the bucket and keep in a cool (**35 to 50 degrees**) place that does not freeze. Or bury the bucket deep enough that it will not freeze. One disadvantage to damp storage is that when the temperatures warm in the early summer, the nuts will sprout and die since they cannot reach sunlight.

Hickory mast was a significant food source in early historic years, and the archaeological record suggests they were a significant resource to prehistoric people in eastern North America for thousands of years. (Mast is nuts from forest trees that accumulate on the ground.)

(Read "Peterson Field Guides: Field Guide to Eastern Trees" by George A. Petrides. Read the excellent book "Botany in a Day: The Patterns Method of Plant Identification" by Thomas J. Elpel.)

 Forage **oak trees for acorns** (fagaceae family; quercus genus) **September and October.** Acorns were an important part of the diet for Native Americans and pioneers. Flowers late spring.

Varieties Common in Northeastern United States:
White Oak: (fagaceae family; Quercus alba) Usually grows to about 65-85 feet tall. It grows very wide. Can live up to 600 years. All oaks in the White Oak group have leaves with ends of the lobes rounded and flaky white bark. White oaks can take up to 50 years before they produce acorns. Acorns in the White Oak group need little or no processing. They have **low tannic acid and are naturally sweet**. The **inside of the acorn is hairless**. They produce mast (acorns) every year, requiring only 1 season to mature. **Squirrels eat White Oak acorns first** and bury Red Oak acorns for eating later. Many oak trees in **Appalachia** are white oak.
Red Oak: (fagaceae family; Quercus rubra) It grows straight and tall usually up to 90 feet but can grow to 140 feet tall. Can live up to 500 years. Oaks in the Red Oak group have leaves with ends of the lobes pointed. Acorns in the Red Oak group **have a lot of tannic acid** that must be leached out to make them edible. The **inside of the acorn is hairy.** They produce acorns every other year requiring 2 years to mature.

Storing Unshelled Acorns: If you will not be shelling them right away, spread the acorns on a cookie sheet and roast in the oven at 325 degrees for 30 minutes. This kills moth eggs that may be in the acorns. If not killed, the acorn will be eaten by the larva.

To Make Meal or Flour: Acorns can be made into flour. **Shell acorns** by hitting with a rock/hammer or using a nutcracker. Remove the brown paper beneath the shell too. Crush the acorn meat into smaller pieces. A food processor or mortar/pestle can be used.
Put them into a pot of already **boiling water.** When the water is dark brown (every 10 minutes or so), strain out the acorn meats and switch them to another pot of already boiling water. It is important that the water is boiling in the second pot. Continue until the meats no longer taste bitter. This is to remove the bitter tannins. If you use the boiling

method, save the brown water. It is a **tannic acid solution**. (White Oak acorns do not need to be boiled as long as Red Oak. Some White Oak acorns do not need to be boiled at all.)

The meat can be crushed into a meal that can be used as **flour**. Since acorn flour is heavy, mix with lighter flour such as wheat or corn. The meal can be dried or frozen.

Another way to **leach out tannins** from acorns is to put them in a mesh sack. Leave it in a **running stream** for a week or so.

Medicinal: The tannic acid water is antiviral and antiseptic. It can be used as a skin wash for rashes, skin irritations, burns, poison ivy, cuts, etc. It can be gargled for sore throats or drunk as a mild tea for diarrhea, or used externally on hemorrhoids. Consult your herbalist, doctor and/or health care specialist.

Uses of Tannic Acid: It can be used as a dye for clothing or as a laundry detergent. Put a couple of cups in each load of wash of dark colors only. Your clothes will smell good. **Animal hides** can be tanned by soaking them in tannic acid. Hide tanning is the process of making raw animal skin into a comfortable, durable piece of clothing.

Store jars of tannic acid solution in the refrigerator (preferred) or cool place. If mold forms on top, remove it and reboil to kill the mold and store again.

(Read "Peterson Field Guides: Field Guide to Eastern Trees" by George A. Petrides. Read the excellent book "Botany in a Day: The Patterns Method of Plant Identification" by Thomas J. Elpel.)

Harvest **rosa rugosa** rose hips. The one inch fruit is about the size of a small walnut. It ripens to bright red, orange or purple. A ripe hip (seed pod) is soft. The leathery covering is eaten as for its vitamin C.

Inside the **hips** are seeds and many tiny, sharp hairs that are very irritating if eaten. So remove them and the seeds by slicing each hip in half and scooping seeds and hairs out.

Uses: Can make into tea, jelly, jam, etc. Rose hips used for jellies do not need to be seeded or scraped. Liquefy hip and strain out juice. Some people remove the stem and hairy end before liquefying. A half and half mixture of rose hip juice and apple juice makes a tasty jelly.

To **dry** spread the hips on a clean surface. Let dry until the skin begins to feel dry and slightly shriveled. Then split the hips in half and take out all of the seeds and tiny hairs. Then let the hips dry completely. Do not wait to remove the seeds until hips are completely dry, or it will be hard to remove seeds. They store 2-3 months in the refrigerator. For longer storage freeze it.

To make **puree**, clean and simmer them in water. Then mash and strain.

Some say let hips stay on plant until after the **first hard frost** which turns them orange.

(See April Garden-Plant. For **harvesting leaves**, see April Garden-Harvest.)

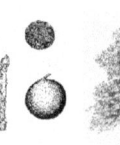Harvest **walnuts** in **late September and October.**

Black walnut tree (juglandaceae family; Juglans nigra) is native to **eastern North America.** Grows 98-130 feet tall. Its nuts are round and dark covered by a **greenish husk.** The husk changes from solid green to yellowish green when ripe. Press on the skin of the walnut with your thumb. Ripe nuts show an indentation. It usually alternates years with lighter and heavier harvests.

Parts of a Walnut Nut: The nut has 3 layers. First is the **outer husk (hull)** that is easy to remove. It is somewhat soft, and green or brown or black depending on age. Next is the **inner shell** that can be hard to remove. Then on the inside is the **nut meat**.

Allow nuts to ripen on the tree. Pick black walnuts off the ground as soon as they fall and de-husk right away (before they turn black) because otherwise they will mold and be harder to remove. The **husks (hulls)** stain hands brown so wear gloves. Do not put husks in a compost pile since they contain toxins.

Removing Husks (Hulls): You can husk the nuts by driving a **car** over them. Or use a hammer to hit the husk sharply, and then peel the nut out. Or use a manual corn sheller that has had some adjustments done to it.

An oily black appearance beneath the husk usually means the nut has been lying on the ground too long and has maggots. Put the nuts in a bucket of water, throwing away any that float. Then drain the water and spread them in a single layer on newspapers or other surface in a warm, dry, and airy environment. Others in layers of 2-3. Turn every 2 days. Let **cure** for 2 months or longer. Check periodically for dryness and throw away moldy ones.

Storing Unshelled Nuts: After dry, you can pack in alternating layers of dry peat and salt if you do not want to crack them yet. Or store unshelled nuts in a well-ventilated area at 60 degrees or less that does not freeze. Cloth bags or wire baskets discourage mold. Keep the relative humidity fairly high, about 70%.

Cracking Nuts (Shells): To crack open nuts, place nut in a **vise or vise-like nutcracker** with its seams parallel to the vise jaws. Then tighten the vise until the shell splits open. If you need extra leverage, you can slip a length of pipe over the vise handle. If the nutcracker does not state it will crack walnuts, then it probably can not crack them. Wear safety glasses when breaking walnuts open. Shells crack and send sharp fragments into the air.

Or **soak** the nuts in hot water for 24 hours before cracking. Then draining and soaking again for two more hours. Then baking nuts at 215 degrees for 15 minutes. Then cracking the nuts.

Remove the **nutmeat** by shaking or prying it loose. The nutmeats usually come out in four neat quarters. Store the nuts in the refrigerator or freezer.

Uses for black walnut husks (hulls): To make a black to sepia (reddish-brown) dye for clothing, baskets or other items. To make ink. Can put powdered husks in soap. To make gritty or abrasive material. Ground hulls are used as a tumbling abrasive for soft metals and jewelry. Do not put in compost because it has a chemical toxic to plants.

Medicinal: Make **black walnut hull tincture.** Put hulls in a 1/2 gallon glass jar. Cover with 40% or more proof alcohol. Let stand for 2-3 weeks and strain. Used as a fungicide externally for problems such as ringworm. Used internally for worms such as tapeworm. Consult your herbalist, doctor and/or health care specialist.

(For **pruning** see February Garden-Maintenance. See April Garden-Plant. For general information about **making a tincture, decoction or infusion,** see April Garden-Harvest.)

(Read "Peterson Field Guides: Field Guide to Eastern Trees" by George A. Petrides. Read the excellent book "Botany in a Day: The Patterns Method of Plant Identification" by Thomas J. Elpel.)

 Greenhouse, Hoop House or Cold Frame  (September)

Mid to late September succession sow annuals in greenhouse:

 Succession sow **through November** hardy plants such as **arugula, claytonia, endive/escarole, mache, mizuna, minutina, mustard, pak choi, parsley, scallion, Swiss chard, tatsoi, and turnip.** (For a **list of hardy vegetables and minimum cold temperatures,** see February Greenhouse. For arugula, claytonia, mache, minutina, mizuna and mustard, see February Greenhouse. For parsley see May Garden-Plant. For Swiss chard see July Garden-Plant. For endive/escarole and turnips, see August Garden-Plant.)

Plant Bull's Blood **beets** for their leaves. Harvest all winter. (For **early beets and sow/grow details,** see March Garden-Plant. See July Garden-Harvest. For **late beets** see July Garden-Plant. For **harvest details** see October Garden-Harvest.)

 Sow **lettuce and radish (summer or winter)** with succession sowings every 7-10 days. Last month to plant radishes. (For **lettuce** see March Garden-Plant. For **summer and winter radish sow/grow details,** see April Garden-Plant. See August/September Garden-Plant. See August Greenhouse.)

Spinach is a productive crop, sow a lot. Will be good until **late March**.
(For **sowing/growing details** see March Garden-Plant. See June Garden-Harvest, August Garden-Plant, August Greenhouse.)

 Sow **creasy greens, sorrel and watercress.** (For **sorrel** see May Garden-Plant. For **creasy greens** see March Garden-Plant. For **sowing watercress,** see April Greenhouse.)

Farm Animals (September)

 **Sheep and goat** breeding season begins. Pregnancy for both is 5 months. Goats usually give birth to twins (kids) though 1, 3 or 4 are possible. Sheep usually give birth to 1 or 2 lambs but can give birth to 3 or 4. (For **details and other animals,** see February Farm Animals.)

 **Open paddocks** previously planted with **brassica forage crops** such as kale, kohlrabi, mangel beet, rutabaga and turnip. Let animals graze here for part of the day only. Sow July through September for grazing September to March.
(See February/November Farm Animals. See **Fall and Winter Pasture: Special Forage Crops** in "Garden Tips".)

 **Clean** barn, chicken coop, turkey/duck pen, and nest boxes. Put diatoms (diatomaceous earth) in nest boxes to kill mites and lice. Put gypsum or lime on floor. Treat for mites. (For **maintaining animal bedding,** see January Farm Animals. For **mites, lice and diatomaceous earth,** see March Farm Animals.)

 **Chickens, ducks and turkeys molt** (replace old feathers with new feathers) in the fall usually **late September and through October** as the day length shortens. Molting usually occurs after a long, intensive laying period. If hens have not been laying long, they may molt next fall or at another time. **Egg production goes down** a lot or stops when hens molt. Males molt before females. Molting starts around the head and neck and moves down the body with the tail last.

Better layers usually have a far more catastrophic molt, and regrow feathers quickly, and may even start laying again before winter. Molting and feather regrowth takes 2-3 months with molting and feather regrowth taking place at the same time. Poorer layers molt over a much longer period.

Hens referred to as **late molters** lay for 12 to 14 months before molting, while others, referred to as early molters, molt after only a few months in production. Late molters are better laying hens and have a more ragged and tattered covering of feathers. The early molters are poorer layers and have a smoother, better-groomed appearance. Early molters drop only a few feathers at a time and may take as long as 4 to 6 months to complete the molt.

Feed: Feathers are 85% protein so require a lot of **protein** to produce (hence the slowing of egg laying) so give birds high protein treats (20-22% protein) such as dried cat food or game bird feed. Also black oil sunflower seeds and/or cod liver oil (1/2 teaspoon per week per bird) help. Reduce the amount of corn or chicken scratch you feed them since it is low in protein.

The less **stress** they are under, the faster they grow back feathers. Make sure they have enough **calcium** from oyster shell, hardshell, egg shells, or other sources. (You can feed egg shells back to hens by crushing them first.)

Honey Beekeeping in September

There is **not much pollen or nectar collection** this month. The first hard freeze kills all the flowers. The drones (male honey bees) may begin to disappear. The **hive population is dropping**. The queen's egg laying is dramatically reduced.

Harvest honey from supers (boxes). Harvest may be August through October. You get about **30-100 pounds of honey per hive**. Leave the colony with at least 60 pounds of honey for winter. Check for the queen's presence.

Feed and medicate towards the end of the month (only the first 2 gallons is medicated). Add Apistan (controls Varroa mites) strips (stay in hive for 42 days). Also add menthol for mite control. Continue **feeding** until the bees take no more syrup. Feed 2 parts sugar to 1 part water.

Watch out for honey robbing by other bees.

Propolis is the glue that holds all the pieces of the hive together. Every time you open the hive, you break the propolis seal. If you do this when warm days are over, the propolis will never seal again so the hive can be blown apart by strong winds.

Read the books "The Backyard Beekeeper: An Absolute Beginner's Guide to Keeping Bees in Your Yard and Garden" by Kim Flottum and "Natural Beekeeping: Organic Approaches to Modern Apiculture" by Ross Conrad and Gary Nabhan.)

(For an **overview of beekeeping,** see January Farm Animals.)

Local **honey** is ready to buy. Stores for a very long time, even a lifetime.

It has been used as a **folk remedy** in cultures around the world for thousands of years. Honey diluted with water supports the growth of bacteria that are helpful to humans while killing off dangerous strains. Never feed honey to a child under 1 year old because it may contain harmful bacteria that their immune system can not handle.

Ale yeast that causes fermentation is used to make honey diluted with water into the **alcoholic drink mead** (honey wine). It is 8-18% alcohol. Beer is 4-6%.

Medicinal Uses

Honey is **antibacterial and antimicrobial**. It is anti-inflammatory reducing both swelling and pain. It reduces scarring. Consult your herbalist, doctor and/or health care specialist.

It is good to put on **wounds**. It breaks down bacteria and does not allow it to grow again. It has a direct nutrient effect as well as drawing lymph out to the cells by osmosis. The honey prevents the dressing from sticking to the wound, so there is no pain or tissue damage when dressings are changed. Other topical antiseptics cause tissue damage.

Honey is effective in treating inflammation of the eyelid, and some types of conjunctivitis (inflammation of membrane covering the surface of the eyeball. It is used to treat athlete's foot and other **fungal infections**.

It is more effective than silver sulfadiazine and polyurethane film dressings for the treatment of **burns**. It has been proven to be more effective in preventing **infection** in wounds than hydrogen peroxide.

Honey is soothing to a **sore throat**. Add honey to tea. Make a gargle or mouth rinse with honey by combining 2 tablespoons of honey with 4 tablespoons of cider vinegar, and add a pinch of salt.

For a **cold or flu**, mix 8 ounces hot water with 2 ounces lemon juice and 1 tablespoon honey. If you have eucalyptus oil or ginger root, add it. This drink helps with congestion.

It can be used as a **sleep aid.**

Honey helps with **stomach** discomfort. People with infections or inflammation of the gastric system have shown improvement when they ate honey. Honey helps fight bacteria in your **intestines**. Honey is also proven to help control diarrhea and constipation.

Honey is healthy for **skin**. A mixture of honey and ground almonds is an excellent scrub for your face. For a moisturizing rinse, mix 2 tablespoons honey with 2 teaspoons whole milk. Apply to your face and neck. Allow it to sit for 15 minutes. Rinse off.

For shiny **hair**, mix 1 teaspoon honey with 4 cups hot water. Lather throughout hair and rinse.

How to Store Honey: Keep in a cool, dark location out of the sun in a tightly sealed container. It does not need to be refrigerated. It can last for decades. Stored honey can crystallize but it is still good to eat. To remove crystals, gently heat jar (if glass) in a pan of hot water.

Real Honey: Food Safety News found that the vast majority of so-called honey sold at grocery stores, big box stores, drug stores, and restaurants do not contain any pollen, which means they are not real honey. It is best to **buy your honey from a local beekeeper** so you know you are getting the real stuff.

Build trap boxes for **meadow voles** (Microtus pennslyvanicus) **and other rodents.** Voles look like chubby, furry mice. They breed rapidly and live in underground burrows. They are a **serious pest for fruit growers** because they eat the bark of young trees. They **eat vegetable crops** and are especially fond of potatoes and sweet potatoes. They chew row cover fabric. They especially like greenhouses.

Voles must be **trapped all year** to keep them under control. They are fruit and vegetable eaters so they are not attracted to baits such as cheese and meat that catch rats and mice. Place the traps in the voles' line of travel which means at their burrow entrances, against walls and on their surface trails. **Proper placement is the key.** You do not even need any bait.

Meadow vole **burrow entrances** do not have any dirt mounding around them. Their travel lanes above ground are about 1 1/2 inches wide. Set box so hole is at entrance.

They love to enter small holes. So **build a wooden box** 12 inches long by 8 inches wide by 6 inches deep with a bottom piece of wood. Make a removable top. At one corner of each of the 8 inch ends drill a mouse-size hole at floor level. Place a snap trap in the box next to each hole. It is a very good trap. Make a lot of them. These boxes also trap mice.

You can make a bigger box with bigger holes for rats. Though rats are not as curious about things as mice are so they are harder to trap.

These boxes are good if you have cats or small children, since it is harder for them to get caught in a snap trap.

Predators such as coyotes, foxes, owls and skunks eat rodents. A barn owl with chicks eats 3,000 rodents in one breeding season. Add an owl nest box to your farm.

(See **Large Animal Pests** in "Plant Health" at the back of this manual.)

(Read "The Winter Harvest Handbook" by Eliot Coleman for **box building and use**.)

 # OCTOBER
October 1 daylight is 11 hours 48 minutes. Oct 31 daylight is 10 hours 45 minutes.

First frost around mid-October to early November. Abundant harvest. Dry season.
Early October: Oak leaves start turning brown.
Mid October: Peak fall colors. Most leaves still on trees. Leaves starting to fall a lot.
Late October: Peak falling leaves.
Very Late October: Most leaves have fallen except off of oaks and a few other trees.

First killing frost in fall, North Carolina: Oct 10-Nov 21

Garden- Maintenance

 Cut **asparagus** plants to the ground when leaves turn yellow. Then mulch. (For **fertilizing** see March Garden-Maintenance. For **planting crowns and seeds,** see April Garden-Plant. For **salt and production,** see July Garden-Maintenance. See April Garden-Harvest.)

 Log trees for **firewood** in dormant season from **late autumn (October) to late winter (February).** At this time there is the least amount of water in the tree. This reduces the time needed to season (dry) the firewood. Green wood has about 35-50% moisture. Firewood should be between 15-20% moisture to burn properly.

Folklore says to cut timber during the last quarter of the moon (week before full moon). It dries better and will not be eaten by worms.

Hardwood comes from deciduous (leaves fall in autumn) trees with broad leaves such as elm, hickory, locust, maple and oak. Hardwood is more dense, burns longer, and produces more heat than softwood. Poplar is a soft hardwood.

Softwood comes from conifers (trees that bear cones and evergreen needle-like leaves) such as cedar, fir, juniper, pine and spruce. Softwood is better for starting fires and as kindling. Softwood creates more creosote (tar deposits in chimney). Tar deposits lead to chimney fires.

After felling the tree, **chainsaw the trunk into lengths** (rounds) that fit into your stove or fireplace. Use a **dull axe or maul** (heavy, long-handled hammer used for splitting a piece of wood along its grain). The axe or maul is meant to split the fibers apart, not to cut the wood. A maul handle is heavier than an axe handle, and the head is heavier because most of the work is done by the weight of the maul, not by the energy put by you into the hit. For wood splitting a maul is easier to use than an axe.

Even better to use is a hydraulic splitter (see third image above).

It is easiest to split wood if it is green or slightly dry. Very dry wood is harder to split. Split wood dries faster. Store wood so it does not get wet when it rains. It should have good air flow.

Always burn well seasoned firewood. This means split wood should **air dry for at least 12 months**. Burning freshly logged or green firewood produces less heat, more smoke and more tar deposits in the chimney.

A **full, true or bush cord** is a stack of wood 4 feet tall, 8 feet long and 4 feet (48 inches) wide (128 cubic feet).

A **face cord** is 1/3 the size of a full cord. It is 4 feet tall, 8 feet long and 16 inches wide (42.6 cubic feet).

The weight of a cord varies by species. Up until 100 years ago wood was the major source of heat energy in the United States.

 Prepare for **freezing temperatures** for unprotected water pipes/hoses and bowls/buckets:

    Disconnect all garden hoses. Disconnect and drain sprinklers, cover with tin cans.
    Turn on water heater tapes and heat bulbs for water in barn, greenhouse and other buildings.
    Turn off water to pipes. Remove waterers not protected from freezing. Use electric heated water buckets.

Divide **mature lovage and other perennial plants** every 3-4 years. (For lovage, see September Garden-Plant.)

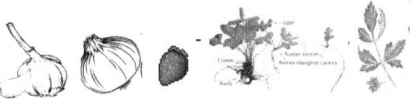 October or November deep **mulch perennials** such as asparagus, garlic, herbs, lovage, onions, rhubarb, stinging nettle, strawberries, etc to protect from freezing weather. Use leaves, hay, straw, etc..

 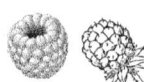 **Prune Bramble fruit such as blackberries and raspberries:**

    For **pruning** non-brambles, see February Garden-Maintenance.
        For more about pruning blackberries and raspberries, see August Garden-Maintenance.
    For **planting** all fruit/nut trees/bushes, see April Garden-Plant.
    For **propagating** blackberries and raspberries, see April Garden-Plant.
        For more about propagating all raspberries, see February Garden-Maintenance.
        For more about propagating black/purple raspberries, see July Garden-Maintenance.
    For **harvesting** blackberries and raspberries, see July Garden-Harvest.

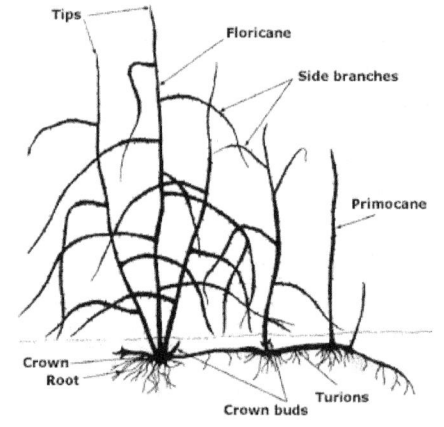

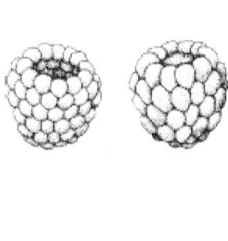

Above images are both Raspberry and Blackberry.
Image above in middle: Basic structure of both **blackberries and raspberries.**
Image above on right: **Raspberry** on left: fruit receptacle (core of the berry) remains on plant when you pick fruit.
                              Blackberry on right: the receptacle is part of the fruit you eat.

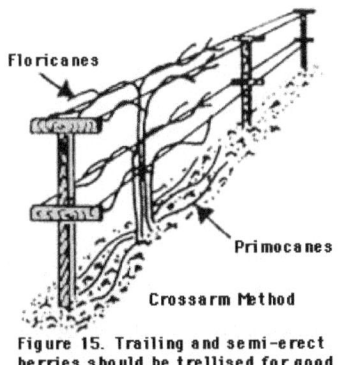 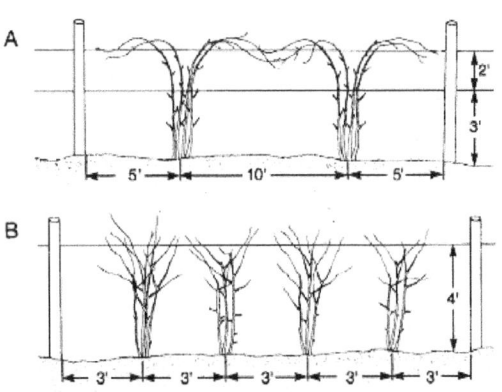 

Figure 15. Trailing and semi-erect berries should be trellised for good sunlight exposure.

Image above on left: **Trailing and semi-erect blackberry. (For trellis,** see grapes in February Garden-Maintenance.) Primocanes are new, 1-year old. Floricanes are old, 2-year old. Notice double-wire trellis.
Images above in middle: "A" is **trailing blackberry**. "B" is **erect blackberry**.
Image above on right: **Erect blackberry** before and after pruning.

 Prune Blackberry in October: Bears fruit on last years canes (**biennial fruit production**).

There are **4 types**- thorny erect (wild), thorny trailing (includes dewberry), thornless erect, and thornless trailing (includes Doyle). Most wild berries in **western North Carolina** are erect blackberries with thorns.

1. Erect Blackberry (includes wild)- thorny and thornless:
Spring (February) or Fall (October) Pruning while dormant: Thin canes to 6-10 inches apart. Remove weak canes. Before bud break, side branches are cut to 12-18 inch long. If needed, shorten canes to 4-5 feet tall.

Summer Pruning and Maintenance: Canes that grow too tall or long can be cut to manageable length. Can cut very long canes shorter to get more lateral (side) growth.

After plants bear fruit in **July or August**, remove all old (2 year or floricane) canes. The primocanes that grew this summer are now lying on or close to the ground. Cut them to 4-5 feet tall or whatever height to prevent berries from laying on ground, or build simple trellis such as horizontal wire 3 feet above ground. If needed, tie the canes to the trellis/wire with plastic or cloth.

New shoots (new primocanes) appear next summer at base of plants.

Can also create new plants through tip rooting. (For **planting and propagating** blackberries and raspberries, see April Garden-Plant. For propagating all raspberries, see February Garden-Maintenance. For more about propagating black/purple raspberries, see July Garden-Maintenance.)

2. Trailing Blackberry (includes Doyle)- thorny and thornless:
Plants should be 6-8 feet apart in rows 6-10 feet apart. Same as blackberries except a **grape-type trellis** is best where a major wire goes horizontal to the ground at 3 feet and 6 feet. (For grape trellis, see February Garden-Maintenance.) (See images above for how to trellis.)

Summer Pruning and Maintenance: The first year tie all canes on one side of the trellis, next year tie on other side. Do not weave canes in trellis (too hard to get out later). Several canes go in each direction along the wire. Cut canes when they go several inches above top of trellis, though it is best not to cut if possible. Trim canes from adjacent plants when they overlap several feet.

Remove all side canes that are in the bottom 3 feet of the vertical primary canes. Side canes above 3 feet should be cut to 2-4 inch long (stubs).

For **Doyle blackberries** do not wait until end of summer to tie up new canes (primocanes) since they will be too hard to tie. Though in very cold climates keeping them on the ground covered with mulch protects them in winter.

After all fruit has been harvested in **July or August,** cut out all 2-year (floricane) canes.

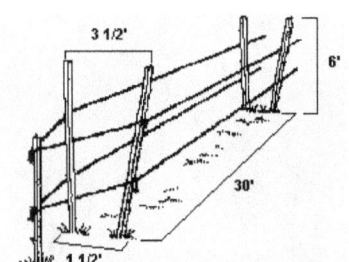  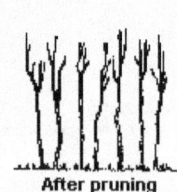

*Image above on left (Figure 2): **Raspberry** yields can be increased by using a **V-trellis** that separates the fruiting canes (floricanes) from vegetative canes (primocanes). In early spring floricanes are tied to wires on the outside of the V. Primocanes are allowed to grow in the middle of the V. Harvesting and pruning are easier because floricanes are pulled to the outside. Primocanes in the middle forces lateral (side) growth outward.*
*Image above on right: **Red Raspberry** pruning.*

*Image above on left: **Black Raspberry** pruning. Image above on right: **Purple Raspberry** pruning.*

  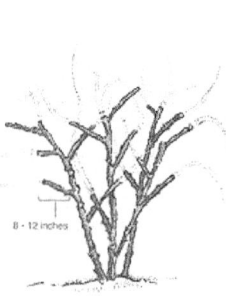  

*Above image on left: Prune **purple/black raspberries** by removing primocane tips in summer, and thinning in winter.*
*Above image on right: Pruning **red raspberries**.*

Prune Black & Purple Raspberry in October:

Spring (February) or Fall (October) Pruning while dormant: Thin to 5-10 primocanes per plant. Side branches should be cut to 8-10 inches for black raspberries and 6-10 inches for purple raspberries. Cut primocanes to 4-5 feet tall or whatever height to prevent berries from laying on ground or build **trellis** such as horizontal wire 3 feet above ground.

Summer pruning: Cut old fruit bearing canes (floricane) to ground after harvest berries in **July and August**. Floricanes are woody, silvery brown with light green leaves. Primocanes are green with darker green leaves.

Propagation: Canes should not be cut if you want to form new plants from cane tips rooting in soil. Black raspberries also grow new canes from crown of plant but do not spread as much as red raspberries. (For **planting and propagating raspberries**, see April Garden-Plant. For more about propagating black/purple raspberries, see July Garden-Maintenance. For **general propagating,** see "Garden Tips" in back of this manual.)

Prune Red Raspberry in October:

Red raspberries bear fruit on canes (floricane) grown previous summer. The middle 60% of plant has most fruit. Also bears some fruit on tips of primocanes.

Spring (February) or Fall (October) Pruning while dormant: Thin primocanes so 1 cane every 4-6 inches. Keep thicker canes. Cut primocanes to 4-5 feet tall or whatever height to prevent berries from laying on ground or build simple **trellis** such as horizontal wire 3 feet above ground.

Everbearing red raspberries such as Heritage produce fruit on the tips of primocanes the first year in late summer through early fall. The second year they produce berries on the lower sections of floricane that did not fruit last year. Therefore, Heritage can be pruned by removing all canes if you want.

Summer pruning: Cut old fruit bearing canes (floricane) to ground after harvest berries in **July and August**. Floricanes are woody, silvery brown with light green leaves. Primocanes are green with darker green leaves.

Propagation: Red raspberries spread by growing new shoots (suckers) away from crown. They can be transplanted to new places. Keep beds 12-18 inches wide. (See April Garden-Plant. For more about **pruning raspberries**, see August Garden-Maintenance.)

Fruit Trees in October:

 **Mow**, ideally with a flail mower, the grass and weeds around fruit trees and berries. (**Flail mowers** have a number of small blades on the end of chains attached to a horizontal axis. The ax-like heads strike the grass. They are used on rough ground and tougher vegetation such as brush/scrub).

 In late October **propagate elderberry step 1.** Take a 9-12 inch hardwood cutting with 1 or 2 nodes (where leaf forms) from a good 1-year-old stem. Fall cuttings are wrapped in plastic to hold moisture and stored in a cool, dark place **until spring** when they are planted. They can also be cut in spring and planted right away.

(For **pruning** see February/July Garden Maintenance. See April Garden-Plant and August Garden-Harvest. For details about herbaceous, softwood, semi-hardwood and hardwood cuttings, see "Garden Tips" in back of this manual.)

(**Propagation:** For propagation with hardwood **step 2**, see March Garden-Maintenance. For propagation with semi-hardwood, see July Garden-Maintenance.)

 In October **propagate kiwi** by taking hardwood cuttings.

(For **trellis types and pruning,** see February Garden-Maintenance. See April Garden-Plant. For **more pruning** see June Garden-Maintenance. For **propagation** see July Garden-Maintenance. See September Garden-Harvest. For details about herbaceous, softwood, semi-hardwood and **hardwood cuttings**, see "Garden Tips" in back of this manual.)

Garden- Plant (October)

Vegetables: Parsley and parsleyroot.
Cover Crops and Grains: Barley, rye and wheat.

 Sow **parsley and parsley root** (umbelliferae family; Petroselinum crispum) **late fall after a few frosts such as late October or November.** Nature stratifies the seeds and they germinate in the spring. **Very hardy biennials.** (For stratification, see "Garden Tips".)

(**Parsley:** See April Garden-Plant. See August/September/October Greenhouse.)
(**Parsleyroot:** See July Garden-Plant. See October Garden-Harvest.)

 Sow **cool-weather winter cover and grain crops** such as barley, hairy vetch, and winter rye, winter wheat. They can also be sown in **March**. There are many different opinions about when to plant barley, oats, rye and wheat in the fall. Times vary from 1-2 or 10-12 weeks before first frost in fall. (For **oats** see September Garden-Plant.)

(For more **cool-weather cover crops and grains,** see in March Garden-Plant. **Barley, rye and wheat** are below. For **hairy vetch** see September Garden-Plant. For more about **cover crops**, see "Garden Tips" in the back of this manual. For **general grain harvesting**, see May Garden-Harvest.)

 Sow **barley (winter)** (poaceae/gramineae/grass family; pooideae subfamily; triticeae tribe, Hordeum vulgare). Barley and oats should be planted about **5-10 days earlier than the dates shown for wheat in the chart below in the winter wheat section. Sow barley and oats 10-12 weeks (2 1/2 to 3 months) before first fall frost. Some suggest sowing 2 weeks before first fall frost.** (For **oats** see March Garden-Plant and July Garden-Harvest. See **Garden Seeders and Broadcast Spreaders** in "Garden Tips" in back of manual.)

Grown and harvested similar to **wheat**. Can be sown **spring (March) or fall (October).** Spring-planted barley ripens in 60-100 days while fall-planted ripens about 60-100 days after spring growth starts. Flowers and matures about a week earlier than wheat.

Sow 60-90 pounds per acre, or 1/2 pound per 100 square feet, or 20-25 seeds per square foot. A plot 100 square feet yields 5-15 pounds of grain. Sow 1 to 1 1/2 inches deep. It germinates rapidly sometimes in 24 hours. Will germinate in temperatures as low as 36 degrees.

Grows 10 inches to 2 feet tall. With a 30 inch tall plant having leaves 12 inches long.

Care: Likes full sun. Likes soil pH 6 or higher. Likes well drained soil. It will do OK in poor soil but likes phosphorus and potassium. Fertilize similar to wheat. It likes cool weather. Drought tolerant. More salt tolerant than other cereal grains. Does not reseed well.

Varieties: 2 botanically distinct types: six-row and two-row which is number of rows of grain on the seedhead.
   1. Six-row barley is more common. It is smaller-grained, less starchy, has more enzymes, and is used to make American-style high-adjunct beers. It is divided into 3 families:
      Malting Barley is grown in the upper Midwest. It is tall and bearded (bristles). It is spring planted.
      Coast Group is grown in California and Arizona. It is fall planted.
      Tennessee Winter Group is grown east of the Mississippi as livestock feed.
   2. Two row barley is grown in the Pacific Northwest and on the northern Great Plains. It is plump and starchy and used to make all-malt brews. It is spring planted and used for feed and malting.

Types based on Seed:
   Bearded barley has a slender bristle (awn) about 3 inches long attached to each seed. Hard to hull.
   Beardless do not have bristles. Used for forage. Better for small farms since do not have to hull grain.

Diseases: May be bothered by the fungus smut. Seeds can be treated with fungicide before planting. Plant resistant varieties. (See "Plant Health" in back of this manual.)

Uses: It is the fourth major **cereal grain** in the world after corn, rice and wheat. It has 3% more protein than corn.
   It is an important **animal fodder** but animals like wheat better because barley hulls are tougher. You can sprout barley and feed it to poultry. Or grind or crack for easier digesting. The stalks and leaves make good straw.
   It is used to make malt for **beer** and other distilled beverages by sprouting it.
   Barley seed is grown into a short grass (6-8 inches tall) and then juiced. The juice is drunk.
   It is used as a **cover crop** or green manure.

Hulling (removing beards): With oats and barley hulling is more complicated because the grain is completely enclosed by the husk in bearded varieties. There is no easy way to do this on a small scale. It requires a lot of threshing. Hulless (naked, Hordeum vulgare L. var. nudum) varieties are available and are best for the small farmer.

(See May Garden-Harvest. For **cool season cover crops**, see March Garden-Plant. For **general grain harvesting**, see May Garden-Harvest.)

 Read "Homegrown Whole Grains" by Sara Pitzer, "Small Scale Grain Raising" by Gene Logsdon.)

 Sow **rye (winter)** (poaceae/gramineae/grass family; pooideae subfamily, triticeae tribe, Secale cereale). It is not the same as annual ryegrass. (For **ryegrass**, see September Garden-Plant.)
    Can be sown in fall or spring (March) but best planted late summer to late fall. It is better to plant winter wheat a little earlier than winter rye. Once established it can tolerate temperatures down to minus 30 degrees. (See **Garden Seeders and Broadcast Spreaders** in "Garden Tips" in back of manual.)

    Sow between 2 weeks before and 4 weeks after first killing frost in fall so that means **early October to mid November**. The North Carolina Extension Service suggests sowing it in the mountains from **September 25 through November 1.**
    Seeds germinate at 33 degrees and higher. Sow 1/2 pound per 100 square feet. Sow 1 to 1 1/2 inches deep. If growing as a grain, plant seed 5 inches apart, or broadcast and rake in.
    Care: Likes soil pH 5.0-7.0. Needs full sun. Soil must be well drained. It is frost hardy.
    It grows slowly all winter. Grows when 40 degrees and warmer. It grows taller than wheat up to 5 feet tall.
    Matures over a 4-6 month (120+ days) period. Rye matures next summer about 1 week earlier than wheat. If planted in October, **harvest in May or June**. Winter rye is hardier than winter wheat and withstands drought better.

    Uses: Grown as **cover crop, pasture (forage) or for grain**.
    (See May Garden-Harvest. For **cool season cover crops**, see March Garden-Plant. For **general grain harvesting**, see May Garden-Harvest. For more about **cover crops**, see "Garden Tips" in back of manual.)
    Read "Homegrown Whole Grains" by Sara Pitzer, "Small Scale Grain Raising" by Gene Logsdon.)

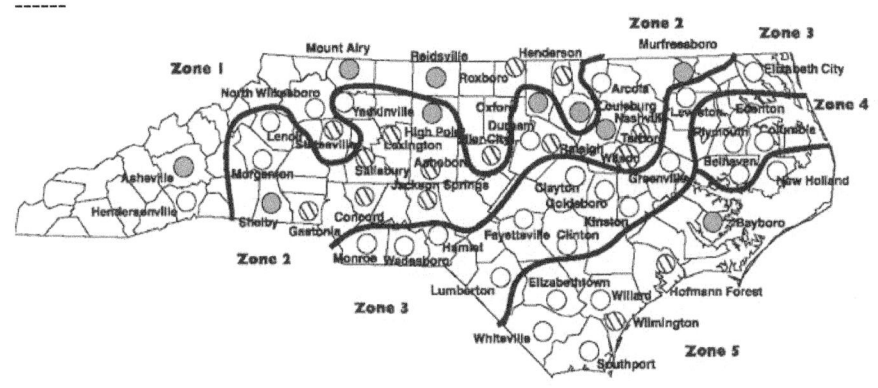

Chart when to sow wheat.

Zone 1	Zone 2	Zone 3	Zone 4	Zone 5
○ 9/29 to 10/15	○ 10/10 to 10/27	○ 10/20 to 11/8	○ 10/24 to 11/10	○ 10/26 to 11/14
● 10/5 to 10/22	● 10/13 to 10/31			● 10/31 to 11/16
◐ 10/7 to 10/24	◐ 10/16 to 11/3			◐ 11/3 to 11/21

 Sow **wheat (winter)** (poaceae/gramineae/grass family; pooideae subfamily, triticeae tribe, Triticum aestivum, cereal wheat) in **September or October. Sow at least 6 weeks before your first fall frost. Can sow 8-10 weeks before first fall frost. Or sow based on above chart.** There are many different opinions about when wheat should be planted. Can also plant **triticale** (a wheat/rye hybrid) at this time. (See **Garden Seeders and Broadcast Spreaders** in "Garden Tips" in back of manual.)

    Plant 1 1/2 to 2 bushels **per acre.** (A bushel of wheat weighs 60 pounds.) Or plant 75-90 pounds per acre if drill, and 90-110 if broadcast. Plant 1/2 pound of wheat to 100 square feet. Or plant seed 5-6 inches apart.
    Plant 1 to 1 1/2 inches deep. Can use lawn roller to firm down the seeds or walk on with a plank of wood.
    It sprouts before freezing weather, then becomes **dormant** until soil warms in spring. Needs **8-12 weeks before the ground freezes.** In the spring you get more yield if you again firm down the seeds with a lawn roller or plank.
    Care: Wheat does not like acid soil. Likes soil pH of 7.0-8.5. Needs full sun and well drained soil. Make sure soil has enough phosphorus so if needed add bonemeal or rock phosphate. Do not add too much nitrogen.
    Matures in about 17-19 weeks (4-5 months). Grows 2-4 feet tall. Warm spring weather causes **rapid growth,**

and harvest within 2 months. **Harvest in May or June**.

Types of wheat: Wheat is the third major cereal grain grown in the world after corn and rice.
1. Hard wheat (Triticum aestivum) has a higher gluten protein content than other wheat. Used for baking yeast breads. **Planted in spring (March) or fall.** Can be red or white. Hard red spring wheat is grown in the upper midwest and plains states.
2. Soft wheat (Triticum aestivum) is used for cake flour, quick bread, and pastry. Contains less protein and gluten than hard wheat. **Planted in spring (March) or fall.** Red soft wheat is grown east of the Mississippi. White soft wheat is grown in western and northern parts of the United States. **Most wheat grown in North Carolina is soft wheat.**
3. Durum (Triticum durum), the hardest wheat, is used for making pasta. High in protein. Not much gluten. **Durum wheat is spring-planted (March).** Grown in plains states.

Uses: Grown as cover crop, pasture (forage) or for grain.
(See September Garden-Plant but details are here. See May Garden-Harvest. For **cool season cover crops**, see March Garden-Plant. For **general grain harvesting**, see May Garden-Harvest. For more about **cover crops**, see "Garden Tips" in back of manual.)

(Read "Homegrown Whole Grains" by Sara Pitzer, "Small Scale Grain Raising" by Gene Logsdon.)

 **Garden- Harvest**  (October)
Before Frost: Sorghum, sweet potatoes, tomatoes, valerian, and winter squash/pumpkin.
After Light Frost: Beets, broccoli, brussels sprouts, cabbage, celeriac, horseradish, leeks, potatoes, rutabaga, spinach, and turnips.
After Hard Frost: Carrots, echinacea, kohlrabi and parsleyroot.
After Frost and Throughout Winter: Various greens, corn, and Jerusalem artichokes.

 Before Frost (usually early October)

 Harvest **sorghum (sweet)** (poaceae/gramineae/grass family; sorghum bicolor) in **October or September** when seeds are dark brown or red, and hard. When the seed plumes (feathery part of seedhead that aids dispersal by wind) begin to appear, cut a stalk near the ground. Peel off outer skin of stalk, and chew the inner pith. If it is somewhat sweet, then it is **ready to harvest.**
Sweet sorghum yields 200-250 gallons of syrup per acre. Leaves and seed heads are removed before processing and fed to livestock. Can remove leaves while still planted in the ground. Then **cut stems** close to the ground. Remove the seed clusters.

Uses: It is a tall, **warm weather crop** that is the third most popular grain in the United States and fifth most popular worldwide after corn, rice, wheat and barley. It is also grown as a fodder crop for animals. It is high in sugar so molasses/syrup, drinking alcohol, and bio-fuel are made from it. It is used to make items such as brooms and for weaving. Blairsville, Georgia holds an annual Sorghum Festival to celebrate the grain.
Sweet sorghum produces small, bitter grains and a thick, juicy stalk. Sorghum syrup is made by running the stalks of sweet sorghum through a mill, then simmering the cane juice. Sorghum molasses is this same syrup further cooked to a thick consistency. Sorghum syrup production has been a part of **Appalachian culture** for a long time. It was a staple food 150 years ago. (See May Garden-Plant. For **general grain harvesting**, see May Garden-Harvest.)

 Dig **sweet potatoes** when vines start to turn yellow or later. Try to dig as late as possible because most of the tuber size is grown in **September and October.** Dig immediately after frost causes leaves to go limp and black. Dig and handle gently. Can let air dry half a day to let soil dry so easier to remove. Eat damaged and very small potatoes right away. Do not wash.

Cure for 2 weeks in hot (80-85 degrees), 90% humidity location. A good place is near a wood stove with a damp towel over them. Curing is important and is needed to harden the skins. Then wrap each sweet potato in newspaper or other paper, and store in room 50-60 degrees with 85% humidity that is dry, airy and away from rodents.

(For **growing slips/sprouts to plant**, see March Greenhouse. See June Garden-Plant.)

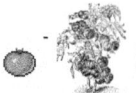 Harvest **tomatoes** before frost.

Mature or almost mature Long Keeper variety of tomatoes can be stored in a cool, dark place for several months or even up to 6 months. When mature they are a reddish yellow.

Regular mature **green tomatoes** can ripen indoors, others will not ripen.

(See June Garden-Plant. For **details about harvesting including green tomatoes,** see August Garden-Harvest.)

 Harvest **valerian** roots in fall (September or October) **before first frost.**
Dig up roots (rhizomes) of plants 2 years and older. Keep roots out of sun while you harvest.

Chop up roots into 1-2 inch pieces. Dry at room temperature out of the sun for 2 weeks to several months. Or dry in an electric/solar dehydrator. They are brittle when dry. Then powder the root (rhizomes) with a mortar/pestle or food processor.

Uses: When preparing use a method that does not drive off the **medicinal oils**. Boiling a long time is not good. A tincture is good. Consult your herbalist, doctor and/or health care specialist.

It has been used as pain killer, sedative, anti-depressant and tranquilizer for over 1,000 years. It is an antispasmodic so is used to relieve stomach cramps and digestive system disorders. Can use for headache.

The **leaves** can be made into a sedative tea. For **tea from the root,** use 1 teaspoon of root to 8 ounces of water. Do not use all the time. Use in moderation. For some people it is stimulating rather than relaxing. Cats and rats love it.

(For **sowing or dividing,** see May Garden-Plant. For **transplanting daughter plants**, see August Garden-Maintenance. For **harvesting leaves**, see June/August Garden-Harvest with details in August. For **how to make a tincture, decoction or infusion**, see April Garden-Harvest.)

 Harvest **winter squash and pumpkin:**

Harvest before frost preferably before temperatures drop below 50 degrees. Frost is bad for pumpkins and squash. The longer a squash or pumpkin sits outside in frosty weather, the more damage that is done.

Harvest winter squash and pumpkin **when stems begin to shrivel and dry** and/or leaves yellow and wither. Or when seeds have matured fully. Or when winter squash is a deep, mostly solid color. The rind gets dulls. Or when skin is hard enough that you can not cut it with your fingernail. **Pumpkin skin** remains soft though.

Bring inside as soon as they are mature. Do not drop, stack or throw them because it will bruise them and cause disease and rotting.

Leave 2 inches of vine on squash or pumpkin except remove **stem** on butternut and hubbard. Do not carry by stem because it weakens it and becomes a place for disease to start. Do not wash before storing. Just gently brush off excess dirt. Or wash them and then bathe in a 10% chlorine bleach solution. (One part bleach to 9 parts water.)

Storage: Winter squash can be stored through winter. Most benefit from a curing stage except for **acorn, sweet dumpling and delicata.** Acorn squash should be immediately stored at 45-55 degrees. Mature acorn squash should be green on the outside.

Cure other squash by keeping it at room temperature (about 70-80 degrees) for 10-20 days in a well ventilated, dry place. This toughens the skin. Do not cure in direct sun. A garage or outbuilding with windows is good. Fans speed curing.

Then **store** in cool, dark place around 45-60 degrees with humidity of 50-70%. An unheated room or basement that does not freeze is good. Do not store in refrigerator. Do not store near apples, pears or other fruit. Do not let them touch. You can wrap them in newspaper or line shelves with newspaper or cardboard in case some rot. Turn over monthly or more frequently when checking for rot. Always handle gently.

    Stores 1-2 months: Acorn.
    Stores 2-3 months: Banana, Buttercup, Butternut, Delicata, Spaghetti, Turban, and most Pumpkins.
    Stores 4 months: Sweet Dumpling.
    Stores 5-6 months: Cushaw, Hubbard, and Sweet Meat.

Uses: Immature or bruised squash should be eaten first or fed to animals. Immature fruit is not as sweet.

Freshly picked, mature pumpkin and squash have high starch levels and high vitamin A. They are at a good time for processing, canning and freezing. As they age, enzymes turn starch into sugar. Some people prefer their squash when it is a month past maturity. Over time they lose water. Skin color changes just before they are too decayed to eat.

Cooking: Winter squash can be baked, steamed or roasted. Usually it is cut in half or into smaller pieces and baked with skin/rind still on. Put in baking pan with rind/skin side up with a little water in the pan.

All winter squash **seeds** are edible raw or cooked. They can be toasted. Spread seeds on cookie sheet in single layer. Let sit for 2-3 days (optional) stirring occasionally to prevent mold. Then sprinkle with 1 tablespoon oil per cup of seeds. Bake in a 250-360 degrees oven for 10-60 minutes. Stir every 10-15 minutes. They are done when light brown and crispy. They can also be fried in a pan. To eat, crack the hull between your teeth and spit it out. (The hull is edible but may be tough.)

(For **winter squash and pumpkin including varieties and storage times,** see May Garden-Plant.)

Harvest After Light Frost (October)

Mid October before hard frost:

 Dig **beets** (late or storage). Beets for eating can be harvested at any time. But **for storage harvest beets before potatoes and carrots.** The **best size for red beets** is usually between 1 1/4 to 2 inches in diameter but some varieties can be picked larger. Large red beets tend to be fibrous. The **best size for mangel beets** for people to eat is 3-4 or up to 6 inches. Animals can eat any size mangel beet. The **best size for sugar beets** is 2 inches in diameter. You can harvest after that but they contain less sugar the larger they get.

Beets can take **light frost** but harvest or mulch before hard freeze. If the roots freeze, they will rot.

Leave a 1-2 inch stem. Eat the greens now or freeze. Do not cut off bottom of root tip. Handle gently so they store better. Beets stored in the refrigerator last about 1-2 weeks.

Storage: Store beets in layers of very damp sand, leaves, peat moss or sawdust near 32 degrees (up to 40 degrees) and 90-95% humidity. Do not cut off roots but leaves do need to be removed. Beets store well in **root cellar**, almost as good as potatoes and carrots. Can also store in barrel, trench or clamp. (A clamp is a hole lined with straw, filled with vegetables, then covered with straw.)

Red beets store 2 months in average storage conditions, and 4-5 months in good conditions.

Mangel beets store for 6-10 months, however after 5 months they have lost a lot of nutrients. Old timers used to store mangels in the field in pits covered with mulch. Good for animal feed over winter.

(For **early beets and sowing/growing details,** see March Garden-Plant. See July Garden-Harvest. For **late beets** see July Garden-Plant. See September Greenhouse.)

Late planted **broccoli** is ready to harvest. Tastes better after light frost. Tastes best when flower buds are tightly closed, firm and green. Those are also the ones best for storage. Harvest **before hard frost**. Leaves are edible.

Cut broccoli head before the flowers open when the head is 4-6 inches across. You get a second harvest of smaller heads below where you cut off the first head. To store it is best to freeze or can it.

(For **sowing/growing**, see March Greenhouse. See June Garden-Plant.)

Brussels sprouts taste much better after several light frosts. **Very cold hardy (down to 0 degrees)** and easily survive a hard frost and snow. Sometimes lasts outside until **November**.

The small sprouts or buds form heads 1-2 inches in diameter. Pick or cut off the stem when firm and about 1 inch. Lower sprouts mature first. Lowermost leaves should be removed when those sprouts are harvested. Harvest before leaves yellow.

To store keep at 32-40 degrees at 90-95% humidity in perforated plastic bags. Will store 3-5 weeks. (See July Garden-Plant.)

 Harvest late or storage **cabbage.** It can take **light frosts**. Red cabbage stores better than green varieties. The best cabbage for storing is picked a little bit **before full maturity.**

Cabbage can be harvested any time after the head has formed. Heads should be firm and solid. **Save tight heads for long term storage.** Harvest before heads split or crack. (Splitting is caused by too much water.) Loose wrapped cabbage does not store as well. Loose cabbage can be eaten right away or made into sauerkraut.

Cut off head with as little stem as possible. (New small cabbage heads will form after the main head is cut off.) Or harvest with roots on. Be gentle so do not bruise. Remove large, loose leaves.

Storage: Do not wash cabbage before storing. Cabbage will store in the refrigerator for up to 2 weeks. For long term storage keep at 32-40 degrees at 90-95% humidity. Can store in root cellar, barrel, clamp or trench. (A clamp is a hole lined with straw, filled with vegetables, then covered with straw.) In a root cellar place heads on shelf several inches apart. Or can wrap in newspaper or layer heads in hay or straw.

Sauerkraut Recipe from "Nourishing Traditions" by Sally Fallon: You need 1 medium cabbage cored and shredded. Also 1 tablespoon caraway seeds, 1 tablespoon sea salt, 4 tablespoons whey or if not available an additional 1 tablespoon salt. Mix cabbage with caraway seeds (or other flavorings), sea salt and whey. (**Whey** is the sour liquid left over after clabbering milk. For **clabbering milk**, see February Farm Animals.)

Pound with a wooden pounder or meat hammer for 10 minutes to release juices. Put in quart mason jar and press down firmly. The top of cabbage should be at least 1 inch below top of jar. Cover tightly and keep at room temperature for 3 days before transferring to cold storage. Can eat right away if you want to. It improves with age. This is **lactic-acid fermentation** that is very good for you. Can do same process with other vegetables. The sauerkraut in grocery stores is not processed this way.

(For **early cabbage** see March Greenhouse. For **sowing/growing details,** see April Garden-Plant. See June Garden-Harvest. For **late cabbage** see June Garden-Plant.)

 Harvest **celeriac** when root is 3-4 inches across. Flavor improves with **light frost**, but harvest before hard freeze. Cut greens off and keep 1-2 inch stem. Cut off some of the finer roots but do not cut too close to main root. Cover with damp sand, leaves or sawdust. Store at 32-40 degrees at 90-95% humidity. Will store 2-3 months in root cellar, barrel, trench or clamp. (For **sowing/growing details,** see May Garden-Plant. See June Greenhouse.)

 Dig up **horseradish** after frost kills the leaves for better taste. Can also be dug any time through **winter or in spring.** Can mulch to keep ground unfrozen longer. One-year-old plants have the best flavor so best to dig it up and replant it.

The **main root** is harvested and one or more large offshoots of main root are replanted 6 inches deep, 2 1/2 to 3 feet apart. Or harvest bottom of root and replant top with some root attached. Spreads via underground shoots and can be invasive.

Storage: Store roots in root cellar, barrel, clamp or trench in sand or sawdust like carrots at 32-40 degrees, 90-95% humidity. Stores well for months. (For **carrot storage**, see September Garden-Harvest.)

Uses: Homemade horseradish is stronger than grocery store types. Take an 8-10 inch long piece of root. Peel the skin off. Chop into pieces. Put into a food processor. Add a couple tablespoons of water. Process until well ground. Keep away from eyes. Strain out water if mixture is too liquidy. Add a tablespoon of vinegar and a pinch of salt. Pulse (blend briefly) to combine. Transfer to a jar. It will keep for 3-4 weeks in the refrigerator.

Horseradish vinegar is made by taking a medium size root and grating it. Put it into 4 cups of vinegar. Let sit for 2-3 weeks before using. Store in cool, dry place.

Medicinal: It is anti-bacterial. The root contains a lot of **mustard oil** that is used in aromatherapy because of its healing qualities. It is used to relieve the symptoms of flu, sore throat, and bronchitis. It can be made into vinegar, tea or syrup. It can be applied as a poultice to improve blood circulation, and to reduce muscle spasm/pain.

For infections drink one cup of **horseradish tea** 3 times a day. Add 1 teaspoon grated horseradish to 1 cup boiling water. Let sit for 10 minutes and then strain. Consult your herbalist, doctor and/or health care specialist.

(See March Garden-Plant but details are in this section. See April Garden-Harvest.)

 Harvest long-season **leeks** September through light frost. (For **sowing/growing details,** see February Greenhouse. See May Garden-Plant. For **storage** see September Garden-Harvest.)

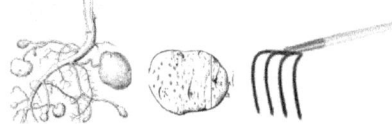 Dig up late **potatoes** (white, red or purple).

Potatoes are planted April through June, and harvested June through October.
(Planting: For when to plant early/mid/late season potatoes, see April Garden-Plant. For sprouting and then planting in April, see March Greenhouse.)
(Harvest: For mid season potatoes see August Garden-Harvest. For late/storage potatoes see September Garden-Harvest but most information is in this section.)

Let **frost** kill potato vines of late/storage potatoes. Then remove dead above-ground plants to help reduce moisture in soil. Let stay in ground 2 more weeks unless problems with slugs or rodents. Or can dig before frost if vines are brown.

Digging: Dig with a long-handled, 3-prong grape (graip or spading fork) or a **potato hook** which is similar to the graip but has tines at a 90 degree angle to the handle (see above image). Place fork just outside where potatoes are and lift/leverage the fork so you do not stab any potatoes.

Harvest on a warm, dry day. Dig and handle carefully to prevent cuts and bruises so stores better. Leave on top of soil several hours to harden skin. But after that keep out of the sun.

Eat small and damaged potatoes first. Do not eat green parts of potatoes because they contain solanine and are poisonous. Under ideal conditions **harvest** is 25 pounds per 1 pound planted. But usually it is more like 10-12 pounds per 1 pound planted but that depends upon variety and other factors.

Storage: Store only mature potatoes that have no damage or disease. Potatoes are mature if the skin stays on when you rub them. If some potatoes are damaged but more than you can eat right away, set aside to be eaten first. Do not wash before storing.

For the **first 10-14 days** store in a dark, well ventilated area that is 50-65 degrees and 85-95% humidity so cuts and bruises can heal. Then move to a dark, well ventilated area with high humidity (95%) and temperatures 38-45 degrees. Do not let freeze. Storage places can be in **root cellar**, unheated room, barrel, trench or clamp where it is dark. **(A clamp is a hole lined with straw, filled with vegetables, then covered with straw.)** You want it **dark** so the potatoes do not form sprouts.

In storage area, cover with newspaper, hay, straw, towel or burlap. If light gets on the potatoes, the skin turns green which is poisonous. (If there is only a little green on the skin, it can be peeled off before eating.) Store in small piles rather than large piles. They can be stored in plastic milk crates or cardboard boxes stacked on each other. Needs some air circulation.

Store on higher shelves in less cold parts of root cellar. Do not store with apples or onions. Check every few weeks to remove bad potatoes. They can **store up to 9-10 months**. Potatoes with sprouts are still good to eat. Just do not eat the sprouts.

When stored for some time, cold temperatures convert starch to sugar, giving potatoes a sweet taste. This sugar caramelizes during cooking and produces brown potatoes with an off flavor. To help potatoes get back some of their starch, take them out of storage and keep in a dark place at room temperature for a few days.

Saving Tubers for Planting: Save some potatoes for planting in the **spring**. However, some professionals say saving your own seed potatoes can lead to a buildup of viruses and diseases. They suggest buying new seed potatoes each year that are certified free of viruses and diseases. If you save your own, only save those that are disease free. Some people save the smaller potatoes for planting next year. Other people save the largest ones because they store better. (For details about **planting potato seeds rather than tubers**, see April Garden-Plant.)

Potatoes from the grocery store will usually not grow because they have been sprayed with sprout inhibitor.

Rutabaga is less frost hardy than turnips. But they do not get pithy, wooden or hollow if left in the ground past their maturity date. Stores very well in root cellar, barrel, trench or clamp for 2-4 months. Remove leaves. Bury in damp sand, leaves or sawdust the same as with carrots. Can wax with beeswax. (For **carrot storage**, see September Garden-Harvest.)

Uses: The roots are large and sweet. Can eat leaves too. Eat raw or cooked. Simmer 20-25 minutes. Or bake at 350 degrees for 45 minutes. (See June Garden-Plant.)

Harvest late **spinach.** The whole plant can be cut or just outside leaves can be removed. In either case the plant continues to **grow until frost.** Can survive temperatures down to 15 degrees. Harvest when not frozen.

Leaves that are 3-4 inches long and 2-3 inches wide are large enough to harvest. Can freeze or can for later.

(For **early spinach and sowing/growing details**, see March Garden-Plant. See June Garden-Harvest but no information there. For **late spinach** see August Garden-Plant. See August/September/October Greenhouse.)

Turnips are more frost hardy than rutabagas but less frost hardy than carrots. A light frost improves the flavor. Dig up **before heavy freeze.** Harvest turnips when bulb is between 1-3 inches in diameter. The greens are edible. Keep 2-3 inch stem on root.

Storage: Roots will keep in the refrigerator for up to 7 days. For long term storage keep in damp sand, leaves or sawdust at 32-40 degrees, 90-95% humidity, the same as carrots. (For **carrot storage**, see September Garden-Harvest.)

Will store 4-5 months. Can **harvest outside through winter** if mulched with straw, hay or leaves.

Uses: Good frozen or canned. Some people turn it into a kraut. Can be used as **winter forage** for animals. (For how to make **kraut**, see cabbage above. See pasture in "Garden Tips" in back of manual.)

(For **early turnips,** see March Garden-Plant. For **late/storage turnips and sowing/growing details**, see August Garden-Plant.)

Harvest After Hard Frost (October)

Late October, or dig after medium/hard frost:

Harvest storage **carrots.** Mature carrots can **store 4-5 months.** (For early carrots see March Garden-Plant, June Garden-Harvest. See August Greenhouse, December Greenhouse. For **late carrots and sowing/growing details,** see June Garden-Plant. **For harvest and storage details**, see September Garden-Harvest.)

Harvest **echinacea roots** (Echinacea purpurea, coneflower) for medicinal purposes. It is best to wait until plants are **3-4 years old**. Harvest roots in **fall** when flowers have gone to seed and there have been a few **hard frosts**. Can cut part of the root so plant continues to grow. Roots are 8-24 inches long.

Drying Roots: Cut roots into 1 inch or smaller pieces. They are very hard when dry so you may want to cut it in somewhat small pieces. Dry in an airy, dry spot out of direct sunlight. Or dry in a electric or solar dehydrator or oven at lowest setting (preferably 85-110 degrees).

Storing: The roots stay good for about a year depending upon storage conditions.

Uses: Echinacea helps your immune system fight infection. You can make a **tea** with the roots or flowers. Do not use all the time. Only use occasionally when you need to help your immune system.

Put 1-2 teaspoons of dried herb or 2-4 teaspoons of fresh herb in a pot with 1 cup of water. Bring to boil and then steep for 15-20 minutes. Consult your herbalist, doctor and/or health care specialist.

(See April Garden-Plant. For **harvesting leaves and flowers**, see July Garden-Harvest. For **how to make a tincture, decoction or infusion**, see April Garden-Harvest.)

 For storage, harvest **kohlrabi** when stems are 2-4 inches in diameter. Bulbs get tougher as they get bigger than that. The bulb should not be cracked. Only store **summer planted** kohlrabi, not spring planted. (Spring planted will not store well.) Trim off leaves if storing. Keep roots on.

Store at 32-40 degrees at 90-95% humidity. Stores 3 months in root cellar, barrel, trench or clamp. (A clamp is a hole lined with straw, filled with vegetables, then covered with straw.) Or can mulch plants for **harvest outside through winter.**

Uses: It is a member of the **cabbage family**. (See "Plant Families" in back of manual.) Leaves can be eaten and taste somewhat like turnip greens. Can be eaten raw or cooked. When raw the root is crisp and sweet. When cooked the root tastes like nutty cabbage. May need to peel more mature roots. Good **food for animals** over winter.

(For **early kohlrabi** see March Greenhouse. For **late kohlrabi and sow/grow details,** see August Garden-Plant. For **animal forage**, see Pasture in "Garden Tips" in back of manual.)

 Harvest **parsleyroot (hamburg). Tastes better after frost.** Dig after first medium/hard frost.

Store in sand or sawdust in root cellar, barrel, trench or clamp like carrots. (For **carrots**, see September Garden-Harvest.) Or overwinter with mulch outside. Some say do not leave outside over the winter

Uses: Can eat leaves but grown for roots. The root is about 6 inches long. It looks like a beige carrot. Can leave in ground over winter if it is mulched. Eat like carrot, raw or cooked. Usually cooked.

(See July/October Garden-Plant with **sow/grow details in July**.)

 **General Harvest** (October)

Harvest after frost and throughout winter:

 **Collard greens, creasy greens, kale, kohlrabi,** turnip, and other frost hardy vegetables. Can **extend harvest** by putting frost blanket, mulch or sheet over vegetables on frosty nights.

(For a list of **frost hardy vegetables and minimum cold temperatures**, see February Greenhouse.)

 Mid to late October local **corn (field)** and other grains are available to buy. (See May Garden-Plant. For **sweet corn** see August Garden-Harvest. For **field and flint corn** see September Garden-Harvest with most harvest details in September.)

 Dig **Jerusalem artichokes** (sunchokes) **after several hard frosts or dig up in spring**.

Dig after plants have turned brown. It is best to wait until at least **2 weeks after the first frost**. The skin is thin so handle carefully. Can continue to **harvest throughout winter as long as ground is not frozen.**

They can be left in the ground for years before you dig them up to eat or replant (they are perennials). The longer they are kept in the ground, the bigger the tubers get. So you could have a **2-3 year rotation** where you harvest the second or third year, letting year 1 or 2 continue to grow.

Harvest is 5-15 tons or even up to 25 tons per acre. (Potatoes are 8-35 tons per acre. A ton is 2,000 pounds so that is 16,000-70,000 pounds per acre.) A 25 square foot area can produce more than 100 pounds of tubers. Variations depend on irrigation/rainfall, soil fertility, pests and other factors.

How to Use: When raw it tastes somewhat like water chestnut. When cooked it is like potatoes except nutty. Can be eaten peeled or unpeeled. To prevent blackening when cut, you can soak in water that has some lemon juice or vinegar.

Unlike most starchy vegetables, the main carbohydrate in Jerusalem artichokes immediately after harvest is **inulin** (a fructose carbohydrate) rather than starch. For most people this is **hard to digest** and leads to gas. But if allowed to go through hard freezes especially if wait until spring before harvest, much of the inulin has changed to a more digestible carbohydrate.

If you have not had hard freezes yet and want to eat some, then put them in the freezer for a few days or weeks. Cooking for a long time such as 12-24 hours at low temperatures such as 200 degrees also reduces the inulin. And cooking it and then eating it the next day can help.

Storage: Can store in root cellar, barrel, clamp or trench. They do not store as well as potatoes. They need a lot more moisture than potatoes otherwise they shrivel up. Put in damp sand or dirt. Make sure it stays moist to very moist.

Cold storage should have high humidity (85 to 95%) with temperature 33 degrees or a little higher. They store at eating quality for several months or longer. Will store longer at a quality good enough to plant.

Even better is to dig them up to eat as you need them (in both spring and fall) unless the ground is frozen. To eat them in the middle of winter, you will have to store them.

For those you want to **replant**, the best method is not to store them at all. In the fall dig up and immediately re-plant. Or leave in the ground until you want to replant in **spring**.

(See March/April Garden-Harvest but details are here. See May Garden-Plant, July Garden-Maintenance.)

Greenhouse, Hoop House or Cold Frame  (October)

 Continue to **sow annuals arugula, claytonia, minutina, mizuna, parsley, spinach and perennials such as creasy greens and sorrel.** Can transplant **sorrel** from outside to greenhouse. Harvest all winter.

(For individual plants, see February Greenhouse. For creasy greens see March Garden-Plant. For sorrel see May Garden-Plant. For a list of **frost hardy vegetables and minimum cold temperatures**, see February Greenhouse.)

Dehydrate vegetables in electric or solar dehydrator, greenhouse, sunporch or car. (For **how to make a solar dehydrator**, see May Garden-Harvest.)

Farm Animals (October)

Worm goats, sheep, horses, cattle, donkeys, pigs, dogs, cats and other animals. Give garlic water to chickens, ducks, turkeys and pigeons. (For **garlic water,** see January Farm Animals. For **mites and lice**, see March Farm Animals. For **natural wormer recipes**, see April Farm Animals.)

Chicken, turkey and duck egg production is down. (Peak **egg** production is April through May.)

  **Honey Beekeeping in October**

There is **not much activity at the hive**. The queen is laying only a few eggs. It is time to **prepare for winter.** Watch out for **robbing.** Keep honey covered as much as possible so bees and other insects can not get to it. Install inner cover wedges for ventilation. Install mouse guard at entrance of hive.

Bee colonies **stay warm** by clustering together. They will stay warm enough if they do not get wet. Place insulite boards (thermal insulation) under hive cover to help keep colony dry. Setup a wind break if necessary. You may need to put concrete blocks on hives to prevent them from being blown over.

Finish **winter feeding of sugar or honey.** Remove Apistan strips (assuming they have been in for 42 days).

(For an **overview of beekeeping,** see January Farm Animals. For **harvesting honey**, see September Farm Animals.)

 # NOVEMBER November 1 daylight is 10 hours 44 min. Nov 30 daylight is 19 hours 59 min.

Early November: All leaves have fallen except for oaks.
Mid November: All leaves have fallen.

 Appalachian Folklore: Prepare for a **hard winter** when there is an unusually large crop of nuts or acorns.

 Garden- Maintenance

Clean driveway ditches, drain pipes, gutters, etc. Clean and oil tools, tiller blades, etc. Clean engines such as tillers and chainsaws. Clean air filters. Change the oil. If not using machine for several months, then drain fuel and remove spark plugs. Run outdoor pumps if they have not been run in a few months.

Log trees for **firewood** in the dormant season from **late autumn (October) to late winter (February).** (For details about **logging and splitting firewood,** see October Garden-Maintenance.)

Mulch fruit/nut trees and bushes with leaves, straw or hay. Mulch and mark location of **overwintered** (to be dug in spring) **root vegetables** such as carrots, salsify, and parsnips.

Propagate **sea kale** (step 1) by root cuttings or thongs in **November or December.** Select a healthy plant at least 3 years old. Dig up part of plant. Cut off side roots of about pencil thickness. Cut into pieces 3-6 inches long. Mark which end is up and which down. Store them in sand in the refrigerator until **March.**

By March **buds** will be on the shoots. Rub off all but the strongest central bud. Plant the cutting 1 inch deep, 15 inches apart. (See March and May Garden-Plant.)

In greenhouse or other enclosed area, cover **turkey fig** and other frost sensitive plants with sheet or frost blanket for frost protection. Turkey fig is hardy to 10 degrees. (For figs see April Garden-Plant.)

 Garden- Plant (November)

 Garden- Harvest (November)

  After frost harvest hardy vegetables such **as** beets, burdock root, carrots, collard greens, creasy greens, kale, kohlrabi, onion/garlic greens, sunchokes, turnip, etc. At the **end of November still picking outside** catnip, chicory, comfrey, dandelion, onion greens, parsley, salad burnet, sage, sorrel, thyme, and perennial fennel.

Greenhouse, Hoop House or Cold Frame  (November)

 Can continue to plant **arugula, claytonia, mache, minutina, mizuna** and **spinach.** (For arugula, claytonia, mache, minutina and mizuna, see February Greenhouse. For a **list of cold hardy vegetables and minimum cold temperatures**, see February Greenhouse.)

(For **early spinach and sowing/growing details,** see March Garden-Plant. See June Garden-Harvest. For **late spinach** see August Garden-Plant, October Garden-Harvest. See August/September Greenhouse.)

 ## Farm Animals (November)

 Give animals extra **kelp, alfalfa meal** and other supplements. Recommended use of **kelp meal** for animals is 2% of the feed ration or up to 5% where there are deficiencies or stress. Use as 1% of feed ration for poultry or young animals. Kelp is high in trace minerals.

Feed **soaked beet pulp to goats, sheep and other animals** throughout winter. Feed soaked beet pulp or alfalfa hay to **pigs** just before and a few days after giving birth. (Roughages are used as a laxative close to farrowing (giving birth) and as food source when weight gain should be kept to a minimum.) You can feed unsoaked beet pulp but soaked is better.

(For more about **kelp and alfalfa meal**, see fertilizers in "Garden Tips". For more about **beet pulp**, see Pasture in "Garden Tips".)

Feeding **granite grit** to poultry not only helps them grind their food in their gizzard but it gives them trace minerals.

 **Open paddocks** previously planted with **brassica forage crops** such as kale, kohlrabi, mangel beet, rutabaga and turnip. Let animals graze here for part of the day only. Sow July through September for grazing September to March.

(See February Farm Animals. See **Fall and Winter Pasture: Special Forage Crops** in "Garden Tips".)

 November or December **butcher hogs, cattle and other large animals.** Some people feed hogs only corn for 4-6 weeks before kill. This is called **finishing.** It is supposed to improve the meat. Other people do not make any changes to the feed. Some people finish them with acorns. Make sure they have been off any medicated feed for weeks before slaughter.

Ideal butchering weight for **hogs** is about 240-260 pounds with 300 pounds if you like a lot of fat. Of course, this depends on the breed of pig. But this is usually the most efficient weight per amount of feed given.

It is best if there is a maximum temperature of **40 degrees outside** when kill. Ideally below freezing all day and night so no bacteria problems. And partially frozen meat is easier to cut with a meat band saw. It is best to vacuum pack meat before putting in freezer. Can dry, smoke or freeze meat.

Consult an expert or your County Extension office.

 Read the book "Basic Butchering of Livestock & Game" by John J. Mettler.)

 # DECEMBER December 1 daylight is 9 hours 58 min. December 31 daylight is 9 hours 51 min.

December 21 or 22 is the **Winter Solstice- shortest day of the year.**
Early December: Only a few outdoor greens (unprotected) are still good to eat.

Garden- Maintenance

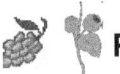 Log trees for **firewood** in the dormant season from **late autumn (October) to late winter (February).** (For details about **logging and splitting firewood,** see October Garden-Maintenance.)

Propagate blueberries and grapes mid December with **step 1.**

Cut dormant shoots that are about pencil size in diameter. Remove top portion of shoot and cut remaining branch into 5 inch pieces for blueberries, longer for grapes. Bottom cut on each should be just above vegetative bud. Make note which is top and bottom. Wrap in paper towel, put in plastic bag in refrigerator until **March**.

(For **step 2 for blueberries and grapes,** see March Garden-Maintenance. For **step 3 for blueberries,** see June Garden-Maintenance. For **pruning,** see February Garden-Maintenance. See April Garden-Plant. For general details about herbaceous, softwood, semi-hardwood and hardwood cuttings, see "Garden Tips" in back of manual.)

Garden- Plant (December)

 Garden- Harvest  (December)

  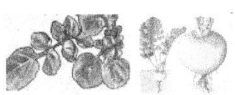 **Harvest after frost and throughout winter:** collard greens, creasy greens, kale, kohlrabi, leeks, parsnips, sunchokes (Jerusalem artichokes), and turnip. Before ground freezes dig up most **parsnips** to store in root cellar. (For details about harvesting parsnips, see September Garden-Harvest.) By mid December most **frost hardy greens outside** are no longer good enough for eating.
(For a **list of cold hardy vegetables and minimum cold temperatures,** see February Greenhouse.)

Greenhouse, Hoop House or Cold Frame  (December)

Frost hardy greens still good to eat in unheated greenhouse, hoophouse or cold frame.
(For a **list of cold hardy vegetables and minimum cold temperatures,** see February Greenhouse.)

In greenhouse late December sow **carrots** for harvest in May. (For **sowing/growing details** see June Garden-Plant.)

Harvest **kale** in greenhouse after hard frost. (For sow/grow and some harvest details, see March Garden-Plant.)

 Grow **sprouts** all winter such as **alfalfa, barley, broccoli, buckwheat, cabbage, clover, fenugreek, lentil, mung, mustard, radish, sunflower and wheat** in house to eat right away.

To make sprouts, soak seeds overnight in water in mason jar or other container. For a cover for the jar use a plastic lid with small holes sold at health food stores, nylon stocking, cheesecloth or similar material with small holes. Once or twice a day, rinse with water and then drain out water.

Shake the jar so the seeds are evenly distributed along its length. Then lay it on its side. Most seeds sprout in 3-6 days. Good eaten raw or cooked. Very nutritious.

 ## Farm Animals (December)

 Early December or late November **butcher** hogs, cattle and other large animals. (See November Farm Animals.)

 Winter Solstice- lowest **chicken, turkey and duck egg** production. Peak egg production is **April and May**. Good egg production is **February through July.** Lowest egg production is **September through December.** (The longest day is in June. Molting is in September and the shortest day is in December.)

(For **egg storage methods,** See September Garden-Harvest. For **molting** see September Farm Animals.)

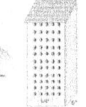 Create a home for the **blue orchard mason bee,** a nonsocial (no hive) native bee (Osmia lignaria) that **pollinates spring fruit trees, flowers and vegetables** especially fruit trees. Of all food people eat two-thirds must be pollinated by insects. Improve garden production by making **habitat for pollinators.**

Life Cycle: This **gentle, blue-black metallic bee does not live in hives**. It nests in hollow stems, woodpecker drillings and insect holes in trees or wood. They are active only a short period of the year. They are not aggressive and one may observe them at very close range without being stung. They **do not produce honey.**

In **spring (usually March)** adult males emerge from tubes first and wait for the later appearance of females to mate. This often coincides with the redbud (cercis) bloom about **April**. When all bees are out of the **wooden block**, clean it to kill parasitic wasps and pollen mites that prey on the bees.

Females alone, begin new nests in holes. Females collect pollen and nectar and lay eggs. Their short foraging range is about 100 yards from the nest. Activity continues 4-6 weeks and then adults die. During the **summer (usually July)**, larvae develop inside the nests, make cocoons, and become new adults resting in the cells. In fall adults go into hibernation. These bees require cold temperatures before spring in order to break dormancy.

Building Bee Homes: The native eastern species of Orchard Mason Bee (found in **North Carolina**) nest in holes drilled in a **wooden block.** Blocks can be made from any shape or type of wood except treated. Can use 4 x 6 or 4 x 8 inch lumber. Drill smooth holes on 3/4 inch centers, 4-8 inches deep. A 5/16 inch diameter hole is important. A smaller hole encourages higher production of male bees.

Drill blocks on 1 side with shallower and deeper holes. Do not drill completely through. Attach a roof for protection from the midday sun and rain. Outside surfaces may be painted or stained, but do not use wood preservatives.

Using Homes: Face nesting blocks **southeast to catch morning sun** and affix firmly so it does not sway in the wind. It should be at least three feet above the ground.

Do not move the blocks in the **spring** during the weeks of active nesting. Once all nesting activity has stopped, the nesting block may be moved to a shelter such as a shed or unheated garage. Be gentle. This gives the bees added protection from predators and parasites, yet gives them exposure to cold temperatures they need to break hibernation.

Other Homes: Orchard Bees are also reared in **cardboard tubes, hollow reeds, or straws**. Cardboard tubes and straws need more protection from weather and parasites. Paper straws allow better inspection and manipulation. Plastic straws hold moisture and let mold develop and are not recommended. Bees may be purchased commercially.

Vary diameter of drilled holes to **attract different species** of tube-nesting bees or nonsocial, beneficial wasps.

Read the books "Pollination with Mason Bees: A Gardener and Naturalists' Guide to Managing Mason Bees for Fruit Production" by Margriet Dogterom and "Pollinator Conservation Handbook: A Guide to Understanding, Protecting, and Providing Habitat for Native Pollinator Insects" by Shepherd/Buchmann/Vaughan/Black.)

 # Garden Tips

Topics in this Section
General Garden Tips
Appalachian Folklore
How Much is a Bushel
Companion Planting
Garden Seeders and Broadcast Spreaders

Seeds and Seedlings
How to Stratify Seeds
How to Make Potting Mix
How to Make and Use Soil Blocks for Seedlings
Inoculating Seeds
How to Prevent Damping Off of Seedlings
Thinning Seedlings and Plants
Hardening Seedlings Before Planting Outside

Soil and Crops
How and Why to Fertilize
Rock Dusts
Soil pH
Crop Rotation and Cover Crops
Pasture
Special Forage Crops
Weeds and What They Say about Soil Type
Prevent Fungal Diseases

Propagation
Propagation by Tip Layering/Rooting
Propagation by 4 Types of Stem Cuttings
Make Your Own Rooting Hormone

General Garden Tips

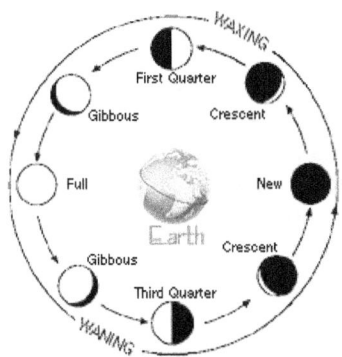   

Appalachian Folklore

    The best time to **plant crops with harvest above ground** is when the **moon is waxing (1st and 2nd quarters) (new moon)** (illuminated area of moon is increasing, going from new moon to full moon). Do not plant on the day of the new or full moon. Do not plant on changing quarters. Graft trees just before the sap flows.

    The best time to **plant crops with harvest below ground** or flowering bulbs is when the **moon is waning (3rd and 4th quarters) (old moon)** (illuminated area of moon is decreasing, going from full moon to new moon). The third quarter is especially good. Do not plant on the day of the new or full moon. Also best time to turn the soil, pull weeds, or kill weeds/trees. A waning moon is good for harvesting most crops, canning and preserving vegetables and jams. Slaughter livestock in the 4th quarter before the new moon.

The **12 zodiac signs for the moon** last a little over 2 days each. The cycle repeats every 28 days. The **earth signs** are Taurus, Capricorn, and Virgo. The **air signs** are Gemini, Libra, and Aquarius. The **fire signs** are Leo, Sagittarius, and Aries. The **water signs** are Cancer, Scorpio, and Pisces.

Use a calendar or almanac (Farmers Almanac) that lists the signs of the moon for each day.

Aries (head)- Good for cultivating the ground, planting beets and onions, and hunting. Bad for planting and transplanting other crops.
Taurus (neck)- Good for planting all root and above ground crops. Good for hunting and fishing.
Gemini (arms)- Good for planting all crops, also for preserving jellies and pickles.
Cancer (breast)- Best for planting above ground and root crops. Good for cooking and fishing.
Leo (heart)- Good for hunting. Bad for planting or transplanting.
Virgo (belly, bowels)- Bad for planting. Good for trading.
Libra (reins, kidneys, lower back)- Good for planting above ground crops and flowering plants.
Scorpio (secrets, loins, groin)- Best for planting flowers and above ground crops. Good for all other crops, fishing and hunting.
Sagittarius (thighs)- Good for hunting, trading, baking and preserving. Bad for transplanting.
Capricorn (knees)- Best for planting root crops. Good for planting flowers and above ground crops.
Aquarius (legs)- Good for planting above ground crops. Good for social events.
Pisces (feet)- Good for planting and transplanting above ground crops, trees and bushes. Good for fishing, and weaning babies and animals.

Other Seed Planting Folklore: Short-germinating and extra-long germinating seeds (that take about 1 month to germinate) are planted 2 days before the new moon and up to 7 days after. Long-germinating seeds are planted at the full moon and up to 7 days after.

Best Fishing Days: One hour before and one hour after low or high tide. Good if wind is coming from the west but not from the north or east. Good when barometer is steady or rising. Good between the new and full moon.

The **Farmers' Almanac** has been published each year since 1818. It has a monthly calendar with moon signs, good fishing days, when to plant/harvest, and astronomical information. It also has weather predictions, and gardening/cooking articles.

(See **Appalachian Folklore** in September Garden-Harvest.) (Read the book "Raising with the Moon: Complete Guide to Gardening and Living by the Signs of the Moon" by Jack Pyle and Taylor Reese.)

 ## How Much is a Bushel?

1 U.S. bushel = 4 pecks = 35.24 liters = 8 dry gallons = 9.31 liquid gallons.

42-48 pounds of apples	48 pounds of barley	28 pounds of beans (pole)
30 pounds of beans (snap)	60 pounds of cowpeas	33 pounds of eggplant
56 pounds of corn (shelled)	70 pounds of corn (in ear)	48 pounds of cucumbers
30 pounds of lima beans (unshelled)	50 pounds of muscadine grapes	32 pounds of oats
57 pounds of onions	25 pounds of peas (field)	50 pounds of peaches
60 pounds of potatoes	50 pounds of rutabagas	20 pounds of spinach
50-55 pounds of sweet potatoes	45-53 pounds of tomatoes	42-54 pounds of turnips (no tops)
42 pounds of white flour		

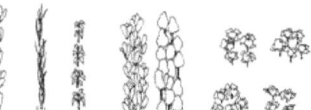 ## Companion Planting

Companion planting is when different types of plants are grown near each other so they **benefit each other** in nutrient uptake (such as nitrogen fixation), pest control (such as insect trap crops), beneficial insect habitat, pollination, wind breaks, and other ways to improve productivity. This has been done for centuries in Europe and Asia.

It brings **diversity and ecological balance** to a garden or farm.

(For **a list of plants that repel unwanted insects and how to make sprays**, see Products to Control Insects in "Plant Health".) (Read the book "Carrots Love Tomatoes: Secrets of Companion Planting for Successful Gardening" by Louise Riotte.)

Plant	Companions (good to grow with plant)	Do not grow with plant
Asparagus	Basil, nasturtium, parsley, tomato	Garlic, onion
Bean, bush	Beet, cabbage, carrot, cauliflower, celery, chard, corn, cucumber, eggplant, leek, marigold, parsnip, pea, potato, radish, rosemary, strawberry, sunflower	Basil, fennel, kohlrabi, onion
Bean, pole	Carrots, cauliflower, chard, corn, cucumber, eggplant, marigold, pea, potato, rosemary, strawberry	Basil, beet, cabbage, fennel, kohlrabi, onion, radish, sunflower
Beet	Bush beans, cabbage, corn, leek, lettuce, lima bean, onion, radish	Mustard green, pole bean
Broccoli	Beet, bush bean, carrot, celery, chard, cucumber, dill, kale, lettuce, mint, nasturtium, onion, oregano, potato, rosemary, sage, spinach, tomato	Lima bean, pole bean, bush bean, strawberry
Brussels sprouts	Beet, bush bean, carrot, celery, cucumber, lettuce, nasturtium, onion, pea, potato, radish, spinach, tomato	Kohlrabi, pole bean, strawberry
Cabbage	Beets, bush bean, carrot, celery, cucumber, dill, kale, lettuce, mint, nasturtium, onion, potato, rosemary, sage, spinach, thyme, tomato	Pole bean, strawberry
Cantaloupe	Corn	Potato
Carrot	Bean, Brussels sprouts, cabbage, chives, leaf lettuce, leek, onion, pea, pepper, radish, rosemary, sage, tomato	Celery, dill, parsnip
Cauliflower	Beet, bush bean, carrot, celery, cucumber, dill, kale, lettuce, mint, nasturtium, onion, potato, rosemary, sage, spinach, tomato	Pole bean, strawberry
Celery	Bush bean, cabbage, cauliflower, leek, parsley, pea, tomato	Carrot, parsnip
Corn	Beet, bush bean, cabbage, cantaloupe, cucumber, morning glory, parsley, pumpkin, squash	Tomato Potato, sage
Cucumber	Bush bean, cabbage, corn, dill, eggplant, lettuce, nasturtium, pea, radish, sunflower, tomato	
Eggplant	Bush bean, pea, pepper, potato	None
Kale	Beets, bush bean, cabbage, celery, cucumber, lettuce, nasturtium, onion, potato, tomato	Pole bean
Kohlrabi	Beet, bush bean, celery, cucumber, lettuce, nasturtium, onion, potato, tomato	Pole bean
Leek	Beet, bush bean, carrot, celery, onion, parsley, tomato	Bean, pea
Lettuce	Almost everything, especially carrot, garlic, onion, radish	None
Lima Bean	Beet, radish	None
Onion	Beet, cabbage, carrot, kohlrabi, leek, lettuce, parsley, pepper, spinach, strawberry, tomato, turnip	Asparagus, bean, pea, sage
Parsnip	Bush bean, garlic, onion, pea, pepper, potato, radish	Carrot, celery
Pea	Bean, carrot, celery, chicory, corn, cucumber, eggplant, parsley, radish, spinach, strawberry, sweet pepper, turnip	Onion
Pepper	Carrot, eggplant, onion, parsnip, pea, tomato	Fennel, kohlrabi
Potato	Bush bean, cabbage, corn, eggplant, marigold, parsnip, pea	Cucumber, pumpkin, rutabaga, squash, sunflower, tomato, turnip
Pumpkin	Corn, eggplant, nasturtium, radish	Potato
Radish	Bean, beet, cabbage, carrot, chervil, corn, cucumber, lettuce, melon, nasturtium, parsnip, pea, spinach, squash, sweet potato, tomato	Hyssop
Rutabaga	Nasturtium, onion, pea	Potato
Spinach	Cabbage, celery, any legume, lettuce, onion, pea, radish, strawberry	Potato
Squash	Celery, corn, dill, melon, nasturtium, onion, radish	Potato
Strawberry	Bean, borage, lettuce, onion, pea, spinach	Cabbage
Tomato	Asparagus, basil, bee balm, bush bean, cabbage, carrot, celery, chives, cucumber, garlic, lettuce, marigold, mint, nasturtium, onion, parsley, pepper	Dill, fennel, pole bean, potato
Turnip	Onion, pea	Potato

Garden Seeders and Broadcast Spreaders (no large machinery)

If you have a large area to sow seed, it helps a lot if you have a **garden seeder** (see first image above) that you can push rather than planting each seed by hand. Corn can be planted with a **corn planter** (see fourth image above). (See corn in May Garden-Plant.)

Crops with small seeds such as **clover, fescue, oats and rye** can be broadcast sown by hand or with a **broadcast spreader.** The broadcast spreader can either be pushed (see second image above) or hang on your shoulders while you turn a crank (see third image above). The push type has 2 wheels and evenly spreads small seeds as you push it. It can also be used to spread fertilizer.

One brand is the **EarthWay Precision Garden Seeder** that digs a row in already tilled soil, correctly spaces the seeds, plants at the proper depth, covers the seed, and marks the next row with precision, all in one pass.

It has 6 standard **seed plates** that plant asparagus, bean, beet, broccoli, cabbage, carrot, cauliflower, corn, cucumber, endive, kale, leek, lettuce, lima beans, mustard, okra, onion, pea, radish, rutabaga, spinach, Swiss chard, tomato, turnip and others. For **wheat** use the beet seed plate. For **sorghum** use the carrot plate.

Seeds and Seedlings

 ### How to Stratify Seeds to Improve Germination

Many seeds from **perennial/biennial plants and hardy trees/shrubs** need to be stratified. Stratification is pretreating seeds to simulate winter conditions that a seed needs before germination. Many seed species have a **dormancy period.** They need **moisture and cold temperatures** for a few weeks to a few months before they will sprout (dormancy is broken).

If you stratify these types of seeds, they germinate faster with higher germination rates.

Method 1: Place seeds in a plastic bag. Add somewhat moist vermiculite, sand, or paper towel. If use soil, sterilize it first. (For **how to sterilize soil,** see Potting Mixes below.) You can add a little kelp or some water that kelp was soaking in. It contains hormones that encourage growth. (For **where to buy kelp,** see fertilizers below.)

Use 3 times the amount of vermiculite as seeds. Close bag and refrigerate for 2 or more weeks. How long depends upon the type of seed. Do not put in freezer.

Method 2: Sow the seeds **outside late fall through late winter**. You can cover them if they need it, or just broadcast on the surface. Frost heaving will cover some of the seeds. They sprout in the spring.

Or sow the seeds in an unheated greenhouse, hoophouse, or cold frame. Make sure the soil stays moist. The seeds can freeze.

 ### How to Make Potting Mix

Supplement for the below mixes: Add **kelp meal** (seaweed) to potting mixes because it helps improve germination rates and plant growth. It has growth hormones. It is high in potassium and nitrogen. It has 60 trace minerals.

Add 1 pound kelp to 100 square feet of garden soil outside. Or add 1/4 cup per 10 square feet. For potting soil add 1/2 cup kelp per 8 gallons of soil. Or 1 teaspoon for a 6 inch pot.

(For more about fertilizers and where to buy kelp, see fertilizers below.)
All seedling mixes should be finely sieved or crumbled.

No-Soil Potting Mix for Seedlings and Cuttings (Experiment to see what works best for you.)
Soil-less Seedling Mixes
    Soil-less Mix A: ProMix/ProMix BX is used by professionals. Can add a little perlite for rooting cuttings.
    Soil-less Mix B: 8 parts sphagnum peat moss, 1 part perlite and 1 part vermiculite.
    Soil-less Mix C: 2 parts peat moss, 1 part perlite, 1 part vermiculite, 1 teaspoon dolomite lime.
    Soil-less Mix D: 4 gallons peat or coir, 4 gallons perlite, 1/2 lb bonemeal, 1/4 lb lime, 1/4 lb bloodmeal.

Soil-less mixes are usually mostly **sphagnum peat moss**. Sphagnum peat is lightweight and inexpensive. It drains water well yet retains water too. The peat must be thoroughly moistened before you plant any seeds.

Peat is acidic (pH 3.5-4.0). You can add a little **dolomite lime or wood ash** to decrease the acidity. Seeds do not need fertilizer to germinate.

Pro-Mix BX with biofungicide is composed of Canadian sphagnum peat moss (75-85%); dolomitic and calcitic limestone (pH adjuster); endomycorrhize (mycorise pro, a beneficial fungus); macronutrients; perlite (20-30%, horticultural grade); micronutrients; vermiculite; and a wetting agent.

Other amendments that can be added to the mixture:

Bark is added to improve drainage and increase air space. Bark mixes are better for mature plants that need to dry between waterings. It should be crumbled.

Coir is a coconut fiber by-product that provides good drainage while retaining water.

Leaf mold is rotted leaves. If you do not have peat moss, you can use leaf mold. (You can mix it with grain hulls or rotted sawdust if you have it.) Leaves take about 2 years to rot.

You can make leaf mold by letting leaves decompose over time. It is dark brown to black, has an earthy aroma, and crumbly texture. It looks like compost. It takes 6-12 months for the leaves to rot.

One way to make leaf mold is to pile up leaves about 3 feet tall and 3 feet wide or more. Keep moist. Another way is to fill a large, plastic garbage bag with leaves. Moisten it. Cut a few holes in the bag for air flow. You can speed up this process by chopping the leaves first.

Perlite looks like pebbly white styrofoam but it really is a volcanic mineral. It does not affect nutrient quality or pH. It improves water drainage and retention while increasing air space. It holds 3-4 times its weight in water. It is sterile and pH neutral. (If you do not have perlite, you can use coarse sand.) (See a few paragraphs below about types of sand.)

Vermiculite is silvery-gray flecks that are a mica-type material. The particles soak up water and nutrients and hold them in the mix. Handle gently so do not lose air in the mix. Has almost neutral pH. Use coarse vermiculite also known as horticultural grade. Do not use fine vermiculite. Vermiculite can be mixed with very small seeds to make it easier to sow them.

Soil Potting Mix for Seedlings/Cuttings and Mature Plants (Experiment to see what works best for you.)

To make potting soil for seedlings and mature plants: Sprinkle fine sand or a dusting of peat moss on the surface of the soil after putting mix in pot.

Seedling Mix A: 6 parts compost, 3 parts soil, 1 or 2 parts coarse sand, 1 or 2 parts aged manure, 1 part peat moss (pre-moistened), 1 or 2 parts leaf mold, and 1 part bonemeal or rock phosphate.

Seedling Mix B: 2 parts sterilized loam soil, 1 part peat moss, 1 part sharp sand with 4-6 ounces lime per every 8 gallons of mixture. (Loam is soil with roughly equal parts sand, silt, and clay.)

Seedling Mix C: 5 parts perlite, 1 part sifted compost.

Seedling Mix D: 1 part vermiculite, 1 part sifted compost.

Seedling Mix E by Eliot Coleman: 40 quarts peat moss, 20 quarts coarse sand or perlite, 20 quarts sifted well-aged compost, 10 quarts soil, 1 cup colloidal phosphate (soft rock phosphate), 1 cup greensand, 1 cup bloodmeal (optional), 1/2 cup lime. Mix in wheelbarrow. Good for **soil blocks** (see below).

Various potting soil mixtures for mature plants:

Potting Mix A: 1 part sifted soil, 1 part aged manure, 1/8 part sifted peat, some bonemeal (or rock phosphate) / trace mineral powder / blood meal.

Potting Mix B: 1/3 mature compost or leaf mold, 1/3 sieved garden loam, 1/3 coarse (builders) sand.

Potting Mix C: 1 part sphagnum peat, 1 part peat humus, 1 part compost, 1 part builders sand.

Potting Mix D: 7 parts sterilized loam soil, 3 parts peat, 2 parts sand or grit with 4 ounces fertilizer and 3/4 ounce lime per every 8 gallons of mixture.

Potting Mix E: 8 quarts potting soil with vermiculite or perlite, 1 quart coarse sand, 4 quarts sphagnum peat moss, compost, and/or rotted manure.

Potting Mix F: 8 quarts potting soil, 1 quart perlite, 1 quart vermiculite, 8 quarts sphagnum peat moss, 1 cup greensand, 1 cup gypsum.

Use **coarse sand** (builders sand), not fine sand (play sand). Coarse sand adds air space to the potting mix. Fine sand settles into the air spaces and makes a dense mix that excludes air. Fine sand (1/8-1/4 mm), medium sand

(1/4-1/2 mm), coarse sand (1/2-2 mm). **Sharp sand** has a wide range of sand sizes.

Seedlings started in a mix of 40% vermiculite, 30% peat moss, and 30% composted cow manure had faster growth and less incidence of disease than those started in a 40% vermiculite, 60% peat moss mix.

Fertilizer mix to make for potting soil: 2 cups rock phosphate (or other phosphorus source), 2 cups greensand (or other potassium source), 2 cups blood meal (or other nitrogen source), 1/4 cup kelp meal. Mix with 15 gallons of potting soil. (See fertilizers below in Soils and Crops.)

If using soil from garden, you may want to sterilize soil if there is a history of soil-borne diseases. Use **sterile soil** for germinating seeds, planting seedlings, and starting rooting cuttings. Older plants do better with regular soil.

Ways to sterilize soil:

1. In an **oven**, heat slightly moist soil above 160 degrees but never over 190. A cover should be on the soil. Heat for 20-30 minutes. Or in **summer sun** a 3 gallon black plastic pot will heat up hot enough in a few hours.

2. To use a **microwave** to sterilize soil, put about 2 pounds of moist soil in a thick, plastic bag. Leave the top open and place in center of microwave. Heat it for 2-5 minutes on full power, checking the temperature in the middle of the soil with a thermometer. When it reaches 180-200 degrees, close the bag and put in a cooler to hold heat in the soil.

3. Another **microwave** method: Place dampened potting mix in a container with the cover on loosely. Microwave on high for 8-10 minutes, until there is condensation under the lid. Close lid and let the potting mix cool.

4. In the **summer** choose an area that receives at least 6 hours of sun during the day (8 hours is best). Lay out clear plastic sheeting and cover with a layer of dirt about 4 inches deep. Spray the soil generously with water but not so much that it becomes runny. Cover with another sheet of clear plastic and secure by laying rocks along the edges. Bake the soil for at least 4 weeks in hot, sunny weather and 6-8 weeks in cooler weather.

5. Put a clear plastic tarp on the soil surface for 4-6 weeks during the **hottest part of the year**. It reduces many soil pests including nematodes, fungi, insects, weeds and weed seeds. Good bugs are killed too.

How to Make and Use Soil Blocks for Seedlings

A soil block is **a block of growing medium (potting soil) that has been lightly compressed** and shaped into a form. You plant a seed or several seeds in the block. You do not need plastic or peat pots, or growing trays (flats).

Why Use: Seedlings grow stronger in soil blocks because the roots have more oxygen and the roots stop growing when the soil stops rather than growing around in a circle like they do in a plastic pot. Seedlings get established more quickly when transplanted. They do not experience as much transplant shock.

The **soil mix for the blocks** needs more fiber than regular potting soil to hold itself together.

Blocking mix recipe:
- 30 quarts premium grade standard peat moss
- 20 quarts coarse sand or perlite (see sand above)
- 20 quarts well decomposed compost
- 10 quarts garden soil
- 3 cups fertilizer (Equal amounts blood meal, alfalfa meal or other nitrogen, colloidal phosphate = soft rock phosphate for phosphorus, and greensand for potassium. Optional, can add 1/2 cup azomite for trace minerals. Mix them.)
- 1/2 cup dolomite lime (for calcium and magnesium plus raises soil pH)

Mix together and then moisten thoroughly with 1 part warm water to every 3 parts mix. It has the consistency of soft putty or wet cement. A small amount of water oozes through the block maker when the blocks are made.

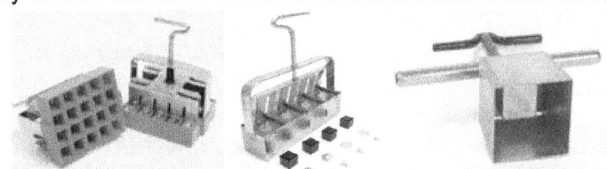

Soil Blocker sizes are 3/4 inch, 2 inch and 4 inch.

The device that makes a soil block is called a **Soil Block Maker or Soil Blocker.** There are 3 types:

1. Mini 20 Soil Blocker makes 20 blocks of the **3/4 inch square**. Good for lettuce, small-seeded flowers and herbs, and seedlings to be transplanted into medium blocks later. Dimples in tops of plungers make small depressions in the tops of soil blocks for seeds.

2. Medium 4 Soil Blocker makes 4 blocks **2 inch square** for all vegetables, large-seeded flowers and herbs. Standard dibbles make small depressions in the tops of soil blocks for seeds. Can also be fitted with long dibbles for large seeds or cuttings, or with 3/4 inch square dibbles to receive mini blocks (3/4 inch square).

3. Large Single Soil Blocker makes one huge **4 inch square** block with a 2 inch square dibble to receive medium size soil blocks. Great for transplanting larger plants like tomatoes, peppers, eggplant, melons, and cucumbers from medium block to large block without transplant shock.

How to make a soil block: Push the above soil mix into a mound that is 1 1/2 inches deeper than the height of the soil blocker. Push the blocker into the mix until it hits the work surface. Twist the blocker a quarter turn and lift. Put the blocker on a plastic tray or other surface, and push down on the handle while raising the blocker. Rinse blocker with warm water and repeat.

Put seed in soil block and cover seed with soil. At first water blocks with a fine mist to prevent blocks from falling apart. Once the seed starts growing, the roots hold it together so a mist is not needed anymore. Do not let blocks dry out.

 For **more about soil blocks and other information**, "The New Organic Grower" by Eliot Coleman.)

 **Inoculating Legume Seeds**

Inoculants improve the nitrogen-fixing process for legumes such as alfalfa, bean, clover, pea, and vetch. They take nitrogen from the air and put it in usable form in the soil that plants can use. Some legumes can fix 300 pounds of nitrogen per acre. Soil bacteria known as rhizobacteria or rhizobium live on the roots of legume plants and create the nitrogen. Most soil contains few if any of these bacteria so the inoculant adds these bacteria. Though **good fertile soil** probably has enough.

Each plant species has a different bacteria species that lives on its roots. There is an inoculant for alfalfa and clover. And another one for fava beans, peas, vetch and lentils.

Buy fresh inoculant at garden centers. Do not leave inoculant packages in the sun. Store in a cool, dark location. Put on the amount the package says or even more. It is not toxic.

Slightly **wet the seed** before putting on the inoculant. Use non-chlorinated water or much better is a mixture of whole milk and molasses (1 cup milk, 1/2 tablespoon molasses). Can use a sprayer to moisten the seeds. Put in the inoculant and mix. Plant right away. Then water the soil.

How to Prevent Damping Off of Seedlings

Damping off is seedling death by any of several fungal diseases, including root rots (pythium, phytophthora) and molds (sclerotinia or white mold, botrytis or gray mold).

Seedlings start to grow and then wither, break off or collapse while they still have their first leaves. It looks as if the seedling has been pinched off at the soil line.

1. Use a sterile potting mix (see above).
2. Sterilize pots with a solution of 1 part bleach to 9 parts water.
3. Do not crowd seedlings. They need good air circulation.
4. Water seedlings from the bottom so leaves stay dry.
5. Do not overwater seedlings.
6. Put a layer of sand on top of the potting mix to keep surface dry.
7. Run a small fan on the seedlings periodically. Make sure room has good air flow.
8. Give seedlings plenty of heat and light.
9. Remove any sick plants immediately.

How to Make Organic Fungicide for Seedlings:
   A strong brew of **chamomile or cinnamon tea.** Use it to water and/or mist seedlings.
   Or 1 tablespoon of 3% solution of hydrogen peroxide per quart of water as a seedling mist.
   (For **growing chamomile,** see May Garden-Plant.)

Bleach (sodium and calcium hyprochlorite) Solution to Disinfect Tools and Cuttings

Use bleach solution to **disinfect greenhouse and garden tools**, potting benches, potting flats, seed starting equipment, and **cuttings**. Dip plant cuttings before rooting in a 10% solution (1 part bleach to 9 parts water). Disinfect tools, traps, cages, and row covers with a 10% solution. Drench seedling soil with a 2% solution (2 1/2 tablespoons bleach to 1 gallon water).

 ## Thinning Seedlings and Plants

Water plants thoroughly prior to thinning. It softens the soil, making the plant come up easier.
Know the proper spacing for the plants. **Keep the largest, healthiest and true-to-type seedlings.**
Loosen the roots from the soil with a knife or small trowel. Then lift out the seedlings.
Try not to disturb remaining plants. If roots are exposed, add more soil.
For small seedlings, use a small scissors and **cut the seedling off at ground level.**
For some plants the thinned plants can be eaten such as chard and spinach.

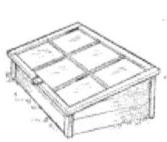  ## Hardening Seedlings Before Planting Outside

Before planting outside, seedlings need to be hardened. This means they need to be **gradually prepared for the colder, harsher climate** than what they had in the greenhouse or other protected area. A **cold frame** can be used. (For image and description of cold frame, see January Greenhouse.)

Or can put in pot outside greenhouse on pleasant days and bring inside at night or bad days. Can put them in a wagon and wheel in and out of greenhouse.

Start about 1 week before planting seedling outside. The goal is to gradually increase the amount of sun, wind and cool air the plant is exposed to. Plants can get sun burned. It is best to transplant them on an overcast day.

Apply **calcium and magnesium** fertilizers such as dolomite lime a few weeks or more before hardening to increase chill hardiness.

Soil and Crops

 Most plants should be fertilized in spring to encourage new growth. Fertilize with manure/ straw from the barn and chicken coop. Add **soil amendments** such as alfalfa meal, azomite, bloodmeal, bonemeal, boron, calcium lime, dolomite lime, kelp meal, granite meal, greensand, gypsum, magnesium sulfate (epsom salt), rock phosphate, sulfate of potash, sulfur, wood ash and/or rock dusts. **The most important fertilizers are manure, rock phosphate, greensand and lime.** NPK= nitrogen, phosphorus, potassium.

Contact your County Extension Office about getting a free soil test. When you are working in your garden, it is important not to walk on your garden beds because it **compresses the soil** and makes it harder for the plants to grow.

Natural Fertilizers

1. **Alfalfa meal** (NPK= 2-1-2) invigorates biological activity in soil. It is also a **protein feed for livestock**. It contains 17% crude protein which translates into 2.75% **nitrogen**. Apply at 2 1/2 pounds per 100 square feet, or 10-25 pounds per 1000 square feet, or 400-800 pounds per acre.

2. **Azomite (montmorillonite clay)** is agricultural grade **trace minerals** (around 100) that is a **rock dust**. (For rock dusts, see below.) It is a highly paramagnetic clay powder. It can be **fed to farm animals** for better health. Apply at 5 pounds per 100 square feet, or 75-150 pounds per acre.

3. **Bloodmeal** (12-0-0) is dried blood and is an excellent source of quickly available organic **nitrogen**. It contains 13-14 percent total nitrogen. It is ideal for heavy feeders such as corn. Apply at rate of 7 and 20 pounds per 1000 square feet, or 300-500 pounds per acre.

4. **Bonemeal** (3-15-0) is considered one of the best natural sources of **phosphorus** because it breaks down rapidly by soil and microbial activity. It also contains calcium and some trace minerals. Apply 5-7 pounds per 100 square feet, or 400-600 pounds per acre.

5. **Boron** deficiency is an uncommon disorder affecting plants growing in deficient soils and is often associated with areas of high rainfall and leached soils. Boron may be present but locked up in soils with a high pH. Symptoms include dying growing tips and bushy stunted growth, and in extreme cases may prevent fruit set. The range between a correct application rate, and a toxic one is not large, so it is easy to apply too much boron. Because of this, it is very

important to get uniform mixing and application. A soil test is needed first. Application rates are 1-3 pounds per acre.

6. Copper sulphate (bluestone) corrects copper deficiency in soil. It is 25% copper. Apply at rate of 10-40 pounds per acre.

7. Granite meal is a rock powder (0-0-5) from soft granite that contains **trace elements and potassium** in a water insoluble, slow release form. Powders from softer granite found in the southeastern United States breaks down more easily than granite powders from the northeast. Hybertite granite from North Carolina is used throughout the United States. It is used at 50-100 pounds per 1000 square feet, or 1000-4000 pounds per acre.

8. Greensand is a mined mineral (iron-potassium silicate or glauconite) (0-1-3) that is a rock dust. Contains **potassium** (7% potash), iron, magnesium, calcium and phosphorus plus 30 other trace minerals. Loosens clay soil and is a soil conditioner. As a soil conditioner, apply 25 pounds per 1000 square feet or 1000 pounds per acre if tilled in. Apply 5-10 pounds per 100 square feet if top dressed. To correct potash (potassium) deficient soils, anywhere from 20 to 100 pounds per 1000 square feet (or 800 to 4000 pounds per acre).

Some say do not apply rock phosphate and greensand at same time since may somehow cancel each other out.

9. Gypsum is a naturally occurring **calcium sulfate** that provides calcium to soils without affecting pH. It lightens compacted heavy soils. It contains 23 percent calcium, 19 percent sulfur, and trace amounts of potassium and magnesium. Gypsum is a **feed grade** material and can be used as a calcium supplement for animals. **Gypsum can be used in the barn to preserve volatile nitrogen in manure.** It reduces the ammonia odor. It works better than lime. Apply 150 pounds per 1000 square feet, or 500-2000 pounds per acre.

(📖 Read "Soil Fertility and Animal Health, The Albrecht Papers" by William A. Albrecht and Charles Walters.)

10. Kelp meal contains **60 trace minerals** in chelated form including zinc, iron, copper, manganese and molybdenum. It has 14 vitamins, plant growth regulators, enzymes, and **growth hormones.** Research shows that kelp **improves seed germination**, root and plant growth, fruit set, and overall health of plants and soil. **Use it when germinating seedlings** in a greenhouse. Makes plants more disease and stress resistant. Increases soil fertility and microbial population.

Unfortunately, it is expensive to use. Another way to use kelp meal is as an **animal feed supplement**. This improves production and health for the animal and enriches the manure for use in compost or on crops. For use on soil, it is generally applied at the rate of 10 pounds per 1000 square feet, or 1 pound per 100 square feet. Add 1 teaspoon to the potting soil of a 6 inch pot.

11. Lime: If you need calcium and magnesium, use dolomite lime. If you need calcium, but your magnesium is adequate, use hi-calcium lime (calcium lime). Do not add too much because the addition of lime to soils increases the rate of loss of organic matter even though it does improve plant production in the short term. Lime raises soil pH. Apply 3-5 pounds per 100 square feet, 30-50 pounds per 1000 square feet, or 500-2,000 pounds (1 ton) per acre.

a. Calcium lime (hi-calcium lime) is a finely ground calcite lime that contains **30% calcium** (calcium oxide and calcium carbonate) and **3-4% magnesium** (magnesium oxide and magnesium carbonate). Use on soils low in calcium but high in magnesium.

b. Dolomite lime is a natural liming material which supplies **21% calcium and 12% magnesium**. Use on soils low in calcium and magnesium. Apply 3-5 pounds per 100 square feet.

12. Magnesium Sulfate (Epsom salt) is used where magnesium (Mg) deficiencies occur in already alkaline conditions. It contains 9.8 percent magnesium, 2.2 percent calcium, and 13 percent sulfur. In acidic soil that needs to have its pH raised, dolomite is a better source of Mg. Apply 10 pounds per 1000 square feet, or 100-200 pounds per acre.

13. Manure is high in nitrogen and phosphoric acid (phosphorus), potassium and other nutrients. For NPK, the nitrogen varies from 1-3, phosphorus from 1-2, and potassium around 1. **Chicken manure** has twice as much phosphoric acid as horse, cattle, goat or sheep manure. **Rabbit manure** has twice as much as chicken. The nitrogen content of manure varies with the type of animal, feed ration, amount of litter/bedding/soil, and amount of urine. **Gypsum** (see above) reduces the nitrogen loss in manure so use it in the barn in stalls or in the coop.

The moisture content of fresh manure is 70-85%. The moisture content of air-dried manure is 9-15%.

Handling affects the fertilizer value of manure. Ammonium (derived from ammonia) is lost to the air, and nitrates are leached by rain. Ammonium losses can be reduced by not stockpiling manure while it is moist, minimizing its handling, and working it under immediately after spreading. Ammonium is lost to the air each time the manure is moved.

Nitrogen compounds in manure are eventually converted to the available nitrate form. Nitrate is water soluble and is moved into the root zone with water. It is the same form available to plants from commercial nitrogen fertilizers.

However, **manure releases its nitrogen more slowly** which is better. The nitrogen in poultry manure is released the fastest with about 90% released the first year. Dry feedlot cattle manure releases only 35% of the nitrogen the first year. So soil builds up its nitrogen over the years.

Manure adds organic matter to the soil that improves soil structure, aeration, moisture-holding capacity, and water percolation. Apply at rate of 30 pounds per 1000 square feet, or 5-10 tons (1 ton is 2000 pounds) per acre depending upon moisture content.

14. Rock Phosphate (0-3-0) is a **rock dust** that is a natural, untreated soft phosphate with colloidal clay containing **minor minerals and phosphorus.** It encourages growth of soil bacteria and earthworms. Some sources say do not apply rock phosphate and greensand at the same time since they may somehow cancel each other out. **Rock Phosphate can be used in the barn to preserve volatile nitrogen in manure.** Apply at 10-40 pounds per 1000 square feet, or 300-2000 pound per acre. (Read "Hands-on Agronomy: Understanding Soil Fertility and Fertilizer Use" by Neal Kinsey and Charles Waters.)

15. Sulfate of Potash (K_2SO_4, potassium sulfate) (**Sul-Po**) (0-0-50) is a natural potash mineral that contains **51% soluble potash (potassium) and 18% sulfur**. It contains trace amounts of calcium and magnesium. Potassium is second only to nitrogen in terms of the abundance needed for plants. Many crops use as much as 250 pounds of potash (potassium) per acre per year. Application rates should be based on soil tests. If you also need magnesium, use Sulfate of Potash, Magnesia (see below). Apply at rate of 1 pound per 1000 square feet. Many crops use 250 pounds of potash per acre.

16. Sulfate of Potash, Magnesia ($2MgSO_4$) is called **Sul-Po-Mag** (0-0-50). It is a natural mineral salt and contains **22% potash (potassium), 11% magnesium and 23% sulfur**. Most eastern states are deficient in magnesium so in **North Carolina** it is better to use Sul-Po-Mag, not Sul-Po unless you add magnesium from another source. Apply at rate of 1 pound per 1000 square feet. Many crops use as much as 250 pounds of potash per acre per year.

17. Sulfur is 90% sulfur for soil application. Useful in **lowering soil pH**. Excellent for blueberries, rhododendrons, azaleas and other acid loving plants. Use according to soil test recommendations—do not over apply. Maximum use is 100 pounds per acre, or 1/4 pound per 100 square feet.

Can feed sulfur to animals. Good to feed to poultry if they have mites or lice. (For **poultry mites and lice,** see March Farm Animals. For **worming goats with recipe with sulfur,** see April Farm Animals.)

18. Wood ash (0-1.5-7) from your fireplace or wood stove is good if you have not burned any pressure treated wood. (You should not burn it anyway.) Wood ash contains **4% potassium, 2% phosphorus and 2% magnesium**. It makes soil more alkaline. It raises soil pH.

Wood ash is put on the ground around plants to deter weevils, rust flies, slugs, snails, wireworms and other pests. Can use as dust for chickens, ducks and turkeys that have mites and lice. (For **mites and lice,** see March Farm Animals.)

Of course, you do not use all of these at one time. You need to **get your soil tested** by your local County Extension Service for free. Then add what your soil needs. They also offer free gardening and farm advice. (For more about **County Extension services**, see March Garden-Plant.)

Some of these **soil amendments** you can get at your **local feed store or hardware store**. Some can be special ordered by your feed store. Others such as azomite, diatomaceous earth, kelp, greensand and rock phosphate can be hard to find. **Countryside Natural Products** (www.countrysideorganic.com) in Fisherville, Virginia carries these products and other natural/organic amendments and feed such as 50 pound bags of buffered **sodium bicarbonate** for ruminants. It is too expensive to ship by UPS but you can order 44 bags of 50 pounds each on a pallet and have it delivered by a trucking company. Their web site lists **local distributors** too.

Nitrogen: Alfalfa meal, bloodmeal, manure. (Good for leaf growth and green leaves. Promotes rapid growth. You can apply too much and damage plants.)

Potassium: Granite meal, greensand, sulfate of potash, wood ash. (Good for roots. Hard to apply too much.)

Phosphorus: Bonemeal, rock phosphate, wood ash. (Good for flower/fruit development. Hard to apply too much.)

Calcium: Calcium lime, dolomite lime, gypsum.

Magnesium: Dolomite lime, magnesium sulfate, sulfate of potash magnesia, wood ash.

Sulfur: Gypsum, magnesium sulfate, sulfate of potash, sulfur.

Trace Minerals (many): Azomite, granite meal, greensand, kelp, rock dusts, rock phosphate.

Other Minerals (individual): Boron.

General Natural Fertilizer Formulas

#1 Formula: Nitrogen: 2 parts blood meal or 3 parts fish meal. Phosphorus: 3 parts bonemeal or 6 parts rock phosphate. Potassium: 1 part kelp meal or 6 parts greensand. Mix together.

#2 Formula: 17 pounds cottonseed meal, 8 pounds rock phosphate, 45 pounds granite dust. This is a 5-10-10 formula (5% nitrogen, 10% phosphorus, 10% potassium). Mix together.

#3 Formula: 1 part kelp meal, 2 parts alfalfa meal, 4 parts of either cottonseed meal / fish meal / soybean meal (or any combination of them), 1 part rock phosphate. Mix. **Good for perennials.**

Plants that are Heavy Feeders

Artichoke, asparagus, broccoli, cabbage, cauliflower, celery, corn, cucumber, eggplant, kale, lettuce, melon, pepper, pumpkin, spinach, squash, Swiss chard, and tomatoes.

Plants that are Light Feeders

Bean, carrot, garlic, onion, parsley, pea, pepper, potato, radish, rutabaga, sweet potato, Swiss chard, and turnip.

Good sources of organic material: alfalfa meal, coffee grounds, compost, corn cobs (ground or chopped), corn husks, grass clippings, cover crops, hulls from grains, leaves (chopped), manure, oilseed meals, pea/bean pods/vines, peanut shells, peat moss, sawdust, straw, tea leaves, vegetable parings, weeds and others.

Fertilizers can be tilled in or left on top of the soil. Tilling destroys weeds. Till when soil is moderately moist. Do not till when it is very wet because it may cause hardpan. **Hardpan** is a hardened layer, typically of clay, that occurs just below tilling depth that does not allow much water or air flow.

Tilling breaks up clods of soil (crumbs or aggregates) and damages the soil structure that influences how water and air move through the soil. Too much tilling destroys this flow. It increases soil erosion. Try to till only once per crop. It is best to add organic matter when you till.

Or use **no-till methods**. With this method the ground is tilled only the first time the bed is created. In future years tilling is not used. Water and soil organisms bring fertilizer into the soil. Mulch is used to reduce weeds.

(📖 Read the book "Hands-on Agronomy: Understanding Soil Fertility and Fertilizer Use" by Neal Kinsey and Charles Waters. Also read "Fertility Farming" by Newman Turner and "Quality Pasture" by Allan Nation.)

Rock Dusts (Stone Meals)

Plants need organic matter, nitrogen, lime, and rock dusts (minerals) in a well aerated, porous soil. These minerals or micronutrients include boron, chlorine, cobalt, copper, iron, manganese, molybdenum, silica, chlorine, flourine, alumina, potassa, soda, lime, and zinc. All of these minerals were originally created by geological forces. To replace these nutrients one source is **rock dust which is a by-product of the gravel industry.**

Types: Rock dust the size of minus 200 mesh (a powder) is best. The finer the mesh, the more easily it is used by the micro-organisms in the soil. The **best rock dust is glacial gravel, river gravel, ground volcanic rock** such as basalt, montmorillonite (azomite), greensand, and granite. A variety of rocks provides the widest range of nutrients. Some say only apply one type of rock dust per year because they may cancel out each other.

Lime: Soils need lime but if you apply too much lime then the old saying is true: "Manuring with lime makes rich fathers but poor sons." It increases the rate of loss of organic matter in the soil. For lime to work well there must be enough potassa (potash or potassium) and soda (such as muriate of soda, sulphates of soda, and phosphates of soda) in the soil. Potassa and soda are found in rock dusts.

Manure: Manure is high in nitrogen and phosphoric acid. If there is too much of it, then plants grow too fast making them weak and vulnerable to disease and pests. Rock dusts balance out the plant making it stronger.

How Much Rock Dust: Apply a minimum of 2 tons per acre (10 tons is better), or 14 pounds per 100 square feet. You can add 2-20 pounds to every cubic yard of compost. You can sprinkle it in the barn on top of animal bedding.

It is also good to throw some **gypsum** (a sulphate mineral high in calcium and sulphur) in animal stalls because it changes the structure of ammonia so it is odorless and not lost to the air. (See fertilizers above.)

Rock dust is a very low cost and effective way to improve soil nutrition. (You can get 1000+ pounds of **rock dust or slurry at a quarry** for a low price.)

Soil and Plant Health: You make healthy plants by creating healthy soils. For healthy soil, you need healthy microorganisms which process the nutrients in the soil and make them more available to the plants. Rock dusts **remineralize the soil** helping crops and even trees be healthier.

Many people have had great success with adding rock powder to the garden and then having dying plants come back to life especially with tomatoes and strawberries. They also make plants more drought and frost tolerant. Plants need less of the major fertilizers such as nitrogen, phosphorus and potassium.

Cosmic Forces: Rock dusts raise the paramagnetic forces (weak magnetic forces) of the soil and plants. This is the ability to resonate or collect magnetic forces of the universe. Healthy plants have fewer problems with pests and disease. Healthy plants are more nutritious for people to eat.

How to Use: If you do use rock dusts, they are much more effective if you do not use any inorganic chemicals (non-organic or manmade fertilizers) such as ammonium nitrate, ammonium sulfate, potassium chloride, sodium nitrate and triple superphosphate.

It is extremely important that the soil have good levels of organic matter (humus) for the rock dusts to be effective. Lime also needs to be present. (See lime above.) Rock dusts and lime need to be applied every year to fields that grow crops. Pastures need them every few years. But do not over apply lime. Also some rock dusts contain lime. **Ask your quarry about the composition (percent lime, etc.) of their rock dust.**

Rocks: Related to rock dusts is the use of gravel and rocks to help plants grow. Gravel and rocks can be used as **mulch to help preserve soil moisture, reduce erosion**, and very slowly add minerals to the soil as the rocks break down. Rocks and gravel also help **stabilize soil temperature** by keeping heat in winter and keeping soil cool in summer. Chat or small gravel makes great garden paths that keep down weeds. Small stones should be left in the soil.

(Read the books "Stone Mulching in the Garden" by J.I. Rodale, "The Survival of Civilization" by John Hamaker, "Bread from Stones" by Julius Hensel, "Secrets of the Soil" by Peter Tompkins & Christopher Bird, and "The Enlivened Rock Powders" by Harvey Lisle. Also look into Biodynamic farming and energy fields in plants. These books and many others show a new way at looking at plant and soil health that is very expansive in its thinking. Soil health is complex and requires a well thought out plan to have the healthiest plants and then the healthiest people.)

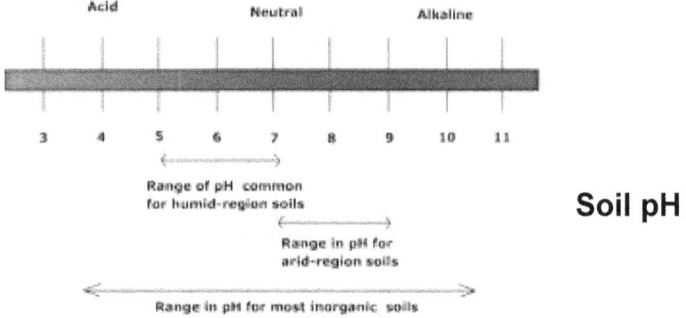

Soil pH

Above image: Soil pH 5-7 is common for humid-region soils. Soil pH 7-9 is common for arid or dry regions.

The acidity or alkalinity of soil is measured by pH (potential Hydrogen ions). **Soil pH in most of North Carolina is usually around 5.5, an acid pH.**

pH is a measure of the amount of lime (calcium) in the soil. Usually, soil in wet climates tend to be acid, and those in dry climates tend to be alkaline. Soil with a pH lower than 7.0 is acid (sour). Soil with pH higher than 7.0 is alkaline (sweet). Neutral is 7.0.

You can take one or several **soil samples** to your County Cooperative Extension office. They test it for free. They test for major plant nutrients such as phosphorus, potassium, calcium, magnesium and sulfur; minor plant nutrients such as copper, manganese and zinc; soil class; pH, humic matter, and other tests. Or you can buy an inexpensive pH test kit at a garden center.

It is easier to make soils more alkaline than it is to make them more acid. To increase pH by 1 point and **make acid soil more alkaline,** add 12 ounces of agricultural lime per square yard in clay soils Correct a high acid soil slowly over several years.

Adding hardwood ash, bonemeal, crushed marble or crushed oyster shells also raises pH. Add sulfur to **make alkaline soil more acid**. To reduce pH by 1 point, mix in 3.6 ounces of ground rock sulfur per square yard to clay soil. Sawdust, composted leaves, wood chips, cottonseed meal, leaf mold and especially peat moss, lower soil pH.

Most garden plants prefer a neutral pH (7.0). Plants such as blueberries like an acid soil.

Crop Rotation and Cover Crops

Crop Rotation

Most of the time, the type of crops grown on a particular piece of land should be changed or rotated. This helps to reduce unwanted insects and disease. Pests live in the soil over winter and if a suitable plant host is grown there in the spring, the pest continues to live. If no suitable plant hosts are there, most or all pests die. This works particularly well with soil borne diseases.

Most pests attack all plants in a particular family. So rotate crops according to family. (See "Plant Families" section.) Crop rotation also balances out the fertility needs of plants. Different plants have different nutrient needs.

A season or a year of **letting land lie fallow (uncultivated)** is a part of crop rotation. During the fallow time weeds grow that bring up nutrients deep in the soil. (For weeds, see below.) It also helps reduce unwanted insects and disease. It encourages biodiversity and a good farm ecology.

One type of rotation is **legumes** that add nitrogen to the soil rotating with **cereals** (poaceae/grass family). For instance alternating soybeans and corn, or alternating fava beans with wheat.

Crops such as lettuce, onions and squash benefit all crops grown after them. Most crops grow poorly when planted after beets, carrots and members of the cabbage (Brassicaceae or Cruciferae) family.

Plant **heavy feeding crops** such as cabbage, corn and tomatoes before **light feeders** such as bulb crops, herbs and root vegetables. Then plant a soil building crop such as clover, oats, mustard, rye or vetch.

For **perennials** such as asparagus and rhubarb, rotate the crops around them.

Sample 2-way rotations: Root crops to leaf crops. Row crops to sod such as fescue. Vegetables to berries. Grains/grasses to legumes (beans, clover).

Sample 3-way rotations: Corn to wheat or oats. Then to clover or timothy hay.

Sample 4- and 5-way rotations: Corn to soybeans or beans. Then to wheat or barley. Then to clover.

Or corn to oats or clover in March. Then to wheat in September. Then to hay. Then to pasture.

Cover crops or green manures should be used in the crop rotation.

Why Use: Cover crops increase soil organic matter (humus), fertility, nitrogen, earthworms, water retention, and beneficial organisms. They reduce erosion by wind and rain, reduce soil compaction, and suppress weeds. They bring minerals deep in the soil to the surface. They provide habitat for beneficial insects. They break up subsoil.

How to Use: When mature, cover crops can be tilled under (**green manure**), or cut and then composted in place. The best time to till under is when 50% of the flowers are in bloom.

Sowing: Sow seeds no deeper than 3 times the width of the seed. If not tilling seeds in, then cover with a light mulch. Keep soil moist. A legume such as clover, pea or vetch has the added benefit of adding nitrogen to the soil. Some cover crops are also grown for grain such as oats, rye and wheat. (See **Garden Seeders and Broadcast Spreaders** in General "Garden Tips" above.)

1. Cool weather cover crops

They are sown **in early spring (March)** when cool, moist soil favor rapid growth. Cool weather cover crops should be at full height before hot weather (85 degrees or more).

Or they can be sown **in the fall usually September and October** so they have about 4–6 inches growth before winter freezes make them dormant. They start growing again in the spring.

Cool weather cover crops include **clovers** such as crimson clover (Trifolium incarnatum), red clover (Trifolium pratense), annual white sweetclover (Melilotus alba), yellow sweetclover (Melilotus officinalis), berseem clover (Trifolium alexandrinum), subterranean clover (Trifolium subterraneum), and white clover (Trifolium repens).

It includes **legumes** such as Austrian winter pea (Pisum sativum subsp. arvense), **faba/fava/bell bean** (Vicia faba), **field peas,** lupin (Lupinus spp.), **vetch (hairy)** (Vicia villosa), and Sericea lespedeza (Lespedeza cuneata).

It includes **non-legumes** such as **barley** (Hordeum vulgare), **fescue, fenugreek** (foenugreek), **flax, kale, mustard** (Brassica nigra), **oat** (Avena sativa), **radish (daikon), rape/canola** (Brassica napus, a type of kale), **rye (cereal)** (Secale cereale), **ryegrass (annual)** (Lolium multiflorum), triticale (a wheat/rye hybrid), and **wheat (winter)** (Triticum aestivum).

The above crops in bold are written about in the months part of this manual.
Mustard is the quickest growing cover crop.
For **general information about cool weather cover crops**, see March Garden-Plant.
For **cool weather cover crops planted in the fall,** see October Garden-Plant.
For **clover, fava bean, fescue, flax, kale, mustard, pea, and radish,** see March Garden-Plant.
For **fenugreek** see May Garden-Plant.
For **kale, oats, ryegrass (annual) and vetch,** see September Garden-Plant.
For **barley, rye (cereal) and wheat,** see October Garden-Plant.

2. Warm weather cover crops

Sow frost-sensitive cover crops **after the spring frost-free date** and at least **6-8 weeks before the fall first frost date**. It is best if soil has warmed to at least 60 degrees in the spring. This is usually May when corn is planted.

Warm weather cover crops include **black eyed pea (cowpea), buckwheat, garbanzo bean (chickpea),** millet, sesame, **sorghum, soybean,** and sudangrass (Sorghum bicolor drummondii). Sudangrass is not a grain sorghum.

For **beans, sorghum, and soybean** see May Garden-Plant.
For **buckwheat** see June Garden-Plant.

Pasture (forage for animals)

Pastures with a lot of different plants withstand diverse weather conditions better, provide a better range of nutrients for animals, and give higher milk yields. Pasture should be a mix of grasses, legumes and herbs.

What to Plant in Pasture:

1. Plant grasses such as meadow fescue, orchard grass (cocksfoot), perennial ryegrass, and timothy. Most animals do not like hard fescue. Bermuda grass and tall fescue are deep rooted so are more drought tolerant.

Hard fescue (festuca durusicula) is a cool season grass that provides an undergrowth of fine, thin leaves. Very hardy. Shade and drought resistant. Sow 15-30 pounds/acre.

Kentucky bluegrass is a cool season grass that grows from March 15 to November 1. The peak months of production are March to May, and September. It grows very little in July and early August. It is very palatable and high in quality. It does well in mountain areas. Good for grazing. Sow 5-20 pounds/acre.

Meadow fescue (festuca pratensis) is a cool season grass that is slow to establish but then very productive. It does well on clay soil but Timothy is better. Sow 15-30 pounds/acre.

Orchard grass is a cool season grass that is deep rooted so is good in drought. It also takes heavy grazing. Orchard grass has higher yield than Kentucky bluegrass. Sow 20 pounds/acre.

Tall fescue (festuca elatior) is a cool season grass that is taller and broader leaved than Meadow fescue. It is deep rooted. It has a higher yield than Kentucky bluegrass. Sow 15-30 pounds/acre.

Timothy is a cool season grass that likes clay soil. It is somewhat slow to establish. A winter hardy perennial grass. Seed at 1/2 pound per 1000 square feet, or 12-15 pounds per acre alone, or 6-9 pounds/acre in mixes.

2. Plant legumes such as alfalfa, alsike clover, sweet clover, white clover, and vetch. (For clover see March Garden-Plant. For vetch see September Garden-Plant.)

Alfalfa is a cool season perennial used as forage, hay, cover crop and green manure. It has an extensive root system. Likes full sun. Likes well drained soil with pH of 6.8-7.5. Drought tolerant. Can grow in a wide range of soils. Usually planted in spring. Sow 1 pound per 100 square feet, or 12-20 pounds per acre. Germinates in 7-10 days.

3. Plant deep-rooted herbs such as caraway (carum carvi), chicory, plantain, salad burnet and yarrow. Deep rooted herbs bring up minerals from deep in the soil. Salad burnet is slow to establish. Chicory and plantain establish quickly. Chicory produces the greatest bulk.

More herbs: Agrimony, angelica, balm, borage, burdock, chamomile, chervil, chickweed, clary sage, cleavers, coltsfoot, comfrey, cowslip, daisy, dandelion, dill, eyebright, fat hen, fennel, fenugreek, feverfew, flax, garlic, goats rue, hawthorn (berry), hogweed, hops, horehound, horseradish, ivy, knapweed, licorice, mallow, marigold, marjoram, meadowsweet, milk thistle, milkwort, mint, motherwort, mullein, nettles, parsley, penny royal, peppermint, primrose, purslane, raspberry, rosemary, rue, sage, sheep's sorrel, shepherd's purse, sorrel, thistle, thyme, vervain, vetch, watercress, woodruff, and wood betony.

Goats and other ruminants like burnet, chicory, sweet clover, kidney vetch, ribgrass plantain and yarrow. **Pigs** like chicory and plantain.

Most herbs can not handle extended grazing so you may want to create a **special paddock of herbs** that you let your animals graze on for short periods of time.

In **western North Carolina** seeds sown for pasture, cover crops and lawns are Austrian winter peas, Bermuda grass, brown top millet, crimson clover, hairy vetch, kenland red clover, Kentucky bluegrass, Kentucky fescue, Korean lespedeza, ladino clover, orchard grass, red creeping fescue, serecea lespedeza, tall fescue, timothy, weeping love grass, and white dutch clover.

Korean lespedeza is an acid-tolerant, drought-resistant, summer annual legume used for pasture, hay and soil improvement. **Serecea lespedeza** is a perennial considered an invasive plant that is hard to get rid of.

(For **pasture management and worms**, see April Farm Animals. For **hay harvest**, see June Garden-Harvest.)

Read the book "Fertility Pastures" by Newman Turner.)

Fall and Winter Pasture: Special Forage Crops

Grow frost hardy crops such as **forage kale and other brassicas** for animals to eat in the pasture in winter. Forage brassicas are high quality, high yielding, fast growing crops that are particularly suitable for grazing by livestock. Leaves and roots are edible. Kale has highest yield of all brassicas and the greatest cold tolerance.

Keep the animals out of the brassicas until fall or winter when pasture growth is slow or dormant. Open up the paddock to let them **graze for short periods** each day.

Sow 4-5 pounds of **kale** seed per acre. Sown July-September, can be **eaten November to March.**

Mangel beets, kohlrabi, turnips and rutabagas are sown mid-July to mid-August and ready to eat 70-90 days after sowing. They are good for **forage September-January**.

Brassicas are very high in crude protein and energy, but extremely low in fiber. Their low fiber content results in rumen action similar to when concentrates are fed. Therefore, supplement with roughage such as hay. Brassicas should never be more than 2/3 of the forage portion of livestock diets.

Forage Feed: Higher than 18% crude fiber. Includes all hay, soybean/almond hulls, ground corn cobs.
Energy Feed: Less than 18% crude fiber and less than 20% crude protein. Includes all cereal grains, wheat/rice bran, fats and molasses.
Protein Feed: Less than 18% crude fiber and more that 20% crude protein. Includes meals derived from soybean / linseed / cottonseed, brewers yeast, fish meal, sunflower seeds, and dehydrated milk.
Beet pulp is the by-product from extracting simple sugars (sucrose or table sugar) from beets. The remaining pulp has little or no sugar. At 10% crude protein and 18% crude fiber, it is somewhere between being a forage and an energy feed. It is higher in calories than any of the forages, and is lower in energy than any of the cereal grains. The energy in beet pulp is primarily derived from both soluble and insoluble fiber (energy which is released relatively slowly so blood sugar remains stable). Beet pulp can replace **up to 50% of the forage portion of a feed ration.** It can be fed dry or soaked but it is better to feed it soaked in water for a few hours or overnight. The above is true for horses and ruminants.

(For **beets and kale**, see March Garden-Plant. For **rutabagas**, see June Garden-Plant. For **kohlrabi and turnips** see August Garden-Plant.)

Weeds and What They Say about Soil Type

A weed is a plant that is in a location where you do not want it. Some people call them wayside plants. Weeds tell a farmer what type of soil it is.

There are 3 types of weeds:
1. **Succulent plants** store water and thereby survive for short periods without rain or irrigation. They include agave, aloe, cactus, jade, and orchids.

Garden Tips p. 15

2. Broadleaf weeds indicate an imbalance between phosphate and potassium (potash). The ratio should be 2 parts phosphate to 1 part potash for crops and vegetables. The ratio should be 4 parts phosphate to 1 part potash for grasses and pasture. They include dock, bindweed, knotweed, plantain, pokeweed, thistle, milkweed and many others.

3. Grasses vary with what they indicate based on type of grass.

Soil Types & Weeds:
Acidic (sour) soil weeds: cress, daisy, knotweed, moss, plantain, and sorrel.
Alkaline (sweet) soil weeds: chickweed, chicory and spotted spurge.
Disturbed Soil: most broadleaf weeds especially lambs quarters and pigweed.
Dry or sandy soil weeds: carpetweed, mustard, nettle, pigweed, sorrel, speedwell, sandbur, thistle, Queen Anne's lace (wild carrot) and yarrow. Ragweed (also means low potassium.)
Fertile or well-drained, humus soil weeds: chicory, dandelion, foxtail, horehound, lambs quarters, pigweed, and purslane.
Good phosphate soil weeds: Cocklebur.
Heavy clay soil weeds: nettle, plantain, and quack grass.
Hard compacted soil weeds: bluegrass, chickweed, cress, dandelion, goose grass, knotweed, morning glory, mustard, nettle, and plantain.
Low calcium soil weeds: foxtail and quackgrass.
Poor or low fertility soil weeds: bindweed, clover, crabgrass, daisy, dandelion, fennel, ragweed, morning glory, mugwort, mullein, plantain, thistle, Queen Anne's lace (wild carrot), and yarrow.
Wet or moist soil weeds: chickweed, crabgrass, ground ivy, Joe-pye weed, knotweed, moss, sedge, spotted spurge, and violets.

Weeds have deep roots that go down into the lower subsoil and bring up **minerals** such as potassa, sodia, lime, magnesia, manganese, iron, silica, alumina, phosphoric acid, sulfur and flourine. Weeds are very good at growing in **compacted soil** making it easier for crops to grow there later. Their roots promote **water capillary action** to bring water to the soil surface for crops with shallow roots. They can be good for a field of crops if the weeds are not too dense.

This is one of the reasons why land should be allowed to lie **fallow (uncultivated)** sometimes. A season or a year of letting land lie fallow is a part of crop rotation. (For crop rotation, see above.)

If **weeds are cut** and allowed to decay into the soil of a pasture or field, it will give the pasture or field the nutrients that it is low in. Succulent plants build up carbonic ions so that the soil holds more water.

Certain types of weeds are often associated with certain **insect pests and diseases** due to the same nutritional problems in the soil. If you have good healthy soil, you have fewer problems with weeds. And the weeds you do have, will be easier to control.

(Read the books "Weeds: Control Without Poisons" by Charles Walters, "Weeds: Guardians of the Soil" by Joseph Cocannouer, and "Paramagnetism: Rediscovering Nature's Secret Force of Growth" by Philip S. Callahan.)

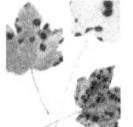

Prevent fungal diseases by having **well drained soil.** Make sure plants have enough space around them for **good air circulation. Remove weeds and garden debris** that harbor insects that transmit diseases. Select growing sites with good air drainage, full sunlight and low humidity. Use drip irrigation to avoid excessive leaf wetness. Take care not to transport the disease by hand or on infected tools and equipment. (See the section "Plant Health".)

Propagation (Reproduction of Plants)

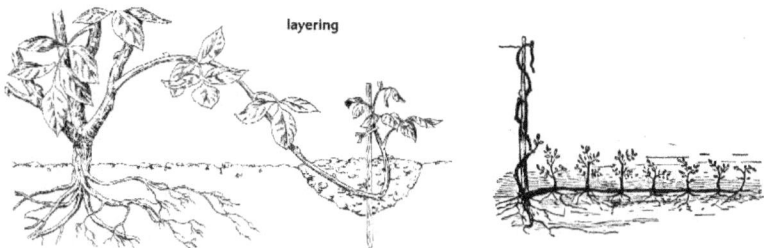

Propagation by Tip Layering / Rooting *(See 2 images above.)*
Tip layering is pulling a branch to the ground and covering it with dirt. You may need to place a rock on it to hold it down. It forms roots. A few months later or in early spring, the plant can be separated from the mother plant and transplanted somewhere else.

A more efficient method of tip layering called simple layering is to cut off the shoot tip to induce lateral (side) branching. During the **summer** dig a 3 inch deep hole and put middle and end of shoot in hole. Cover with 3 inches of soil. By fall new rooted shoots will have developed at each lateral branch. Transplant in **spring**.

Fruit plants that do well with this technique are blackberries, currants, gooseberries, grapes and raspberries. Other plants include azalea, boxwood, climbing roses, forsythia, honeysuckle, rhododendron, and wax myrtle. It is good for most **perennials that have flexible branches** such as bay, lemon balm, rosemary, sage, thyme, and winter savory.

Propagation by 4 types of Stem Cuttings

1. Herbaceous cuttings are made from non-woody, herbaceous plants (a plant that has leaves and stems that die down at the end of the growing season). In **summer** cut a 3- to 5-inch piece of stem from the parent plant. The leaves on the lower 1/3 to 1/2 of the stem are removed. Most root and do so quickly.

2. Softwood cuttings are from succulent, new growth of woody plants, just as it begins to harden (mature). Softwood cuttings should be from wood that can be snapped easily when bent. It still has a gradation of leaf size (oldest leaves are mature while newest leaves are small). For most woody plants, this occurs in **May, June or July**. Soft shoots are very tender so make sure they do not dry out. They root quickly.

3. Semi-hardwood cuttings are from partially mature wood of the current season's growth, just after a flush of growth. This type of cutting is usually made from **mid-July to early fall**. The wood is fairly firm. The leaves are a mature size. Many broadleaf evergreen bushes and some conifers (trees and bushes that bear cones) are propagated by this method.

4. Hardwood cuttings are taken from dormant, mature stems in **late fall, winter or early spring**. Plants are fully dormant with no signs of active growth. The wood is firm and is hard to bend. These cuttings are usually used for deciduous (leaves fall in autumn) shrubs but can be used for evergreens. Examples include fig, forsythia and grape.

How to Root the Cuttings

Early morning is best time to take cuttings. Keep cuttings cool and moist. If there will be a delay in putting cuttings in soil, store in a plastic bag in a refrigerator. Cuttings are generally 4-6 inches long.

Remove leaves from the lower 1/3 to 1/2 of the cutting. On large-leafed plants, the remaining leaves may be cut in half to reduce water loss (see second image above).

Remove all flowers and buds from cuttings. Plants that have been fertilized heavily, especially with nitrogen, may not root well. Usually cuttings from young plants root better than cuttings from older plants. Cuttings from lateral (side) shoots usually root better than cuttings from terminal (top or end) shoots. Putting **rooting hormone** on the cutting helps it root better. (For how to make your own rooting hormone, see below.)

Rooting mediums (potting mixes) used are coarse sand, a mixture of 1 part peat and 1 part perlite, or 1 part peat and 1 part sand (for details about potting mixes, see above). Moisten medium before putting in cutting. Insert cuttings 1/3 to 1/2 their length into the medium. Insert cuttings in same up/down direction as was on plant (buds are pointed up).

Cover cuttings with **plastic** (see above images on right). Place in indirect light. Keep the medium moist until the cuttings have rooted. Rooting is improved if cuttings are misted.

Transplant newly rooted cuttings into a nursery area. Let them grow larger before transplanting to permanent location.

Make Your Own Rooting Hormone with Willow

Why Use: Rooting hormone is needed with plant types that do not readily produce roots on cuttings. The hormone encourages the plant to grow roots when it usually would not. It can be applied either to the cut tip of the cutting, as a foliar (leaf) spray, or watering the soil with it.

Rooting hormone is made with the **willow tree or bush** (salicaceae family; Salix genus, osiers, sallows) that is found in northern climates such as North Carolina. It likes moist soil. There are 400 species with 100 in the United States. They usually have multiple trunks and grow as thickets. They are 25-100 feet tall. The leaves are narrow and elongated.

How to Make: Make your own rooting hormone with any species of willow tree. Cut some pencil size branches from a willow tree. Cut into small pieces and add 2 cups to 1/2 gallon of warm water. Let soak for 1-2 days.

For a stronger brew, use more than 2 cups of cut branches. If you need it right away, use boiling water and soak overnight.

Use soon. It is best to make a fresh brew each time. However, you can put the willow water in an airtight jar in the refrigerator. It will keep about 60 days.

Applying Rooting Hormone: Soak the cut part of the cuttings in willow water overnight. Or if you plant the cuttings in soil or potting mix, just water it with willow water. Two applications should be enough. Some difficult plants may need to have cuttings put directly in a jar of willow water.

(Read "Peterson Field Guides: Field Guide to Eastern Trees" by George A. Petrides. Read the excellent book "Botany in a Day: The Patterns Method of Plant Identification" by Thomas J. Elpel.)

 # How to Save Seeds

Topics in this section:
Open and Hybrid Pollination
Inbreeding/Outbreeding
Pollination Methods
Annual/Biennial/Perennial
Genetic Vigor and Inbreeding
How Difficult Seed is to Save
Harvesting/Preparing Seed
Storing Seed for Planting
How Many Years Seed Remain Viable

There is a lot of seed saving information in "Plant Families".

When saving seed for planting, always save seed from healthy plants that have the characteristics you like. Seed saved at the end of the growing season is usually the best. Only harvest ripe seed. When seeds are mature they usually turn from white to cream colored, or from light brown to dark brown.

Before saving seed, you need to know for each plant the method of pollination, time of seed bearing, whether plant is a hybrid, and how seeds should be collected. Keep written records of your seed saving activities. Label all saved seed with name and date.

All living things are divided into categories. Starting with large groups: Kingdom (plant or animal), Division, Class, Order, Family, Tribe, Genus, Species, Variety. Plants are identified by their **genus and species** such as "Solanum tuberosum" for a potato. A **variety** is a subclassification of a species such as "Solanum tuberosum **var** Desiree". **Cultivar** is a variety of cultivated plants as opposed to a wild plant. (See "Plant Families".)

Seed saving organizations such as "Seed Savers Exchange" at www.seedsavers.org sell open-pollinated and heirloom seeds to the public. You can also join, and then buy, sell or trade seeds with other members.

For saving your own seed, read "Saving Seeds: The Gardener's Guide to Growing and Storing Vegetable and Flower Seeds" by Marc Rogers. For a **simple guide** read "Seed Sowing and Saving" by Carole Turner. For a **detailed, professional guide** read "Seed to Seed: Seed Saving and Growing Techniques for Vegetable Gardeners" by Suzanne Ashworth. Read "Breed Your Own Vegetable Varieties" by Carol Deppe.) (More information about seed saving is in "**Plant Families**" section.)

 ## Open and Hybrid Pollination; Inbreeding and Outbreeding

Plants are either open-pollinated or hybrid.

Open pollination is pollination by insects, birds, wind and other natural methods. Open-pollinated seed can either be pure seed, meaning it has bred with others of its type. Or it can be a hybrid that has bred naturally with plants not of its type. **Heirloom seeds** are open-pollinated seeds that have bred with their own type and have been saved for 50 years or more. **There is more genetic diversity with this method.**

There are 2 types of hybrid pollination.
1. **One occurs naturally among closely related plants** that are not of the same type or variety such as bees pollinating a sugar snap pea (fabaceae family, Pisum sativum var macrocarpon) with pollen from an English pea (Pisum sativum var sativum). If you save the seed, it will not breed true.
2. **The other type is human controlled pollination** where people artificially pollinate a plant.
This is of 2 types:
   a. One type is the most common and **pollination is done by hand** by a person among related plant types. Anyone can do this type of pollination in the garden. This has been done for hundreds of years.
   b. The other is a modern method known as **genetically modified organisms (GMO)** where genetic material from a completely different species even an animal is introduced into the DNA of the plant with special instruments in a laboratory. (**For more about GMO, see corn** in May Garden-Plant.)

In **human controlled hybrid pollination,** plant types are deliberately crossed between 2 different inbred lines creating F1 hybrids. The parents that create the F1 hybrid are very inbred. Human created hybrid plants are **more uniform than open-pollinated varieties.** Seed saved from hybrids, whether natural or human pollinated, does not produce true to type, meaning the exact plant characteristics are unknown until the seed grows. All commercial varieties of corn are hybrid. Most common, commercial crops are hybrids.

If you only have hybrid seeds, you can over time turn them into **open pollinated seeds** by planting many seeds and then removing all plants that are not breeding to the type that you want. You will have to remove most plants but over a few years you will get consistent breeding to the type that you want.

Landraces or folk varieties are heirloom seeds that have diverse genetics with a lot of variability among the seeds collected from the same type of plant. (All of these seeds are not genetically identical the way F1 hybrid seeds are.) These seeds are good because there will usually be some plants that do well in your location during the environmental conditions during each growing season.

Local Breeding: Plants that reproduce through natural means (rather than human controlled hybrids) adapt to local conditions over time, and become reliable performers, particularly in their localities. These become great seeds to save for your location. It is important to remove all poor performing plants before they go to seed so they do not spread their problems to the next generation.

Inbreeding is when 2 plants that are closely genetically related to each other create seed. If plants do not breed in groups, they are inbreeding plants. Inbreeding plants usually have entirely **self-contained flowers** with both male and female parts in 1 flower. The pollen does not leave the plant; they **self pollinate**. These types of plants are genetically predisposed to being tolerant of a small gene pool. An example is beans.

If plants breed in groups, then they are referred to as **outbreeding** plants. These plants are genetically predisposed to needing a large gene pool to stay healthy. If seed is saved from a small population, the vigor and health of the plants decreases. An example is corn which is wind pollinated.

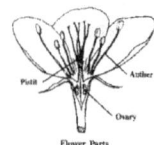 **Pollination Methods**

There are 3 pollination methods used by nature: air/wind-borne, insect and self. To remain genetically pure, plants must be pollinated with pollen from the same variety.

1. For air/wind-borne pollinated plants, there must be no other varieties of the same species and sometimes same family within 1 mile shedding pollen at that time, if you want pure seed for saving. If you are not saving seed, then distance does not matter.

These plants include beets, corn, spinach, Swiss chard and others. If this is not possible, the plants can be put in cages made with window screen or row covers (spun polyester, Reemay) to keep out unwanted pollen. Though this will not work for plants with tiny pollen such as spinach.

2. For insect pollinated plants, there must be no other varieties of the same species and sometimes same family within 1/4 mile if you want pure seed for saving. Some say within 500 feet. If this is not possible, the plants can be put in cages made with window screen or spun polyester to keep out unwanted pollen.

These plants include asparagus, broccoli, brussels sprouts, cabbage, carrots, cauliflower, celeriac, celery, collards, cucumber, eggplant, kale, kohlrabi, melons, mustard, onions, parsley, parsnips, peppers, pumpkin, squash, radish, rutabaga, turnip and others.

3. Self-pollinated plants transfer pollen directly to the stigma within the same flower (male and female parts are all in the same flower). Just to be sure pollination is pure, separate these varieties by a few rows. However, pollination can still sometimes occur by insects. A cloth bag made of spun polyester (Reemay) can be put on flowers of self-pollinating plants to prevent the occasional insect pollination.

These plants include beans, chicory, endive, lettuce, peas and others.

Hand pollination can be done by the gardener usually for plants that need insect pollination and sometimes those that use wind pollination such as corn. Uncontaminated pollen from a male flower is placed on the stigma of a female flower that has been protected from unwanted pollination by screening or row covers.

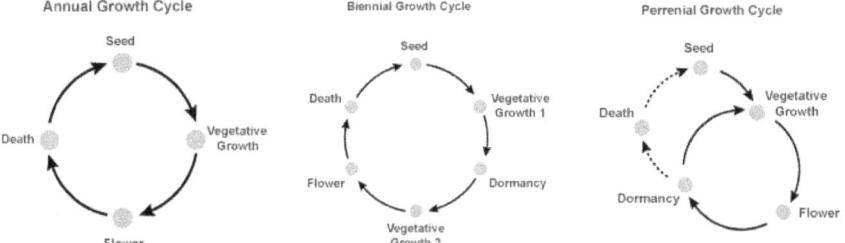

Annual, Biennial or Perennial

Plants are either annual, biennial or perennial (see above image).

1. Annual plants germinate, flower, and die in 1 year or season. For food plants this means it is sown in spring and dies in the fall. These plants include most grown for food such as beans, corn, cucumber, eggplant, pepper, squash and tomato.

2. Biennial plants live for 2 years. The **first year** they only grow leaves, stems and roots. The **second year** they develop their flowers and seeds. Most root crops are in this category. However, most root crops are grown as annuals. (They are dug up and eaten at the end of the first summer or fall.)

To save seed in cold climates, the roots are dug in the fall and stored at 33-45 degrees in sphagnum moss, peat moss or sand. They must not freeze. A root cellar is good for this. Then they are re-planted outside early spring so flowers can grow that summer and then seed is collected.

Biennial plants include beets, brussels sprouts, cabbage, carrot, celeriac, celery, collards, Florence fennel, kale, kohlrabi, leeks, onion, parsley, parsnips, rutabaga, salsify, Swiss chard, and turnip.

3. Perennial plants live for more than 2 years, sometimes decades. Herbaceous perennials grow and bloom over spring and summer, then the leaves die back every fall. In the spring the plants grow from their root-stock rather than from seed as an annual plant does. These plants include asparagus, chives, comfrey, dandelion, lovage, mint, oregano, rhubarb, rosemary, sage, sorrel and thyme.

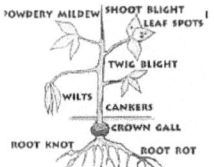

Genetic Vigor and Inbreeding

When saving seed for a particular variety, save seed from as many different healthy plants as possible. Seed should be saved from at least 20 inbreeding (usually self pollinating) plants or 100 outbreeding (usually insect or wind pollinated) plants.

(Read "Seed to Seed" by Suzanne Ashworth and "Breed Your Own Vegetable Varieties" by Carol Deppe.)

Saving diverse seed is not a problem with most **self-pollinating plants** such as beans, since they are naturally inbred with little diversity anyway.

To maintain genetic vigor and prevent inbreeding, these are the recommended **minimum number of plants to be grown together when saving seed for planting:**

    a. 25 cucurbit plants. (Vine crops such as cucumbers, gourd, melons, squash, and watermelon which are all outbreeding.) If you have less than 25 plants, then also hand pollinate between all the plants.

    b. 50-100 brassica, mustard, or radish plants.

    c. 200 corn plants. Corn is particularly sensitive to inadequate population size.

Another way to prevent inbreeding is to **share your seed** with others. Every few years trade the same variety of seeds to increase your diversity. Then plant different groups together and collect new seed.

To prevent plants with unwanted characteristics from pollinating other plants in the same variety or family, you have to remove their flowers or kill them before they flower so they never produce pollen.

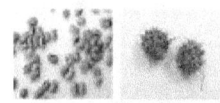
How Difficult Seed is to Save

This system of categorization is used in the section "Plant Families".

      A. Vegetable seeds that are easy to save: Annual plants that are inbreeding and/or self pollinating. They produce seed the same season as planted and are self-pollinating so do not need to prevent cross-pollination. Examples include bean, lettuce, pea, and pepper.

      B. Vegetable seeds that are of medium level to save: Annual plants that are outbreeding or need wind/insects for pollination. These plants produce seed the same season they are planted but require separation to prevent unwanted cross-pollination with other varieties in the same family. Examples include corn, cucumber, muskmelon, radish, pumpkin, spinach, and squash.

      C. Vegetable seeds that are hard to save correctly: Biennial plants that are inbreeding or outbreeding. These plants require more than 1 year for seed production. Most must be separated to prevent cross-pollination. Examples include beet, Swiss chard, cabbage family, carrot, escarole, radicchio, endive, and turnip.

Harvesting and Preparing Seed

More information is available for individual plants where mentioned in **Garden-Harvest for each month.**

1. Flowers:
Harvest seed heads from flowers such as carrots, dill, parsley, and parsnip after they dry but before seed dispersal. This means the flowers are faded and may have puffy tops. If seeds are small or lightweight, put the dry seed heads into paper bags to catch seed as it falls out.

2. Fruit and Fleshy Vegetables:
Seed is extracted from fruit or fleshy vegetables after it ripens but before it rots. (Seed is extracted **after prime picking time for eating**.) This includes cucumber, melon, rose, squash, and tomato. Leave cucumbers, pumpkin, summer squash and winter squash on the vine until after frost. Remove seed mass from fruit or vegetable. Separate seed from pulp, then dry at room temperature. Or can use the fermentation process below.

3. Pod Plants:
Leave pods on vines such as beans and peas until the pod dries. The pod is brown. Harvest before the seed is dispersed. The seed may need to dry a little longer in an airy location out of direct sun.

Fermentation for Fruit and Fleshy Vegetables
Scoop the seed mass out of the fruit or lightly crush fruits. This includes plants such as cucumbers, melon, squash and tomato. Put in a small amount of warm water in a bucket or jar. Let ferment for 2-4 days. Stir daily. The fermentation kills viruses and separates good seed from bad seed and fruit pulp. After 2-4 days, the **good viable seeds sink to the bottom** while the pulp and bad seeds float. Pour off pulp and bad seed.

Can put on window screen, glass, dish, cookie sheet, or plywood to dry. Do not put seeds on paper or cloth because seed will stick to it. Never let seeds go above 85 degrees. Do not dry seeds in the oven because it kills the seeds.

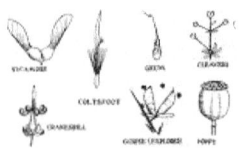
Storing Seed for Planting

When seed is dry, gently hand rub to get rid of any chaff (plant debris). **Seeds should have 5-8% moisture when stored.**

Store in an envelope, barrier pouch, plastic bag or glass jar with rubber gasket lid (an **airtight container** is much preferred) in a cool, dark, dry, rodent-free place such as a **refrigerator, freezer or bottom shelf of a root cellar** (since hot air rises). However, if seed has too much moisture, it will rupture and die in the freezer. It is important that the temperature and humidity be stable.

Label with seed name, variety and date. **Seeds germinate best the following year.** Thereafter, germination declines. Onion, parsley, parsnip, salsify and sweet corn are usually viable for only 1 year. Most seed is viable for 3 years if stored correctly.

Vigor is the seed's ability to germinate rapidly with good disease resistance. The longer a seed is stored, the less its vigor. Good seed vigor is lost long before germination ability stops. Replace saved seed so germination rate is 70% or better.

A **3 year rotation** can be done for most seeds. Seeds need to be dried to **8% or less moisture if stored in freezer**. If greater than 8% moisture, seeds will rupture in freezer and die. To test moisture level, try to bend seed. If less than 8%, seeds will break. To test hard-shelled seeds such as beans and corn, hit with a hammer. If 8% or less it will shatter, rather than mash.

Silica gel that indicates color is a **good way to dry seeds once other types of drying are done.** It is blue from cobalt chloride, a moisture indicator. **It is blue when dry, and pink when moist.**

Place equal weights of seed and silica gel in an airtight container for 6-8 days to completely dry. (The silica gel is in a bag with thin cloth or other porous container.) After seeds are dried, remove silica. This dries seeds from a usual starting moisture of 12% down to 3-5%. Seeds are immediately put in airtight container.

Legumes should not be dried below 6% moisture, otherwise the seed may become irreversibly dormant. This means it may not germinate. Remove legumes from the drying container after 5-6 days.

Silica gel can be **reactivated** by drying in an **oven** for 8 hours at 200 degrees, or 300 degrees for at least 3 hours. When it is dry, the gel turns medium blue. The oven method is preferred.

Or reactivate gel by setting a **microwave** to the medium to medium-high setting. Dry for approximately 3 to 5 minutes. It is dry when gel turns medium blue. If the gel has not dried, stir it with a spoon. Heat for another 3 to 5 minutes. Stir the gel each time. Approximate drying time is 8 to 12 minutes per pound of gel.

How Many Years Seeds Remain Viable under Good Storage Conditions

Asparagus 3 years	Celery 3	Kale 4	Pepper 2	Turnip 4
Beans 3	Chard, Swiss 4	Kohlrabi 3	Pumpkin 4	Watermelon 4
Beets 4	Chicory 4	Lettuce 5	Radish 4	
Broccoli 3	Corn, sweet 2	Muskmelon 5	Rutabaga 4	
Brussels sprouts 4	Collards 5	New Zealand Spinach 3	Salsify 1	
Cabbage 4	Corn Salad (mache) 5	Okra 2	Scorzonera 1	
Cabbage, Chinese 3	Cress 5	Onion 1	Sorrel 4	
Carrot 3	Cucumber 5	Parsley 2	Spinach 2	
Cauliflower 4	Eggplant 4	Parsnip 2	Squash 4	
Celeriac 3	Endive 5	Pea 3	Tomato 5	

How to Save Seeds p. 6

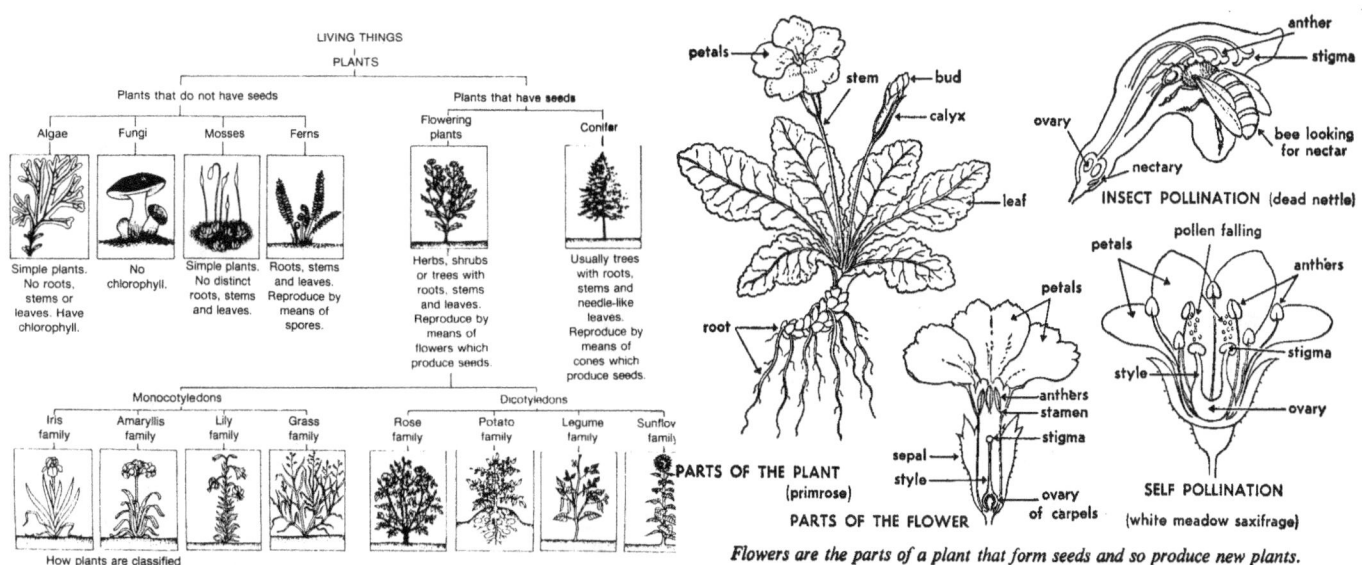

Flowers are the parts of a plant that form seeds and so produce new plants.

Plant Families: Care, Diseases, Pests and Seed Saving

Amaranthaceae, Amaryllidaceae, Brassicaceae/Cruciferae, Chenopodiaceae, Compositae, Convolvulaceae, Cucurbitaceae, Gramineae/Poaceae, Labiatae, Leguminosae/Fabaceae, Liliaceae, Malvaceae, Polygonaceae, Portulacaceae, Solanaceae, Tetragoniaceae, and Umbelliferae/Apiaceae.

All living creatures are divided into groups and subgroups. Each successive group has fewer members. Going from large to small: **Kingdom (plant or animal), Division/Phylum, Class, Order, Family, Genus, Species, and Variety**. There are many genus in one family. There are many species in one genus. There are many varieties in one species.

A **Binomial/Binary name** is given such as "Allium sativum" for all garlic. Allium is the **genus**, and sativum is the **species**. For a particular **variety**, it is for example "Allium sativum var. ophioscorodon" for German Stiffneck garlic. A **cultivar** is a variety of cultivated plants as opposed to a wild plant.

Different varieties of plants within the same species will cross pollinate. Very few outside the species cross pollinate except in certain families such as Cucurbitaceae where members sometimes pollinate outside of their species.

Read the book "Seed to Seed: Seed Saving and Growing Techniques for Vegetable Gardeners" by Ernest Ashworth, Kent Whealy and Suzanne Ashworth for much more information about each species. An excellent book about plant families and identifying plants is "Botany in a Day: Patterns Method of Plant Identification" by Thomas Elpel.)

In the section "How to Save Seeds", **seeds were divided into 3 categories:** A. Easy to save., B. Medium level to save., C. Hard to save. This system is continued in this section.

To help **determine seed saving methods**, each plant is listed as either annual/biennial/perennial and inbreeding/outbreeding. (For definitions of these terms, see "How to Save Seeds" section.)

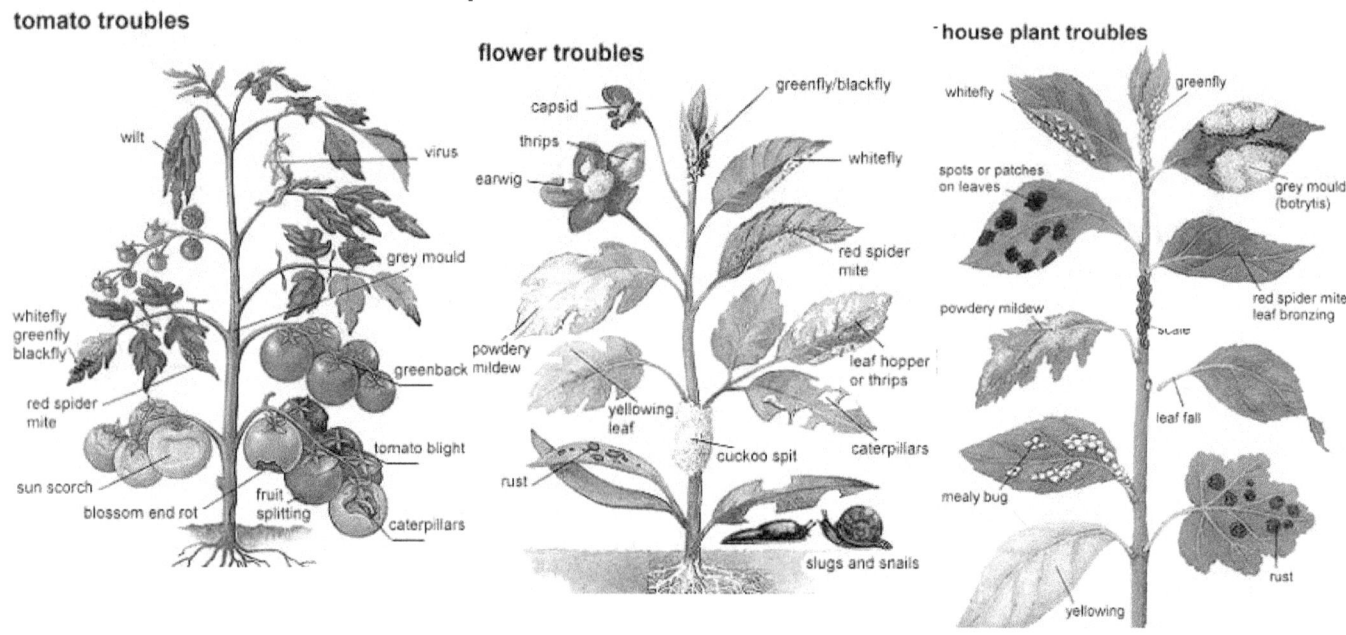

 **Amaranthaceae Family** (B. Medium level of difficulty to save seed.)

Genus- Amaranthus
    Species-   **Grain amaranth** (cruentus, hybridus, hypochondriacus) Outbreeding Annual
    Species-   **Leaf amaranth** (tricolor-black seeds) Outbreeding Annual

    Amaranth was the sacred food of the Aztecs. This family has not been properly divided botanically so there is still confusion about classification of varieties.
    Seed Saving: Varieties of amaranth **will cross pollinate with other varieties in the same species** but not with varieties in other species. All varieties are wind pollinated. They have male and female flowers on the same plant.
    For pure seed the minimum distance between 2 varieties of **leaf amaranth** is 500 feet if there is a barrier such as other tall crops or a building between them. Minimum distance between 2 varieties of **grain amaranth** is 2 miles. You can prevent cross pollination by using corn tassel bags around 5 plants at a time.
    Harvest: Seeds mature gradually from bottom to top. Seeds can be harvested as they mature. If extra drying is need, put in a dry, airy place out of the sun. Wear gloves and rub to release the very small seeds. Amaranth seed will store for many years if stored properly.

    Disease and Pests: Amaranth usually has no major disease or pest problems.
        Amaranth does get fungal diseases but no fungicides are labeled for it.
        Various root/stem rots contribute to lodging (falling over) late in the season if soils are wet in August.

Amaryllidaceae Family (C. Hard level of difficulty to save seed. However, garlic and multiplier onions are easily propagated by saving bulbs.)

Genus- Allium
    Species-   **Leek** (ampeloprasum) Outbreeding Biennial
               Common seed onion, shallot, multiplier onion, topsetting onion (cepa) Inbreeding Biennial
               Bunching onion (fistulosum) Inbreeding Biennial
               Garlic, rocambole (sativum) Propagated by bulbs.
               Common chives (schoenoprasum) Perennial, can propagate by dividing roots/bulbs.
               Garlic chives (tuberosum) Perennial, can propagate by dividing roots/bulbs.

Plant Families p. 2

Seed Saving: Plants in the allium genus are mostly pollinated by insects. Varieties within the same species should be separated by 1-3 miles. Immature flower heads are bagged, and then hand pollination needs to be done every day for 2 weeks. Seeds store for only 1-2 years.

Multiplier/perennial onions and garlic can also be propagated by replanting bulbs. This is easy. There is no cross pollination to worry about.

Fungal Diseases: If plants in the allium genus get a disease, it is usually a fungal disease. (For more about **fungus control**, see "Garden Tips" and "Plant Health" sections.)

1. Onion smut is a fungal disease that attacks the leaf just as it appears. The plant usually survives the first attack, but later it appears as black streaks and blisters in the leaves and bulbs, and it may die. It happens when spring weather is cool and wet. Planting seeds too deep helps the fungus develop.

Onion smut is prevented by planting seeds that have been treated against it or by growing from disease-free transplants.

2. Botrytis leaf blight shows up on leaves as gray/white spots. The spots become sunken/straw-colored. The leaf dies back starting at the tip. It grows when the weather is warm and moist. Botrytis is lessened by removing all plant debris from the area.

3. Downy mildew is an airborne fungus so cultural practices such as crop rotation and keeping the area clean do not help. It is a lavender/purple/gray velvet-looking growth on leaves. It occurs when weather is wet and cool. It can destroy the whole allium crop in 30-45 days.

Grow plants in well-drained soil. To prevent spread of disease, wash tools and shoes with 10% bleach solution (1 part bleach, 9 parts water) after working in the garden. Spray plants with fungicidal soap.

4. Purple blotch is a fungus that attacks plants weakened by other diseases. It appears on older leaves as brown spots with purple centers. It happens when the weather is warm and wet.

5. Pink root, white rot and basal rot are soil-born fungi.

Pink rot turns roots pink or maroon. It occurs in dry conditions. It causes roots to die, and individual plants to become weak and stunted. The bulb is smaller than usual.

White rot appears as yellowing and dieback of leaves. Bulbs and roots develop a white fluffy mold. It is less likely in warm weather. It causes bulbs to rot in storage.

Basal rot is the yellowing of leaves and tip dieback. It happens when soil is very warm, almost hot. It causes the whole plant to collapse and the roots to rot. When the plant is pulled out, the roots stay in the soil.

Insect Pests:

1. Onion root maggot is legless, white, and about 1/3 inch long. It attacks onion plants near the base and burrows into the bulb. It prefers light colored onions. Yellow onions are less likely to be affected than white. Red onions are least likely to be bothered by them. They cause the most damage to young plants.

Cover onions with a row cover as soon as they are planted. Also put a layer of sand or wood ashes on top of the soil. Can plant radishes as a catch crop (companion planting) to lure away the insects.

2. Onion thrips are thin, beige, active and very small. They scrape leaf surfaces and feed off the sap that flows from the injured spot. They can be a problem during hot, dry summers. Control with insecticidal soap.

Brassicaceae or Cruciferae Family (B. Medium level of difficulty to save seed or C. Hard to save seed.)

Genus- Armoracia
    Species- **Horseradish** (rusticana) Perennial, can propagate by dividing roots.

Genus- Brassica
    Species- **Mustard greens** (juncea) Outbreeding Annual, biennial or perennial depending on variety.
    Rutabaga (napus) Inbreeding Biennial
    Broccoli, brussels sprouts, cabbage, cauliflower, collards, kale, kohlrabi (oleracea) Outbreeding Biennial
    Broccoli raab, Chinese cabbage, turnip (rapa) Outbreeding Biennial

Genus- Crambe
    Species- **Sea kale** (maritima) Outbreeding Perennial

Genus- Eruca
    Species- **Roquette / Arugula** (sativa) Outbreeding Annual

Genus- Lepidium
    Species-   **Garden cress** (sativum) Outbreeding Annual
Genus- Raphanus
    Species-   **Radish** (sativus) Outbreeding Annual
Genus- Rorippa
    Species-   **Watercress** (nasturtium) Perennial usually propagated by leaf/stem cuttings.

Seed Saving: All members in this family can cross pollinate with each other if they are in the same species. Sometimes (rarely) they cross pollinate if in the same genus. This makes it harder to save pure seed. The flowers need insects for pollination. For pure seed, plants must be 1/2 mile away from all other in this family that have pollen at the same time. Or you can cage the plants and hand pollinate.

Harvest: Seedpods must mature fully on the plants before seed is collected. As the pods dry, they turn light brown. Seedpods tend to shatter. Seeds mature first near the bottom of the seed stalk, so collect over several weeks as each pod matures but before the pod shatters. Dry the stalks out of direct sun. After dry, break up pods and collect seed to store.

Brassicaceae Family Disease and Pests:

Brassica have some persistent diseases and pests. Problems can be greatly reduced by following a strict 4 year rotation for members of this family with 3 of those years planted with other families.

Insect Pests:

1. Aphids can spread viral diseases. They can be controlled by a short, sharp blast of water from the hose. Ladybirds, lady bugs, Chalcid wasps, hover flies, predatory midges and lacewings eat aphids.

Or squash aphids by hand. You can spray with Sevin. Organic methods are insecticidal soap, pyrethrum or rotenone. They over-winter on brassica stumps so remove these after harvest. Put reflective mulch such as aluminum foil under plants to deter aphids. (See **apple and bramble pests** in April Garden-Plant.)

2. Cabbage root fly maggots eat the root. Infected plants wilt and droop initially. Young plants die, but older plants may live though they do not grow properly. There will be tunnels and/or white maggots eating the roots. The roots blacken and die. In April and May the first generation of flies (looks like housefly) emerge that have over-wintered in the soil. The female flies lay eggs that turn into maggots. After 3 weeks the maggot is fully grown and goes a short distance into the soil. A week later, another generation emerges.

Starting seedlings in sterile soil in pots gives them a good start. (See **Seeds and Seedlings** in "Garden Tips".) But there is no insecticide to kill the flies.

For stemmed brassicas such as cabbage, cauliflower, and broccoli, the best control method is to use a circle or square of barrier material. Cut a piece of carpet underlay, milk carton, or cardboard about 6 inch square, cutting a line to the center. Place it around the stem.

To keep flies away from turnips/swedes, grow under horticultural cloth (row cover) such as Reemay.

Dig the soil over in fall and winter to expose the pupae to birds. All immature stages of the cabbage root fly have natural enemies in the soil such as small beetles. (See **How to Make Traps and Barriers for Crawling Insects** in "Plant Health".)

3. Cabbage white butterfly (cabbage whitefly) lays many eggs, and its voracious caterpillars make leaves skeletal in a few days.

An organic solution is a spray of Bt (bacillus thuringiensis), a parasitic bacterium that kills the caterpillars. Another solution is to cover the crop with row covers to prevent the butterfly from getting to the plants.

Regularly check under the leaves for yellow or white round eggs in clusters. Wash off or crush with your fingers. Use insecticidal soap that blocks the breathing holes of the pest.

Use pantyhose or nylon stocking to cover individual cabbage heads as soon as they start to form. It keeps the cabbage butterfly out. (See **white flies in Greenhouse Pests** in "Plant Health".)

4. Cabbage worms which includes 3 species: imported cabbage worms, cabbage loopers and diamond back moth worms. The adults lay eggs on the leaves. The larvae eat holes in the leaves. There are more problems with fall plantings than spring plantings.

    a. Imported cabbage worms are velvety green caterpillars. The moth is white and is seen during the day hovering over plants.

b. Cabbage loopers (inchworms) are smooth, light green caterpillars. They crawl by making a loop, much like an inch worm. The moth is brown and is most active at night.

c. Diamondback worms are small, pale, green caterpillars that are pointed on both ends. The moth is gray with diamond-shaped markings when the wings are closed. The damage looks like shot holes in the leaf.

Control for all cabbage worms: Handpick worms early morning and kill them. Sprinkle damp leaves with corn meal, rye flour, or a mixture of one part salt to two parts flour. Cabbage worms that eat this coating bloat and die.

Use nylon netting, fine screening or row covers to cover plants. Seal all the edges to the ground.

Parasitic (non-stinging) wasps parasitize the eggs of imported cabbage worms. They can be attracted to your yard by flowers that look like daisies.

Organic pesticides with Rotenone kill cabbage worms.

5. Cutworms feed on roots and kill seedlings by wrapping themselves around the base of the plant stem and chewing right through it. They have smooth brownish or gray bodies. They are about the size of a nickel. They curl up into a defensive position when disturbed.

Put foil, thick paper or cardboard collars around the base of the plants at planting time. (For image see tomatoes in March Greenhouse.)

Go out at twilight with a flashlight and kill cutworms as they emerge from the soil. Toss the worms into a bucket of soapy water to kill them. You can add any nocturnal slugs or snails too.

During the day dig around damaged plants and destroy the cutworm.

Plant a trap crop of sunflower, a favored host, around the garden. Destroy cutworms on it every day.

Bacillus Thuringiensis or BT, is an organic, biological control for all types of caterpillars. Apply to the soil around the plants, rather than to leaves. Reapply after it rains.

Beneficial parasitic nematodes live in the soil and like moist conditions to help them migrate towards their prey. Steinernema carpocapsae and S. feltiaeare are 2 commercially available nematodes that control cutworms.

Birds, ants and parasitic wasps prey on cutworms. Plant sweet alyssum flowers to attract parasitic wasps and flies. (See **How to Make Traps and Barriers for Crawling Insects** in "Plant Health".)

6. Flea beetles are the worst on radishes but can affect turnips and swedes too. (It also likes eggplants.) It is a small beetle that jumps when disturbed. It eats holes (looking like shot holes) in the leaves.

Spray with an organic pesticide that contains rotenone. Or use kaolin clay (Surround brand), garlic or pepper sprays. A non-organic pesticide to use is Sevin. You can try spinosad (Entrust). Also try adding beneficial nematodes to your soil. Wood ashes deter them.

For non-chemical control, row covers are very effective at preventing beetles from reaching the crop, if it is grown in rotation following a non-susceptible crop. It should be placed on as soon as seeds or seedlings are planted. (See **potato pests** in April Garden-Plant.)

Brassicaceae Family Fungal Diseases:

1. Club root (clubroot) is a soil-borne fungal disease that is very serious. If soil becomes infected, it can remain in your garden for many years. It is easily spread by shoes and clothing. The first sign is the wilting of plants, especially in dry weather. Plants do not develop well. Look at the roots. There are swellings and roots that look knobby.

There are no chemical controls to kill this fungus. However, there are varieties that are more resistant to it.

If you have club root, remove the plants, roots and all. Do not put in compost. Take it to a quarantine area in an unused part of your farm or burn.

A great advantage in growing brassicas from seed is that you do not bring the disease into your garden by nursery-bought seedlings. Start from seed in potting soil that is sterile. Add a small amount of lime to the potting soil. (For information about **potting mixes**, see the section "Garden Tips".)

Clubroot thrives best in acid, wet soils. Raise the soil pH to at least 7.5, and 8.5 is good. Before transplanting outside, dig a hole and dust with lime.

Chenopodiaceae Family (C. Hard to save seed.)

Genus- Beta
    Species- **Garden beet, sugar beet, mangel, Swiss chard** (vulgaris) Outbreeding Biennial

Genus- Chenopodium
    Species- **Lambs quarters / pigweed** (album) Outbreeding Annual
               Good king henry (bonus-henricus) Outbreeding Perennial, can propagate by dividing roots.
                Beetberry (capitatum) Outbreeding Annual
                Quinoa (quinoa) Annual (unknown whether out or inbreeding)

Genus- Spinacia
    Species- **Spinach** (oleracea) Outbreeding Annual

Seed Saving: Beet and Swiss chard seed is actually several seeds in one woody covering. Plants in this family are wind pollinated. Pollen can travel up to 5 miles. Beets and spinach can be bagged to maintain seed purity. Spinach and plants in other families such as broccoli, cabbage, carrot, cauliflower, celery, kohlrabi, leeks, lettuce, onion, turnip also have very small seeds. Pollen can sometimes get through bags.

Seeds are harvested off the plants when dry. Then if needed, can dry in a covered, airy location.

Disease and Pests:
Plants in the chenopodaceae family are relatively trouble free although some bolt (form flowers) in hot weather.

Insect Pests:
    1. Aphids (See Brassicaceae Family.)
    2. Beet and spinach/chard leaf miner is the larvae of a small fly. It lays white eggs on the underside of leaves. The larvae leave trails on the leaves. It has little effect on the root but it can effect the leaves.

To control, cover plants with fine netting, cheesecloth or floating row cover (Reemay) to protect them from adult flies. Remove for pollination to take place. Kill eggs and insects by hand and destroy infested leaves. Keep well weeded. Encourage parasitic wasps by planting dill and yarrow. In severe cases use neem oil. (See Solanaceae Family.)

Fungal or Viral Diseases:
    1. Cercospora leaf spot. When watering try not to get the leaves wet. Water in the morning so leaves can dry. Keep plants far enough apart so good air circulation. Keep well weeded. In fall, get rid of all diseased leaves and fruit. Crop rotation reduces or eliminates this disease.
    2. Downy mildew can be a problem on young spinach when soil conditions are too moist. (See Amaryllidaceae family.)
    3. Scab is any of several bacterial or fungal diseases that form crusty lesions on fruit, tuber, leaf, or stem.

When watering try not to get the leaves wet. Water in the morning so leaves can dry. Keep plants far enough apart so good air circulation. (See **potato diseases** in April Garden-Plant.)

Compositae Family (thistle) (C. Most hard to save seed.)

Genus- Arctium
    Species- **Japanese burdock or gobo** (lappa) Inbreeding Biennial

Genus- Chrysanthemum
    Species- **Endive, escarole** (endivia) Inbreeding, usually Biennial but sometimes Annual.
                Chicory (intybus) Outbreeding Biennial
                Chrysanthemum Flower (cinerariaefolium, coccineum and 28 others) Outbreeding Annual

Genus- Cynara
    Species- **Cardoon** (cardunculus) Inbreeding Perennial
                Globe artichoke (scolymus) Inbreeding Perennial

Genus- Helianthus
    Species- **Sunflower** (annuus) Outbreeding Annual
                  Jerusalem artichoke (tuberosus) All seed is sterile. Propagated by tubers.

Genus- Lactuca
    Species- **Lettuce** (sativa) Inbreeding Annual

Genus- Scorzonera
    Species- **Black salsify** (hispanica) Inbreeding Biennial

Genus- Tragopogon
    Species- **Salsify** (porrifolius) Inbreeding Biennial

Seed Saving: Most plants in this family are self pollinating but many need insect pollination. Flowers are produced on a seed stalk. Plants can be caged and hand pollinated. Do not save seed from biennial plants in this family if they go to seed the first year. Seeds are harvested off the plants when dry. Then if needed, can dry in a covered, airy location.

Fungal and bacterial diseases can be greatly reduced by good sanitation and rotating crops every 4 years.
    1. **Downy mildew** is a fungus. (See Amaryllidaceae family.)

Insect Pests:
    1. **Aphids** (See Brassicaceae Family.)
    2. **Armyworms** feed on plants at night. When the food supply is gone, they move en masse to a new area. They can destroy an entire plant in one evening. They are 1 1/2 inch caterpillars. They are pale green when first hatched, then change to olive green with a white stripe. They turn into moths that are gray-brown with a white dot on the wing.

    Look for them on the undersides of leaves and on tender new growth. Rake up leaf debris to eliminate daytime hiding places. Encourage birds to visit your garden. Attract predatory wasps that eat armyworms by planting dill, fennel, coreopsis and brightly colored flowers. Use horticultural oil in July to kill the eggs of second-generation armyworms.

    Spray Bacillus thuringiensis, an organic control for caterpillars, in late afternoon or early evening. Spray during the winter with a dormant-season oil spray to head off recurring infestations.
    3. **Cutworms** (See Brassicaceae Family.)
    4. **Cabbage white butterfly (cabbage whitefly, cabbage worms)** (See Brassicaceae Family.)
    5. **Loopers (one type of cabbage worm)**. (See Brassicaceae family.)
    6. **Slugs and Snails** can be controlled by putting wood ashes, sand, small gravel, diatomaceous earth, or limestone around the plants. (For how to make **slug and snail traps**, see "Plant Health" section.)
    7. **Tarnished Plant Bug** (Lygus lineolaris) is one of the most common garden pests in the United States. They are 1/4 inch long, oval shaped and have black, yellow and brown markings. It has 2-5 generations per year. They overwinter in plant debris so keep good garden sanitation.

    They are hard to find so you can use unbaited, nonreflective, white sticky boards hung low in fruit trees to trap them to see how many you have. They suck sap from stems, buds and fruit, and inject a toxin that deforms flowers and blackens terminal shoots.

    Ladybird beetles, spiders, and parasitic wasps eat them. Natural Insecticides such as sabadilla dust, rotenone and pyrethrum help control them. (See **Sprays and Dusts** in "Plant Health".)

Convolvulaceae Family (Propagated by tubers, not seeds.)

Genus- Ipomoea
    Species- **Sweet Potato** (batatas) Perennial propagated vegetatively by sprouts.

Yams are in the Dioscoreaceae family.

Disease and Pests:
To prevent diseases, plant varieties with multiple resistance. Use certified plants. Use crop rotation.
    1. **Mice, voles and moles** can be a problem. Use traps in tunnels and at entrances. (For **how to make traps**, see September Farm Animals.)

2. Root knot nematodes leave small necrotic lesions in the sweet potato near the top. Use the resistant variety Jewel.

3. Sweet potato scurf is a soil fungus that causes black, necrotic scab which decays rapidly. Control by growing sweet potatoes in acid soil.

   **Cucurbitaceae Family** (C. Hard to save seed.)

Genus- Citrullus
    Species- **Watermelon** (lanatus) Outbreeding Annual

Genus- Cucumis
    Species- **Cantaloupe, casaba, honeydew, muskmelon** (melo) Outbreeding Annual
        Most cucumbers (sativus) Outbreeding Annual
        Teasel Gourd (dipsaceus) Outbreeding Annual

Genus- Cucurbita
    Species- **Some squash- banana, buttercup, hubbard, turban; Some pumpkin** (maxima) Outbreeding Annual
        Some squash- white cushaw, silver gourd; Some pumpkin (mixta=argyrosperma) Outbreeding Annual
        Some squash- butternut, cheese, golden cushaw; Some pumpkin (moschata) Outbreeding Annual
        Some squash- acorn, crookneck, scallop, striped & warted gourds, spaghetti, zucchini; Some pumpkin (pepo) Outbreeding Annual

Genus- Lagenaria
    Species- **Hard shelled gourd** (siceraria) Outbreeding Annual

Genus- Luffa
    Species- **Angled luffa gourd** (acutangula) Outbreeding Annual
        Smooth luffa gourd (aegyptiaca) Outbreeding Annual

Seed Saving: Plants in this family have vines with tendrils and alternate leaves. Most like warm climates. All are insect pollinated.

For plants in most families, cross pollination usually occurs only among members in the same species. However, **some cross pollination between species occurs in the genus Cucurbita** among pumpkins, squash and gourds. Usually even among Cucurbita the species will not cross but sometimes they do. This is only an issue if you are saving seed. If you are not saving seed and just eating them, cross pollination does not matter.

Try to plant them as far from each other as possible. Most of the time 500 feet is enough distance to prevent crossing. The safest way to save pure seed is to hand pollinate. There is disagreement about which species will cross with each other but most say:
    C. pepo will cross with **C. mixta** and **C. moschata**. **C. pepo** will not cross with **C. maxima**.
    C. maxima will cross with **C. moschata**. **C. maxima** will not cross with **C. mixta**.
    Cross pollination does not occur between melons, cucumbers or other species. (For more about **Cucurbita pollination**, see winter squash in May Garden-Plant.)

Sometimes plants must be caged and hand pollinated to prevent cross pollination. Each plant has male and female flowers. Female flowers sit on a small, immature fruit. Male flowers are attached only to a straight stem. Pollen from the male flower is put on the stigma of the female flower. Take pollen from same and different plants of the same variety.

Flowers that are about to open show some color along their seams and the tip of the flower is starting to come apart. Already opened flowers are wilted. Do not pollinate them. Check flowers morning and night. Keep flowers taped shut to prevent unwanted pollination.

Harvest: Fruit must be grown to full maturity before collecting seed. The best time to harvest is 20 days after full maturity (after prime time for eating). (To process seeds, see the fermentation process in "How to Save Seeds" section.)

Disease and Pests: (See **cabbage** in April Garden-Plant. See **melons and watermelon** in April Greenhouse.)

This family has many pests and diseases. It is important to do a 4 year crop rotation. Old leaves and fruit must be discarded and not allowed to rot in the bed. Check the plants for pests and diseases regularly.

Fungal, Bacterial or Viral Diseases:

1. Cucumber wilt. The bacterium that causes wilt in cucumber plants usually overwinters in the belly of the striped cucumber beetle. In the spring, the beetles feed on cucumbers. This spreads the bacteria. It produces blockages in the vascular (water) system.

To find out if you have cucumber wilt, cut the stem and squeeze both ends. A sticky sap oozes out. If you stick these ends back together and then pull them apart again, if it makes a rope-like connection between the two then they have the bacteria. Once cucumbers have wilt there is no saving them.

To control wilt, you have to control the **striped cucumber beetle** (see below). You can use Sevin which must be applied frequently. Or use row cover cloth to keep beetles off the plants.

2. Downy mildew (See Amaryllidaceae family.)

3. Gummy stem blight is a fungus. A wide range of leaf (foliar) symptoms occurs which make diagnosis difficult. There may be a water-soaked lesion on the leaf, dying scorched leaves, or circular lesions.

Cover the plants with a row cover immediately after planting. Consistent applications of neem oil diluted with water can be effective against fungal diseases, but do not spray if it is hot.

4. Mosaic virus. (See Leguminosae Family Disease and Pests.)

5. Powdery mildew is a common disease on many types of plants. Different species of fungi cause the disease on different plants. All infections have a powdery white to gray fungal growth on leaves, stems and heads. Infection can occur on dry leaves. Warm temperatures and shade encourage the fungus to grow.

It usually grows on leaves and stems rather than on the vegetable itself. However, it can affect the flavor of melons and squash and reduce yield. Woody species such as grapes, fruit trees and roses are more badly affected. New growth is often distorted. The young fruit of apples and grapes can develop rough skin.

Select powdery mildew resistant varieties. Plant in full sunlight in a well-drained area. Make sure plants have good air flow. It thrives where lots of nitrogen has been used because high nitrogen promotes tender leaf formation.

Prune off infected parts of plants. If the infestations are severe, remove and destroy entire plants. Disinfect pruning tools in a solution of one part bleach to 9 parts water. Water plants in the morning to give time to dry off.

Organic sprays: Sulfur is highly effective against powdery mildew if used regularly with at least 7 to 14 days between applications. Garlic contains high levels of sulfur and a few cloves crushed in water can be used to make a homemade spray. Begin at the first sign of the disease. Sulfur can damage some squash and melon varieties.

Another way is to spray once a week with a solution of baking soda. Baking soda increases the surface pH of the leaf making it hard for powdery mildew to grow. Be sure to spray the undersides of leaves as well.

Recipe for making spray: 1 teaspoon baking soda, 1 quart water, a few drops of liquid soap. (See **apple trees** in April Garden-Plant. For more about **sprays**, see "Plant Health " section.)

Insect Pests:

1. Spotted and striped cucumber beetles eat plants but their worst threat is they carry bacterial diseases like cucumber wilt (see above) and squash mosaic virus (see above). They can increase the incidence of powdery mildew, black rot and fusarium wilt. They also damage plants directly by feeding on roots, stems, leaves and fruits.

Crop rotation does not control cucumber beetles since the beetles migrate.

Floating row covers exclude cucumber beetles and squash bugs during the seedling stage. Row covers are removed when flowering begins to allow for insect pollination.

Applying botanical pesticides provides season-long protection after row covers are removed. Use yellow sticky traps to catch beetles. (For **making sticky traps**, see April Garden-Maintenance.)

Predators and parasites that prey on cucumber beetles include hunting spiders, web-weaving spiders, soldier beetles, carabid ground beetles, tachinid flies, braconid wasps, bats, entomopathogenic fungi and nematodes.

2. Squash bugs feed on the plant and inject toxic substances making the plant wilt and die. Squash bug eggs are laid on the underside of leaves in tidy clusters. They are brick red. Remove them.

Protect young plants by using a floating row cover as soon as they are planted. Rotenone and pyrethrin offer some squash bug control.

3. Squash vine borer has clear wings and an orange and black abdomen. It lays eggs on the stem of the plant. Its larva makes a hole in the stem and then tunnels inside. The leaves die and the plant is weakened. A sawdust-like residue is found at its entry point.

Individual borers can be removed by slitting the stem lengthwise at the entry hole. Bury the stem so the vine produces roots there. Rotenone at the base of plants gives some protection. Row covers (Reemay) help. Row covers have to be removed once blooming begins so they pollinate.

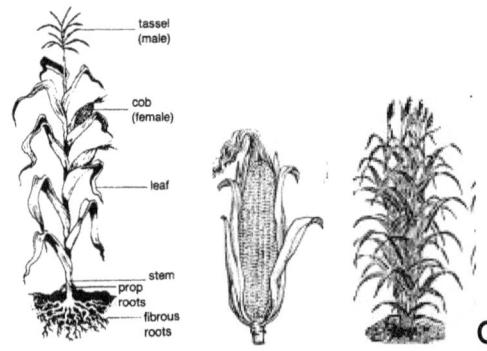

Gramineae / Poaceae Family (B. Medium level difficulty to save seed.)

Genus- Sorghum
    Species-   **Sorghum-grain and sweet, broom corn** (bicolor) Inbreeding Annual

Genus- Zea
    Species-   **Corn or Maize** (mays) Outbreeding Annual

Seed Saving Sorghum: All **sorghums** are self pollinating with some insect pollination. For absolutely pure seed, put corn bags over the tassels as soon as they emerge. Let seeds dry on the plant. Then dry some more in airy location. Seeds can be hit with a hammer to test for dryness. If they shatter, they are ready for storage. Sorghum seeds remain viable for 4 years in good storage conditions. (For **sorghum** see May Garden-Plant.)

Seed Saving Corn is wind pollinated and **easily cross pollinates with other corn varieties**. Pollen is produced by the tassel at the top of each stalk of corn. This is the male part of the plant. The ears and silks are the female parts of the plant. Wind blows pollen from the tassels to the silks. Each silk pollinates 1 kernel of corn.

To maintain pure seed, corn needs to be 2 miles from any other corn that is also producing pollen. Growing corn in blocks rather than long rows helps reduce the amount of outside, unwanted pollination. You should grow at least 200 plants to have good pollination and genetic diversity. Corn is particularly **susceptible to inbreeding from too few plants in a population.** Remove any plants that are not true to type.

Corn can be hand pollinated to have absolutely pure seed. Pollen from the tassel is collected and sprinkled on the silks of corn of another plant. Corn tassel bags are put on the ears before the silks appear and on the tassels when first anthers (part of stamen that has pollen) appear. This is done for 2-3 days. (For **corn** see May Garden-Plant.)

Harvest: (For **corn harvesting**, see September/October Garden-Harvest.) Kernels are removed after the cob and kernels are dry. Kernels should be from at least 25-50 ears of corn from different plants.

Sweet corn remains 50% viable for 3 years if stored in cool, dry, dark conditions. Flint, dent and popcorn remains viable for 5-10 years under good conditions.

Gramineae Family Disease and Pests: (See "Plant Health".)

Corn earworms are the most common pest. Insects and disease are not usually a problem until ears begin to form.

Insect Pests:

1. Corn borers (stem borers) feed on the leaves and internal parts of stalks. It bores into a wide variety of plants with large stems, stalks, and fruits such as bell peppers, snap/lima beans, potato vines, and tomatoes.

You can see insect holes bored into the stalk and droppings on the leaves. Corn borer control is hard because sprays are effective only during the 2-3 day period after eggs hatch and before larvae bore into the stalks. Eggs are white and one-half the size of a pinhead. They are laid in masses that overlap like fish scales.

Uprooting, shredding, and burying infected stalks is the most successful method of destroying the corn borer because it kills the wintering larvae. Removing the borer by hand is the oldest remedy. Split the stem a little below the entrance hole and pick out the worm.

Other methods include attracting the corn borer moth to light traps. Or use parasitic insects such as the ladybug. Spray natural pesticide BTK on the undersides of leaves and into tips of ears after silks wilt.

2. Corn earworms (silkworms) feed on the tip of the ear. To control corn earworms, squirt mineral oil into the top of the ear of corn. This drowns the eggs that have been laid there. Apply it after pollination.

Pirate bugs and ladybug beetles are a good natural defense against earworms. Sevin dust is very effective when applied directly on the silk, or dusted in the air. Remove affected corn stalks. (See **corn** in May Garden-Plant.)

3. Japanese beetles congregate on tip of the ear and feed on silks. This reduces pollination and yields. Eggs laid mid-summer hatch into larvae that eat plants and roots before over-wintering. Grubs start eating plants in July. Adult activity is most common mid-July through August. (See **How to Make Traps for Flying Insects** in "Plant Health.")

Plant varieties of flowers that attract Japanese beetles, but kills when ingested by the insects. This includes white geranium, red and dwarf buckeye, mirabilis and larkspur.

Dry weather kills the eggs.

Apply Neem oil to plants and soil as a repellent once weekly during July and August.

Add a squirt of hand or dish washing liquid to water in a bucket early in the morning. Hand-pick the insects from your plants and drop them in the bucket.

Pour 1/4 cup hot water into a plastic milk jug. Add 1/4 cup sugar, cap the jug and shake. Smash up half of a ripe banana or other fruit, and put it in the jug. Shake well. Cover half the spout with tape to reduce the hole size. Leave the cap off. Place it away from your food plants. Beetles are attracted to the smell. They get in the jug but can not get out. (See **bramble pests** in April Garden-Plant.)

Fungal, Bacterial and Viral Diseases:

1. Corn fungus develops at the ear. It is a blackish purple glob that grows in rainy weather. Remove and destroy the plant. Do not put it in your compost pile.

2. Corn smut is a fungus that is one of the most common diseases to affect corn. Sweet corn and popcorn are more susceptible, but it can affect dent corn too. Corn smut is a mass of greasy, powdery black spores (gall) surrounded by white, smooth corn tissues. It usually forms on ears, nodes and tassels. It can be controlled by removing the galls and destroying them by burning or burying.

Insecticides used to control corn borers can prevent many instances of corn smut. In the fall stalks should be plowed under where the spores will die over winter.

 **Labiatae Family** (A. Easy to save seed.)

Genus- Ocimum
    Species-   **Basil** (basilicum) Inbreeding Annual

    Seed Saving: Basil is an inbreeding plant. Insects sometimes pollinate it. Keeping varieties 150 feet away from each other should keep seed pure. Seeds mature starting from bottom of the stem to the top. When the bottom seeds start to turn brown, the seed stalk can be cut and put in an airy place not in direct sun.
    Harvest: Each flower contains 4 seeds that are hard to remove from the pod. You can rub over a wire screen to release seeds. Seeds remain viable for 5 years when stored in a dry, cool, dark location.

    Labiatae Family Disease and Pests:
    Basil is sometimes bothered by aphids, cutworms, grasshoppers, Japanese beetles, nematodes, slugs, spider mites, and whiteflies. (See Brassicaceae and Gramineae Families.)
    1. Fusarium wilt is an infection caused by fungus. It stunts growth and causes wilting. One way to prevent Fusarium wilt is to plant resistant varieties like the cultivar Nufar. Fusarium can infect seeds and live in soil. It also attacks potatoes, tomatoes, eggplant, and pepper.
    Sterilize garden tools with one part bleach to 9 parts water. Control garden insects such as cucumber beetles, which spread the disease. (See **Cucurbitaceae Family** above.) Remove all weeds from the garden. The biological fungicide Mycostop controls wilt caused by Fusarium.
    Preventative measures include crop rotation, soil solarization, and proper sanitation, aeration and drainage. If basil does succumb, pull and discard the infected plant. Do not throw it in the compost pile. (See **Leguminosae Family** below.)
    To solarize soil, leave a clear plastic tarp on the soil surface for 4-6 weeks during the hottest part of the year. It reduces many soil pests including nematodes, fungi, insects, weeds and weed seeds. Good bugs are killed too. (For more about soil solarization, see **How to Make Potting Mixes** in "Garden Tips".)

 **Leguminosae or Fabaceae Family** (A. Easy to save seed.)

Genus- Arachis
    Species-   **Peanut** (hypogaea) Inbreeding Annual
Genus- Cicer
    Species-   **Garbanzo / chick pea** (arietinum) Inbreeding Annual
Genus- Glycine
    Species-   **Soybean** (max) Inbreeding Annual
Genus- Lens
    Species-   **Lentil** (culinaris) Inbreeding Annual
Genus- Phaseolus
    Species-   **Runner bean** (coccineus) Some inbreeding, some outbreeding. Annual
                Lima / butter bean (lunatus) Inbreeding Annual
                Common bean (vulgaris) Inbreeding Annual
Genus- Pisum
    Species-   **Garden pea, edible pod pea** (sativum) Inbreeding Annual
Genus- Vicia
    Species-   **Fava / broad / bell bean** (faba) Inbreeding and outbreeding, Annual
Genus- Vigna
    Species-   **Moth bean** (aconitifolia) Inbreeding Annual
                Adzuki bean (angularis) Inbreeding Annual
                Mung bean (radiata) Inbreeding Annual
    Species-   **Cow pea** (unguiculata) Inbreeding Annual

Seed Saving: Legumes are **self pollinating most of the time** with occasional insect pollination. The larger the flower, the more likely it is to be pollinated by an insect. But if there are more desirable nectar sources for insects, they tend to avoid the food plants.

Harvest: Save seed from healthy plants with good bean/pea or nut production. Seeds in this family are dried on the plant. If needed, they can be dried more in a dry, airy location. Very dry pods split along the sides, and seeds can be removed by hand. It is **easier to split pod** if done when somewhat dry rather than waiting months until it is very hard.

Storing: Stored seeds are prone to damage by bean weevils. To kill weevils, put the seeds in a container and freeze the dry seeds for 5 days. When you take it out of freezer, let it come to room temperature with the container still closed so condensation does not collect on the seeds.

To make sure seeds are dry enough to freeze (so they do not rupture), put some seeds on a hard surface and hit with a hammer. Seeds that shatter are dry enough to freeze.

Seeds in this family will store for 4-5 years if kept in good storage conditions.

Fungal, Bacterial or Viral Diseases: (See "Plant Health" section.)

To prevent fungal diseases, plant in well drained soil that is free from disease. Remove infected plants. If grown in a container, wash and bleach it before adding a new plant. Clean dead material from the garden such as leaves and rotted fruit/vegetables because virus and fungi overwinter on plant debris. Apply fungicide at the beginning of the growing season before legumes are planted.

Trellising provides more sun and air to plants, so it is a good preventive measure. Plant varieties with resistance. Virus diseases can not be controlled once plant is infected. (For **powdery mildew**, see Cucurbitaceae Family.)

1. Mosaic virus is caused by a variety of viruses. One type is **enation virus**. It is spread by various methods. Mosaic virus damage first appears as green leaves that are mottled or distorted. Often these leaves are curled upward or growth has been stunted. Usually these leaves have yellowish spots. Aphids and cucumber beetles spread the disease. Mosaic virus can badly affect the taste of fruits and vegetables.

There is no chemical control for mosaic virus. Plants must be removed and destroyed promptly if infected. To control spread of the disease by cucumber beetles and aphids, you need to control these insects with a diazinon containing insecticide repeating the application in seven day intervals. (For **aphids** see Brassicaceae Family. For **cucumber beetles,** see Cucurbitaceae Family.)

2. Leaf blight is caused by fungi pathogens, and usually attacks during wet weather when the air is humid. Help prevent it by not letting fruits and vegetables rot on the ground which attracts fungi.

3. Vascular wilt is one of the more serious fungal diseases. The plants ability to absorb water and nutrients is severely damaged. This makes the plant appear dry and the shoots fall off. Eventually the plant dies.

Verticillium wilt and **Fusarium wilt** are 2 types of wilt. There are no cures for these soil borne diseases. Sulfur applications applied weekly once symptoms have been noticed may help but prevention is the key. (See **Sprays and Dusts** in "Plant Health".) Choose resistant varieties. Remove diseased growth or entire plant. Sterilize tools. Solarize the soil before planting again in the same location. Crop rotation does not work well because many crops are susceptible to the disease fungi. (For **Fusarium wilt**, see Labiatae Family. For **soil sterilization**, see Potting Mixes in "Garden Tips".)

4. Root Rot: Fungi attack the roots, crown or stem of legumes causing stunted growth, yellowing of fruit and leaves, wilting, leaf drop and sometimes death. Black, mushy roots indicate root rot. In early stages, outer roots rot while the main root remains intact. Root rot thrives in wet, soggy soil that does not drain properly. Do not overwater. Make sure plants have good air flow.

5. Phytophthora root rot is a fungus that causes significant losses during seedling establishment in wet, poorly drained soils. It can be controlled by selecting a well drained site, planting resistant varieties, and following sound plant hygiene. (For **preventing damping off in seedlings**, see "Garden Tips".)

Insect Pests:

1. Aphids (See Brassicaceae Family.)

2. Mexican bean beetles can be a problem. They look like large ladybugs but are a caramel color. Their yellow grubs eat the leaves. Pick off and discard bean beetles. Bt can be used to control them at the grub stage.

 **Liliaceae Family** (B. Medium level of difficulty to save seed.)

Genus- Asparagus
    Species- **Asparagus** (officinalis) Outbreeding Perennial

Genus- Hemerocallis
    Species- **Day Lily** (fulva, sometimes put in xanthorrhoeaceae family) Outbreeding Perennial

Saving Seed: Asparagus is an outbreeding plant. Flowers are either male or female. They are pollinated by insects. Should be separated by 2 miles from varieties in same family that currently have pollen. Caging can also be used.

The best seed is produced if the edible part of the plant is harvested moderately. Female plants produce red, round, 3/8 inch berries that have 6 seeds in each berry. Ripe berries should be collected before they drop to the ground. Asparagus seeds remain viable for 5 years under good storage conditions.

Insect Pests:
    Aphids. (See Brassicaceae Family.)

Fungal, Bacterial and Viral Diseases:

1. Asparagus rust (Puccinia asparagi) is a fungus. It can be a problem if cultivars are not rust-resistant. It causes leaves to fall off and death of ferns, reducing yields and increasing incidence of root or crown diseases. Symptoms usually occur after harvest season. Small yellow-to-orange spots first appear on the ferns. In the second stage, dusty brick-red pustules appear on both the shoots and ferns; later they turn black. Rust is most severe during times of heavy rain.

Increase planting distance between rows and orienting rows toward the prevailing summer winds. Have good air flow between plants. Remove any infected ferns and burn them. Jersey Giant and Martha Washington varieties are rust resistant.

2. Bacterial Soft Rot (Crown Rot) is caused by a bacteria. The entire plant can quickly become infected and turn into a bad smelling, mushy mess. Your soils might not be infected but newly purchased plants that are infected might bring the problem to your garden. Inspect all plants before planting them. If you suspect a plant might have something wrong, place it in a quarantine bed to observe.

One way to way to prevent problems is to soak all new plants in a solution of 25% chlorine bleach to 75% water for 20 minutes then rinse off before planting.

If you suspect a plant has soft rot, dig the plant up immediately and cut off the infected parts. Soak the rest of the plant in the bleach solution. Treat any cut areas with garden sulphur. Sterilize the soil before planting anything back in the same area. (For **soil sterilization**, see Labiatae Family and Potting Mixes in "Garden Tips".)

3. Fusarium wilt. (See Labiatae Family.)

4. Needle blight (Cercospora asparagi) kills leaves during periods of heavy rain and warm temperatures. A bad infection severely reduces yield the following season. Symptoms appear in June, when affected ferns develop buff to gray, somewhat round spots surrounded by a thin purple band.

Make sure plants have enough air space between them. Remove any diseased ferns.

 **Malvaceae Family** (B. Medium level of difficulty to save seed.)

Genus- Abelmoschus
    Species- **Okra** (esculentus) Inbreeding Annual

Seed Saving: Okra is an inbreeding plant that is self pollinating. It can be pollinated by insects especially since the flowers are large and attractive. Different varieties in the same family need to be separated by 1 mile when both have pollen at the same time. For pure seed the plants can be caged or flowers bagged.

Harvest still green but fully mature pods. Let dry in airy place out of sunlight until they split open. Then break each pod open to release the seeds. Seeds remain viable at 50% for 5 years under good storage conditions.

Malvaceae Family Disease and Pests:
Okra (gumbo) is not usually bothered by disease and pests.

1. Ants can be very destructive to okra pods. They want the moisture and sugar. To control ants, spread diatomaceous earth around the plants. Diatomaceous earth is a non-toxic mineral derived from fossilized shell remains of algae. It is approved as an organic insecticide for crawling insects. (See **fertilizers** in "Garden Tips".)

Ants are always an indication of a high population of aphids. To control aphids, see below.

2. Aphids (See Brassicaceae Family.)

3. Cabbage worms which includes 3 species: imported cabbage worms, cabbage loopers and diamond back moth worms. (See Brassicaceae Family.)

4. Powdery Mildew: (See Cucurbitaceae family.)

5. White Flies: (See Brassicaeae family under "**Cabbage White Flies**". See **Greenhouse Pests** in "Plant Health".)

Polygonaceae Family (Best propagated by root cuttings or dividing plants.)

Genus- Rheum
Species- **Rhubarb** (rhabarbarum) Perennial propagated by root division.
Sorrel- Garden (acetosa) Perennial (in/out unknown) but can be propagated by dividing plant.
Sorrel- Mountain (alpinus) Perennial (in/out unknown) but can be propagated by dividing plant.
Sorrel- French (scutatus) Perennial (in/out unknown) but can be propagated by dividing plant.

Rhubarb seeds do not produce plants that are true to type so root cuttings are used instead.

Sorrel plants are easily divided so seed does not have to be saved. For saving seed the varieties will cross pollinate so use caging to keep seed pure.

Genus: Fagopyrum
Species- **Buckwheat** (esculentum) Outbreeding Annual

Fungal, Bacterial or Viral Diseases:

1. Botrytis blight or gray mold is a fungus disease which infects a wide array of plants. These fungus infections occur more in cool, rainy spring and summer weather around 60 degrees.

Look at any brown or spotted plant material to see if there are masses of silver-gray spores on the dead or dying tissue. These spores are readily liberated, and may appear as a dust coming off the plant. Some species of Botrytis form tiny black structures that may be seen on dead tissue in late summer.

The best way to manage this disease is by inspection and sanitation. Remove blighted flowers, leaves, or entire plants if infected at the base. It is best to do this when it is dry. Drip irrigation is best. Overhead watering is not recommended. To promote rapid drying of plants, space them to allow good air circulation.

In the fall remove plant debris, cut stalks at or below ground level, and destroy/discard plant debris.

Fungicide sprays may help protect plants from infections. Apply these when spring weather is continuously cool and wet or if Botrytis blight was a problem the previous year. Use fungicides with the active ingredients chlorothalonil, neem oil, Bacillus subitlis, or potassium bicarbonate. (See "Plant Health".)

2. Powdery Mildew (See Cucurbitaceae Family.)

3. Phytophthora root rot (See Leguminosae Family.)

 **Portulacaceae Family** (Readily self seeds.)

Genus- Claytonia
    Species-   **Miners lettuce** (parvifolia) Inbreeding Annual
Genus- Portulaca
    Species-   **Purslane** (oleracea) Inbreeding Annual

    Claytonia (miners lettuce) tends to be self seeding. It is an inbreeding plant that is self pollinating. Not much is known about true varieties. Seeds are hard to collect. Seed remains viable for 5 years when stored in good conditions.
    Purslane is an inbreeding plant and is self pollinating but is pollinated by insects too. Tends to be self seeding. Seeds retain viability for 5 years under good storage conditions.

    Portulacaceae Family Disease and Pests: Very few diseases or pests.

 **Solanaceae / Nightshade Family** (A. Easy to save seed.)

Genus- Capsicum
    Species-   **Sweet and chili pepper** (annuum) Inbreeding Perennial grown as Annual.
                 Tabasco, squash pepper (frutescens) Inbreeding Perennial grown as Annual.
Genus- Lycopersicon
    Species-   **Tomato** (lycopersicum) Inbreeding Annual that cross pollinates.
                 Currant tomato (pimpinellifolium) Inbreeding Annual that cross pollinates.
Genus- Nicotiana
    Species-   **Cultivated tobacco** (tabacum) Annual (in/outbreeding unknown but is self pollinating)
Genus- Physalis
    Species-   **Tomatillo / husk tomato** (ixocarpa) Inbreeding short-lived perennial grown as annual.
                 Cape gooseberry (peruviana) Inbreeding short-lived perennial grown as annual.
                 Wild tomatillo (philadephica) Inbreeding short-lived perennial grown as annual.
                 Strawberry tomato / dwarf cape gooseberry (pruinosa)
                    Inbreeding short-lived perennial grown as annual.
                 Downy ground cherry / yellow husk tomato (pubescens)
                    Inbreeding short-lived perennial grown as annual.
                 Purple ground cherry (subglabrata) Inbreeding short-lived perennial grown as annual.
Genus- Solanum
    Species-   **Garden huckleberry** (melanocerasum) Inbreeding Annual
                 Eggplant (melongena) Inbreeding perennial grown as annual.
                 Common nightshade (nigrum) Inbreeding Annual
                 Potato (tuberosum) Propagated by tuber cuttings.

    Seed Saving: All flowers of this family have 5 united or partially united petals. The flowers are **self pollinating** but insect pollination sometimes happens.
    Harvest: Seeds are harvested from fully ripe fruits. Put fruit in buckets where they are chopped and crushed to get out the seed. Water is added and then stirred vigorously. Good seed sinks to the bottom. Repeat several times. Then put through strainer. Lay seeds on a glass, plastic or ceramic surface to dry. Stir the seeds 2 times a day so all parts dry evenly.
    Never dry seeds in direct sunlight or in oven. Store dry seeds in an airtight container in cool, dark place or freezer.

This family has a lot of pests and diseases. Potential pests include aphids (see Brassicaceae Family), cabbage loopers (see Brassicaceae Family), Colorado potato beetles (see below), corn earworms (see Gramineae Family), cucumber beetles (see Cucurbitaceae Family), European corn borers (see Gramineae Family), leafhoppers (see below), leafminers (see below), potato tuber worms (see below), tomato hornworms (see below), pepper maggots, nematodes (see **Insects and Nematodes** in "Plant Health"), and flea beetles (see Brassicaceae Family).

Crop rotations of 4 years, with 1 year for members of the Solanaceae Family, and the other 3 years with other families are strongly recommended. Consistent watering is an effective disease deterrent for blossom drop and blossom end rot. Blossom end rot can be prevented by making sure the soil has enough calcium. Plant disease resistant varieties. (See **tomatoes** in June Garden-Plant.)

Insect Pests: (See **potatoes** in April Garden-Plant. See **tomatoes** in March Greenhouse.)

1. Colorado potato beetles are the worst potato pest. Adults are size of ladybugs but yellow with black stripes. Larvae are red grubs with 2 rows of black dots and a black head. Yellow eggs are laid on underside of leaves.

Controls include covering with a row cover as soon as it is planted, hand picking of insects and eggs, and/or spraying larvae with Bt. They attack all members of this family.

2. Leafhoppers are any species from the family Cicadellidae with 20,000 species. They are tiny and hop when disturbed. The nymphs and adults feed on sap of above-ground stems or leaves of plants. They overwinter in plant debris. They transmit viral diseases.

Signs of damage include stunted growth or leaves that are curled, stippled, or look burned. Can cover plants with row covers to keep them out. Predatory flies and parasitic wasps eat leafhoppers. (See **blueberries** in April Garden-Plant.)

3. Leafminers are the larvae of many different species of insect that eat leaves from the inside in a feeding tunnel. Most are moths (Lepidoptera), sawflies (Symphyta) and flies (Diptera).

They are hard to control with sprays since they are protected by the leaf. Spinosad, an organic insecticide, controls it when they eat it. Two or three applications may be needed in a season but do not spray on bees. Plant trap crops (companion planting) such as lambsquarter, columbine and velvetleaf that attracts leafminers and keeps them away from crops. (See Chenopodiaceae Family. See **apple tree pests** in April Garden-Plant. For **companion planting**, see "Garden Tips".)

4. Potato Tuber Worm/Moth (Tobacco Splitworm) turns into a small, slender moth with narrow, gray forewings and dark brown spots. The moths are most active at dawn and dusk. Eggs are placed on undersides of leaves. The larvae eat leaves and tubers causing papery, grayish blotches on leaves which become brown and brittle. Tuberworm injury is usually on the older, lower leaves. They also attack eggplant leaves.

They overwinter in the soil. Rotate crops with other families. Remove infested parts of plants or whole plants. Apply Sevin to foliage (leaves) for general insect control especially early in season. Bt may be used if applied at the right timing to kill the worms. It will not kill the moths.

5. Tomato hornworms are up to 4 inches long, green with white stripes and have a horn on the back end. They eat the leaves of tomatoes and other members of the family.

They are easily removed by hand or killed by spraying Bt. (Bt works for other worms and caterpillars as well.) Hornworms have natural predators such as the brachonid wasp. Rows of white eggs on a leaf should be left to provide a new generation of brachonid wasps.

6. Spotted and striped cucumber beetles (See Cucurbitaceae Family.)

Fungal, Bacterial and Viral Diseases: (See **potatoes** in April Garden-Plant. See **tomatoes** in March Greenhouse.)

Potential diseases include anthracnose, bacterial wilt, bacterial canker, black leg, blight (see Cucurbitaceae Family), blossom drop, blossom end rot (see tomatoes in June Garden-Plant), fruit rot, internal blackspot, mosaic virus (see Leguminosae Family), rhizoctonia, ring rot, scab (see Chenopodiaceae Family), and verticillium wilt (see Leguminosae Family and **bramble diseases** in April Garden-Plant). (For Botrytis leaf blight, see Amaryllidaceae.)

1. Late blight makes dark brown lesions on stems and leaves, and white fungal growth on lesions especially the underside of leaves, and brown watery pockets on fruit. It is worse in cold, wet weather. It survives over the winter only on living tissue. So remove all tomatoes, potatoes, and plants in the fall. Buy certified, disease-free potatoes.

Tetragoniaceae Family (A. Easy to save seed.)

Genus- Tetragonia
    Species- **New Zealand spinach** (tetragonioides) Inbreeding short-lived Perennial grown as annual.

New Zealand spinach is a vine that does not bolt (form flowers) readily in hot weather. Leaves are high in oxalic acid. It is an inbreeding plant and is self pollinating. Insect pollination is possible.

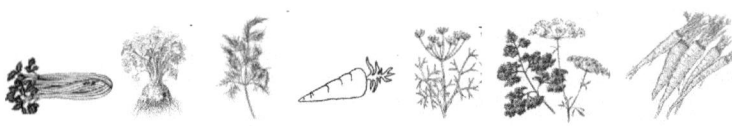
Umbelliferae or Apiaceae Family

(C. Hard to save seed.)

Genus- Apium
    Species- **Celery, celeriac** (graveolens) Outbreeding Biennial
Genus- Anethum
    Species- **Dill** (graveolens) Outbreeding Annual
Genus- Anthriscus
    Species- **Chervil** (cerefolium) Outbreeding Annual
Genus- Coriandrum
    Species- **Coriander** (sativum) Outbreeding Annual
Genus- Daucus
    Species- **Carrot** (carota) Outbreeding Biennial
Genus- Foeniculum
    Species- **Fennel** (vulgare) Outbreeding Annual
Genus- Pastinaca
    Species- **Parsnip** (sativa) Outbreeding Perennial grown as Annual.
Genus- Petroselinum
    Species- **Parsley, parsleyroot** (crispum) Outbreeding Biennial
Genus- Sium
    Species- **Skirret** (sisarum) Outbreeding Perennial

Seed Saving: Plants in this family have **umbrella-shaped flowers.** The main seed stalk forms from the primary umbel (flower). This has the best seed. Additional branches with flowers form below it but the seeds are not as good. The primary umbels mature first.

The flowers of plants in this family are very attractive to beneficial insects. So let some grow flowers if it will not interfere with seed saving plans.

Plants in this family that are grown for their roots are **biennials.** Flowers are pollinated by insects. Plants need to be separated by 3 miles from others in this family that are also shedding pollen. But a problem is wild plants in this family such as Queen Annes Lace (wild carrot) and wild fennel that can cross pollinate with all members of this family.

Caging and hand pollination can be used. It must be done every day for at least 2 weeks. The immature flowers can also be bagged before they open.

Insect pests include aphids (see Brassicaceae Family), carrot rust flies, carrot weevils, cutworms (see Brassicaceae Family), but this family is generally trouble free.

Plantings can be covered with row covers (Reemay) to prevent adult pests from laying eggs at the plants' bases. Crop rotation is the best preventative measure against disease. Adequate calcium and even moisture will help prevent black heart in celery. As with other crops, aphids carry diseases, so control them with insecticidal soap.

Ways to Improve Plant Health

There is a lot about diseases and pests in various families in the "Plant Families" section.

Topics in this Section
Diseases
- Nutrient Deficiency Symptoms
- Disease Symptoms
- Fungi Disease
- Bacterial Disease
- Viral Disease

Insect and Animal Pests
- Insects & Nematodes
- Common Insect Damage Symptoms
- Greenhouse Pests and Disease
- Products to Control Insects
- Large Animal Pests

Methods of Control
- Sprays and Dusts
- How to Make Traps for Flying Insects
- How to Make Traps and Barriers for Crawling Insects
- How to Make Traps and Deterrents for Slugs and Snails
- How to Use Floating Row Covers and Other Physical Crop Protectors
- How to Make Organic Insect or Pest Sprays
- How to Make Organic Fungicide Sprays and Solutions

(📖 Read "Organic Gardener's Handbook of Natural Pest and Disease Control" by Rodale, "Tuning in to Nature" by Philip S. Callahan, and "Bugs, Slugs and Other Thugs: Controlling Garden Pests Organically" by Rhonda Hart.)

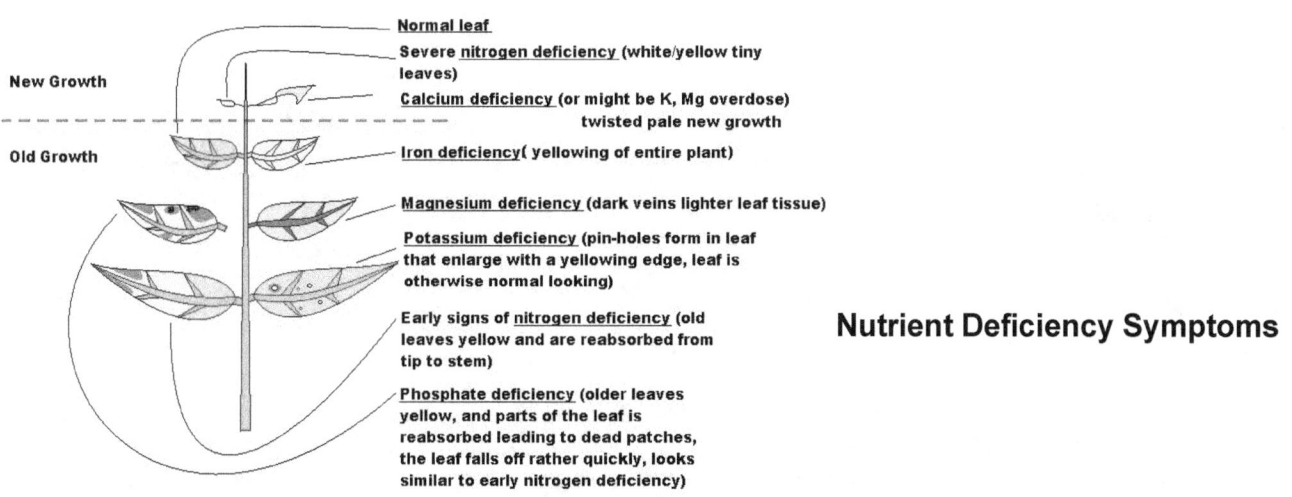

Nutrient Deficiency Symptoms

(**For information about fertilizers,** see "Garden Tips" section.)

Boron (B) *Cure with products containing borax or borate. It is poisonous to plants if apply too much.*
Poor stem and root growth. Buds at ends of branch die.

Calcium (Ca) *Not usually a problem. Deficiency is cured with lime, gypsum, and/or oyster shell.*
New leaves are twisted or hook shaped. The growing tip may die. May lead to blossom end rot in tomatoes, tip burn of cabbage, and brown/black heart in escarole and celery.

Plant Health p. 1

Copper (Cu) *Cure with products that have copper, cupric or cuprous such as copper sulfate (bluestone), copper chelate, or neutral copper.*
Stunted growth. Leaves become limp, curl or drop. Seed stalks are limp and bend over.

Iron (Fe) *Cure with chelated iron or iron sulfate.*
Leaves are yellow but veins are green.

Magnesium (Mg) *Cured with Epsom Salts (magnesium sulfate) and/or dolomite lime.*
Plants have slow growth and leaves turn pale yellow either everywhere or just on the edges. New growth may be yellow with dark spots.

Manganese (Mn) *Cure with manganese or manganous such as manganese sulfate.*
Growth slows. Young leaves turn pale yellow, starting between veins. Develops dark or dead spots. Leaves, shoots and fruit are small size. Failure to bloom.

Molybdenum (Mo) *Cure with products that have molybdate or molybdic such as sodium molybdate. Or use trace minerals such as kelp or azomite.*
Similar to nitrogen deficiency. Older leaves yellow. The rest of the leaves turn light green. Leaves become narrow and twisted.

Nitrogen (N) *Nitrogen is water soluble so gets leached away. Cured with manure, bloodmeal, and/or fish meal.*
Old leaves turn yellow. Younger leaves are light green. Stems may yellow and be spindly. Growth slows.

Phosphorus (P) *Absorption depends on soil pH, best is 6-7. Cure with greensand, bonemeal, rock phosphate.*
Small leaves have reddish-purple tint. Leaf tips look burnt. Old leaves almost black. Reduced fruits and seeds.

Potassium (K) *Cure with greensand, muriate or sulfate of potash, seaweed and/or wood ashes.*
Older leaves look scorched around the edges and/or wilted. Leaves are yellow between the leaf veins.

Sulfur (S) *Happens more in dry weather. Cure with products containing sulfate or pure sulfur.*
New growth turns pale yellow. Older growth stays green. Growth is stunted.

Zinc (Zn) *Is less available in soil with high pH. Cure with products that have zinc such as zinc sulfate. Or use trace minerals such as kelp or azomite.*
Yellow between veins of new growth. Leaves at end of branches form a rosette (rose shaped).

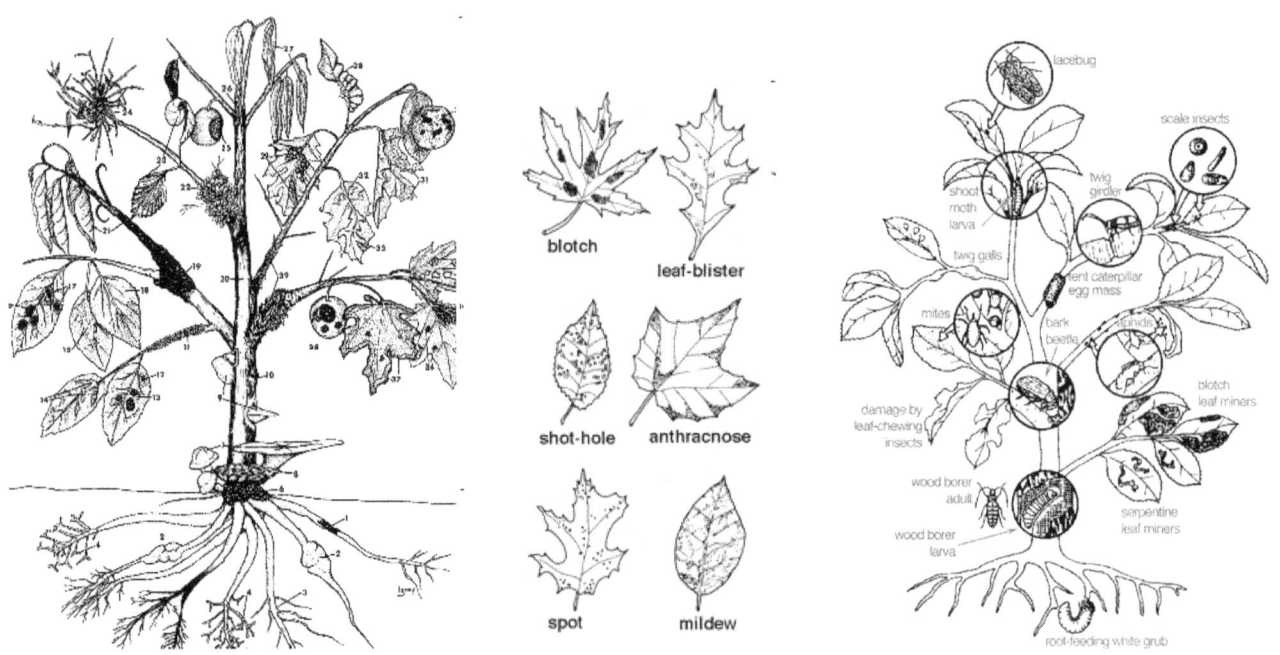

(Image on top left the diseases are: 1. root-lesion nematode; 2. root-knot nematode; 3. root pruning by nematodes; 4. stubby-root nematode injury; 5. root rot; 6. crown gall; 7. fruiting bodies of Armillaria root rot fungus; 8. fruiting body of Ganoderma wood and root rot fungus; 9. fruiting bodies of Fomes wood rot fungi; 10. trunk canker; 11. cedar-quince rust on hawthorn; 12. cedar-hawthorn rust; 13. cedar-apple rust; 14. mosaic; 15. downy mildew; 16. apple scab; 17. leaf spot; 18. powdery mildew; 19. black knot of plum and cherry; 20. wetwood (slime flux); 21. fire blight; 22. American mistletoe; 23. 2, 4-D injury; 24. witches' broom; 25. fruit rot (apple); 26. overwintering canker of fire blight; 27. wilt; 28. leaf curl or blister of peach, cherry, or plum; 29. leaf blister (oak); 30. sooty blotch and flyspeck of apple; 31. leaf blotch; 32. shothole; 33. anthracnose; 34. ringspot; 35. sooty mold; 36. tar spot; 37. leaf scorch; 38. apple scab on fruit; 39. twig & branch canker.)

vegetable troubles

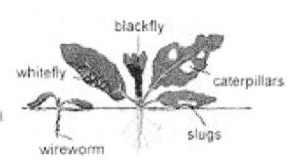

Major Disease Symptoms

Blight
Leaves or branches suddenly wither, stop growing, and die. Plant parts may rot. Some are caused by alternaria blight, bacterial blight and fire blight.

Canker
Cracks, sunken/raised areas of dead or abnormal tissue form on woody stems. Sometimes they ooze.

Galls
Swollen masses of abnormal tissue caused by fungi, bacteria or insects. Cut open a gall and if caused by an insect, it will be in there.

Leaf Blister
Blisters are yellow bumps on the upper surfaces of leaves with gray depressions on lower surfaces.

Leaf Curl
New leaves are pale or reddish and the midrib does not grow properly. The leaves pucker and curl.

Mildew
Downy mildew is white to purple. It grows usually on the underside of leaves and along stems. It turns black later. Powdery mildew is a white to grayish powdery growth usually on the upper side of leaves.

Rot
Rot decays roots, stems, wood, flowers and fruit. They can be soft and squishy, or hard and dry.

Rust
A fungal disease that is a powdery tan to rust color. Cedar apple rust and white pine blister are common. Neal Kinsey, an agronomist, says rust is caused by **copper deficiency** (see copper above).

Wilt
Caused by fungus or bacteria disrupting a plant's water system. Part or all of the plant may die. Looks like blight.

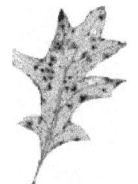

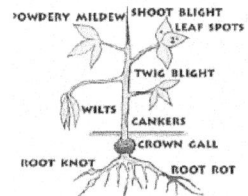

Fungi Disease

(Middle image is galls.)

Fungi are the most common problem in a garden. However, **most fungi are beneficial**. Mushrooms are fungi. Yeasts used to ferment beer and wine are fungi. Mycorrhizal fungus on roots is beneficial.

There are 2 types of fungus. Those that live in the soil and attack roots or crowns of plants. And fungi that disperse spores and damage above ground parts of plants.

Soil sterilization and some pesticides kill both good and bad fungi. (For **soil sterilization**, see potting mixes in "Garden Tips".)

Bad fungi cause apple scab, black knot, botrytis rot, corn smut, cytospora canker, late blight, and scab. Plus:

1. Blight: Early and late blight damage tomatoes and plants in the solanaceae family. It damages or kills leaves and causes rot in fruit or tubers. (For **fire blight** see pears in April Garden-Plant.)

2. Club Root: Common in plants in brassica family. Causes large swellings on roots that stunt/kill plants.

3. Damping Off: Kills seedlings by rotting stem at soil line. (See **damping off** in "Garden Tips".)

4. Fungal Wilt: It disrupts a plant's water system and kills all or part of the plant.

5. Leaf Spot: Caused by many fungi such as alternaria, anthracnose, and septoria.

6. Mildew: Includes downy and powdery mildew. There are spots or white patches on leaves, shoots and other plant parts. Downy mildew can kill plants rapidly. Powdery mildew causes poor growth and low yields. (For more about **powdery mildew,** see the Cucurbitaceae Family in the "Plant Families" section.)

7. Root Rot: Usually attacks older plants killing rootlets leading to stunting and wilting of plant.

8. Rust: Produces orange or white spots on leaves and stems causing poor growth and low yields. Neal Kinsey, an agronomist, says rust is caused by **copper deficiency.**

Prevention
Buy plants that are resistant to fungi infection. Most fungi spores reproduce best when they are wet and warm. So plant in a well drained, airy location in full sun. Prune plants and trees to keep them open so they dry quickly. Use crop rotation. (See "Plant Families".)

General Control
Remove infected parts of plant and destroy by burning. May have to remove entire plant.

    1. Anti-transpirants as fungi preventative such as Wilt-Pruf are sprayed on shrubs and trees to prevent winter damage. It also protects plants from fungal infections.

    2. Biological Fungicides are beneficial microbes that protect plants from fungus such as blight, botrytis, powdery mildew, and pythium. They do not harm beneficial fungus.

    3. Mineral Fungicides kill beneficial as well as unwanted fungus. For example **copper and sulfur based fungicides**. Copper accumulates in the soil and causes problems. Sulfur is a plant nutrient and does not do much environmental damage unless used too much. Sulfur controls brown rot, mites, powdery mildew, rust and scab. Copper and sulfur are usually used in **dormant season.** Use any of them as a last resort. (See **fungicide sprays and solutions** at the end of "Plant Health".)

    4. Neem Oil for fungus, use as a preventive or when the problem is not severe. Can control anthracnose, botrytis, flower blight, leaf spot, and powdery mildew. (See **Oil Sprays** below.)

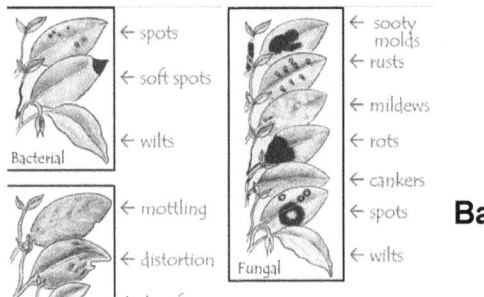

Bacterial Disease

Most bacteria are beneficial and increase fertility of soil. Some bacteria take nitrogen from the air and make it available to plants. They help recycle nutrients in soil. Bacterial disease is usually worse in warm, humid, wet weather.

Bacteria are spread by splashing rain, running water, insects, animals, tools, and moving diseased plants. They usually attack young shoots first.

Some bacterial diseases leave a sticky, gummy material. Cut the stem or leaf stalk and see if there are threads of slime when you pull the pieces apart.

    1. Cankers are bacteria that cause sunken areas that sometimes ooze slime.

    2. Galls grow as a defense against bacterial attacks.

    3. Leaf Spots are bacteria killing plant cells.

    4. Soft Rot are bacteria infecting fleshy fruit, tubers or roots. The cells break down and produce small, mushy, watery spots.

    5. Wilts are bacteria that clog a plants water system. Leaves droop and the plant may die.

Prevention
Buy disease free and disease resistant plants and seeds. Do not stimulate plants in late summer by fertilizing, watering or pruning especially do not add nitrogen. Do not work among plants when they are wet so you do not spread bacteria. Crop rotation helps reduce soil borne disease.

When harvesting handle vegetables and fruit gently. And store where there is enough moisture to prevent shriveling but not more than that.

General Control
Bacterial diseases are difficult to control. Remove and destroy infected plant parts or whole plants if badly infested. Dip pruning shears in a 1 part bleach, 9 part water solution between each cut.

Copper sprays are effective but organic control methods are needed. (See **Sprays and Dusts** below.)

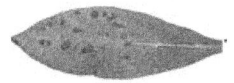 ## Viral Disease

Viral diseases can be difficult to identify. Many insects such as aphids, mites, nematodes, leafhoppers and whiteflies carry viruses to plants. Grafting infected parts to a tree can infect whole tree. Viruses are not usually carried in seeds.

    1. Leaf curling or deformed leaves is caused by viruses.
    2. Mosaic virus causes leaves or fruit to be mottled with patches of light green, yellow or white. (See Leguminosae Family in "Plant Families".)
    3. Ring spot virus show up as pale, yellow spots on leaves.
    4. Rosette is caused by a virus that stunts plants making them short and bushy.

Prevention
Buy virus free and virus resistant plants. Wash your hands after handling tobacco products, so you do not spread tobacco mosaic virus. If possible do not handle plants when they are wet.
Protect plants from insect that carry viruses by covering with row covers.

Control
Remove entire plant if infected with a virus. There are no sprays to cure viral disease.

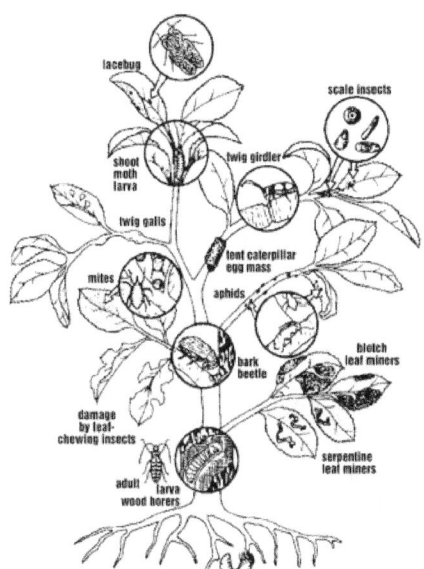

Insects & Nematodes

Nematodes
They are multi-cellular animals with wormlike bodies. They can only be seen with a microscope. Nematodes can transmit viruses. Some are beneficial to plants.
    There are no specific symptoms of nematode problems. Many symptoms look like other disease problems.
    Above ground nematodes can cause leaves, stems and flowers to be twisted and distorted.
    Nematodes in the ground can make a plant yellow, wilt or be stunted. Remove the plant and look at the roots. There will be small galls or lesions, injured roots, root rot or excessive root branching.
    They are spread by swimming in water, and moving in wet soil. Anything that carries soil can carry them.

Nematode Prevention
Use crop rotation. Keep tools clean by dipping in solution of 1 part bleach with 9 parts water. Create a soil high in organic matter so beneficial fungi grow that feed on the nematodes. Buy plants resistant to them.

General Control of Nematodes
You can solarize infested soil. (For **soil sterilization**, see potting mixes in "Garden Tips".)
Spray the soil with a **neem** solution.
Plant **marigolds** in infested soil to lure nematodes. Something in marigolds prevents them from laying eggs. Dig under the marigolds in the fall. (For marigolds, see April Greenhouse.)

Botanical Nematicides contain botanical extracts that protect plants from pathogenic nematodes in the soil. However, they also kill beneficial nematodes.

Beneficial Insects

More than 90% of all insects are beneficial. For instance, insects are needed to pollinate crops. Many beneficial insects prey on unwanted insects.

Beneficial insects that kill unwanted insects include aphid midge, assassin bugs, bigeyed bugs, braconid wasps, damsel bugs, green lacewings, ground beetles, hover flies (flower flies), ichneumon wasps, ladybugs (lady beetles), leafcutter bees, minute pirate bug, parasitic wasps, praying mantid (mantis), predatory mites, predatory wasps, rove beetles, soldier beetles, spined soldier bug, squash bees, syrphid flies, tachinid flies, tiger beetles, trichogramma wasps and many more.

Beneficial insects that are **pollinators** include alkali bees, bumblebees, carpenter bees, honey bees, mason bees, and many others.

Unwanted insects they kill are aphids, cabbage worms, cutworms, gypsy moths, Japanese beetles, mites, scale, stinkbugs, thrips, and others.

The **flowers** of some plants attract beneficial insects. These include alfalfa, alyssum, arugula, bell/fava bean, borage, buckwheat, calendula, caraway, catnip, cilantro, clover, coneflower, coreopsis, daisy, dill, echinacea, fennel, goldenrod, hyssop, lemon balm, lambs quarters, lovage, mustard, parsley, queen annes lace, rosemary, sage, sunflower, sweet clover, thyme, vetch, wild mustard, yarrow, and zinnia. **Try to have some blooming all spring, summer and fall. It is good for bees too.**

Identifying Insects

Insects are attracted to plants that are not healthy. So the first line of defense is to grow healthy plants by having fertile soil, pest-resistant cultivars, and keeping up good garden sanitation.

To control insects you need to be able to identify them. You need a good book or several with color photographs. (Read the books "The Organic Gardener's Handbook of Natural Insect and Disease Control" by Ellis/Bradlye and "Garden Insects of North America: The Ultimate Guide to Backyard Bugs" by Whitney Cranshaw.)

Common Insect Damage Symptoms	Plants Damaged	Caused by Insect Pests
1. Leaf and Foliage Damage		
Small plants cut off at soil line.	vegetables, flowers	cutworms, slugs, snails
Large holes chewed in leaves, no excrement	beans	Mexican bean beetle
	cucumber, corn	spotted cucumber beetle
	vegetables, flowers	Japanese beetles, slugs, snails
	potatoes	Colorado potato beetles
Large holes chewed in leaves, excrement	cabbage family	cabbage looper, cabbageworm, diamondback moth
	tomatoes	tomato hornworm, tomato fruitworm
	vegetables, fruits	woolybear caterpillars
Small, round holes in leaves	cabbage family, potato	flea beetles
	spinach, flowers	flea beetles
Puckered, twisted leaves, sticky honeydew	vegetables, fruits	aphids
Puckered, twisted leaves, no honeydew	vegetables, fruits	thrip, leafhopper, treehopper, spittlebug, nematode
Partial or fully defoliated plants	asparagus	asparagus beetles
	deciduous tree, shrub	gypsy moth, june beetle, cankerworm, tussock moth
	vegetables, flowers	armyworm, blister beetle, cutworm, striped cucumber beetle, rose chafer
	cherry, pear, plum	pear sawflies
	squash family	striped cucumber beetle
Chewed vines and leaves	potatoes	potato tuberworm
	squash family	pickleworm, squash vine borer
Shoot or branch tips wilt and die	fruit trees	oriental fruit moth
	squash family	squash bugs
	vegetable, fruit, flower	tarnished plant bug and other plant bugs

Webbing on leaves, stems, branch tips	apple, rose, trees	leafroller, tent caterpillar, fall webworm
	vegetable, strawberry	garden webworm
Small bag cocoon with leaf bits on branch	fruit tree	bagworm
Tunnels between upper & lower leaf surfaces	apple, rose	obliquebanded leafroller larvae
	beet, chard	leafminer
	nightshade family	leafminer
	cabbage family	diamondback moth larvae
	chrysanthemum	leafminer
Yellow leaves	fruit trees	soft scale, armored scale
	pear, quince	pear psyllas
Yellow and withered leaves	strawberry, brambles	black vine weevil larvae
Russeted (brown, rough) leaves	apple, pear, tomato	rust mite
Small, discolored spots on leaves	vegetables, flowers	garden fleahoppers
White, gray or silvery speckles on leaves	vegetables, fruit trees	spider mite, thrip, lace bugs
Sticky honeydew on leaves	pears, quinces	pear psyllas
	vegetables, fruits	aphids, scale, mealybug, whiteflies
Galls on leaves	oaks, roses	gall wasps
	maples	gall mites

2. Fruit Damage

Early dropping fruit	apple, plum, blueberry	apple maggot, plum curculios
	tree fruit, grapes	mealybugs
	blueberry, currant	fruit flies
Damage around pit of fruit	almond, walnut	navel orangeworms
	cherry	codling moth
Tunnels to core of fruit	apple and other fruit	codling moth, oriental fruit moth, plum curculios
	citrus, fig	navel orangeworm
Large holes or damaged areas in fruit/ears	corn	corn earworm, European corn borer
	squash family	pickleworm
	tomato	tomato hornworm, tomato fruitworm
Distorted, scarred fruit	vegetables, fruits	tarnished plant bug, stink bug, thirp
Russetted (brown, rough) appearance of fruit	apple, pear	rust mites
Holes bored in seeds and pods	beans	bean weevil

3. Stem and Trunk Damage

Holes bored in trunk	apple, fruit trees	roundheaded apple borer, peachtree borer
Holes bored in stems, buds or shoots	corn	European corn borer, southwestern corn borer
	currant, gooseberry	fruit borer
	raspberry	fruit borer
Galleries bored under bark (lots of tunnels)	fruit trees	flatheaded apple borer, shothole borer

4. Root Damage

Holes tunneled in roots	cabbage family	cabbage maggots
	carrot family	carrot rust flies, carrot beetles, carrot weevils
	flower corms	wireworms
	onions, leek, garlic	onion maggots
	potatoes	potato tuberworms, wireworms
	strawberry, grape	strawberry root weevils
	raspberry, other fruit	strawberry root weevils
Knotted, lumpy roots	beans, peas, legumes	beneficial bacteria in roots
	tomato, lettuce, pepper	nematodes

 ## Greenhouse Pests and Diseases

Insecticidal soap will control most soft-bodied pests like **aphids, scale crawler, and spider mites** and is non-toxic to you, your plants and any beneficial insects if used properly. Be sure to spray the underside of leaves. (See **How to Make Organic Insect or Pest Sprays** below.)

1. Whiteflies

Whiteflies are a serious greenhouse pest. They feed on plants, produce honeydew that attracts other insects and sooty mold. Whiteflies can transmit plant viruses. They cause problems with many types of plants and resist insecticides.

Greenhouse tomatoes are especially bothered by whiteflies. The most common whiteflies in a greenhouse are the greenhouse whitefly (Trialeurodes vaporariorum), sweetpotato whitefly (Bemisia tabaci) and the silverleaf whitefly (Bemisia argentifolii).

Several types of **beneficial organisms** are available for biological control of whiteflies. The parasitic wasp Encarsia formosa preys on immature whiteflies and is commonly used for greenhouse whitefly.

Pesticides with neem such as Neemazad and Azatin, insecticidal soap M-Pede, and horticultural oil kill them. Better whitefly control is achieved with thorough spray coverage.

Wider plant spacing and removal of dead lower leaves improve pesticide coverage and pest control. (See Cucurbitaceae Family in "Plant Families" under cabbage white flies.)

2. Fungus Gnats

Fungus gnats are 1/8 inch black flies with long legs and antennae, tiny heads, and clear wings. Females lay tiny yellowish-white eggs. Adults live a week. In a greenhouse, 20-25 days are needed to complete a generation.

Most fungus gnat larvae are scavengers, feeding on decaying organic matter in soil. But others feed on root hairs, roots or stems of plants. Infested plants infested lack vigor and may wilt. Adults are seen running on the leaves before injury caused by larvae is apparent.

Bacillus thuringiensis subspecies israelensis (Bti) provides temporary control and is toxic only to fly larvae such as mosquitoes, black flies, and fungus gnats. Repeated applications are needed. This Bt is a different subspecies from that applied to leaves to control caterpillars. Bt labeled for caterpillars is not effective against fly larvae.

Control gnat populations by reducing water and organic debris and by having good water drainage. Keep compost piles away from greenhouse. Sterilize potting mixes. (For **soil sterilization** see potting mixes in "Garden Tips".) Keep greenhouse clean.

3. Shore Flies

Shore flies are gnat-like insects similar to fungus gnats. They have short antennae, red eyes and heavier dark bodies. They have smoky wings with several clear spots. They are good fliers and can be seen resting all over the greenhouse. They resemble winged aphids, but aphids have two pairs of wings and cornicles (tubes) on their abdomen.

Their life cycle is similar to that of the fungus gnat. The yellow to brown larvae are up to 1/4 inch long. Both larvae and adults feed on algae. They rarely damage plant tissue, but the adults spread soil pathogens.

To control do not overwater. Get rid of algae in the greenhouse. You can use an algaecide. Have good sanitation by keeping everything clean.

 ## Products to Control Insects

(See **Sprays and Dusts** and **How to Make Traps and Barriers for Crawling and Flying Insects** below.)

1. Bacillus Thuringiensis (BT)

This is **bacteria spores** that when eaten by insects with an alkaline pH gut are killed. This includes insects in the Lepidoptera order (butterfly and moth). It usually does not harm beneficial insects except for butterfly larvae.

BTI (BTi), BTK and BTSD are different types of BT. **BTI** attacks larvae of blackflies, fungus gnats, and mosquitoes. **BTK** attacks caterpillars such as cabbage loopers, codling moth larvae, diamondback moths, gypsy moth larvae, cabbageworms, spruce budworms, and tomato hornworms. **BTSD** attacks leaf feeding beetles including black vine weevils, boll weevils, Colorado potato beetle and elm leaf beetles. **BT** will not attack leafminers or slugs.

It comes in liquid, powder and granules. The spray degrades quickly in sunlight, lasting only a few days. Apply late or early in the day. You can add a few drops of liquid dish soap to make it stick better. Pests can develop resistance to BT so use only when you really need it.

2. Botanical Insecticides

They are made from plants and are safer than chemical pesticides, but are still strong and should be used as a last resort. They include **pyrethrin** (see below), **rotenone and sabadilla**. In the long run they can cause more problems than they cure because they upset the natural biological balance. (See **Sprays and Dusts** below.)

3. Milky Disease Spores

Bacillus popilliae and bacillus lentimorbus are **bacteria spores** that are combined to kill grubs of Japanese beetles and other beetles. Does not harm other organisms. The spores continue to reproduce so usually only 1 application is needed. Though in cold climates spores can be killed in winter.

Apply any time the ground is not frozen. Use 7-10 pounds per acre or 10 ounces per 2500 square feet. It is best to apply before rain.

4. Oil Products

This includes dormant oils, growing-season sprays, and spray additives that smother pests such as overwintering aphids, mites, peach twig borer and scale insects. (See **Sprays and Dusts** below. For **applying dormant oil sprays**, see February Garden-Maintenance.)

5. Pyrethrins (a botanical insecticide)

A broad spectrum, kill on contact insecticide that is an extract of chrysanthemum (mum) flower. In 100 years of use insects have not built up resistance to it. Will kill ants, aphids, armyworms, beetles (asparagus, blister, cucumber, Colorado potato, flea, and Mexican bean), cabbage loopers, caterpillars, earwigs, fleas, flies, harlequin bugs, fruit flies, leafrollers, leaf hoppers, mosquitoes, psyllids, thrips, ticks, whiteflies and others.

It is toxic to fish. Kills some beneficial insects. It is safer than chemicals. Breaks down quickly in sunlight. Apply in evening or night. (For **sow/grow chrysanthemums**, see March Greenhouse. See **Non-Oil Sprays** in Sprays and Dusts below.)

6. Insecticidal soap will control most soft-bodied pests like **aphids, scale crawler, and spider mites** and is non-toxic to you, your plants and beneficial insects if used properly. Be sure to spray the underside of leaves. (For **more about insecticidal soap**, see Oil Sprays below.)

7. Herbal Sprays (See **How to Make Organic Insect or Pest Sprays** below. For **how to make a tincture, decoction or infusion**, see April Garden-Harvest. See **companion planting** in "Garden Tips".)

Aphids: mint, peppermint, coriander, chervil, nasturtiums, petunias, dill and rue.
Cabbage moths: oregano, hyssop, mint, peppermint and sage.
Cabbage worms: borage, thyme and geraniums.
Cinch bugs: soybeans.
Colorado potato beetle: flax, horseradish and coriander.
Cucumber beetles: oregano, nasturtiums.
Flea beetles: rue, mint, peppermint, hyssop and sage.
Japanese beetle: geraniums, garlic, soybeans, tansy, rue and white chrysanthemums.
Spider mites: dill and coriander.
Squash bugs: dill, lemon balm and tansy.
Thrips: basil.
Tomato hornworms: borage, basil, marigold, and petunias.
White flies: nasturtiums.

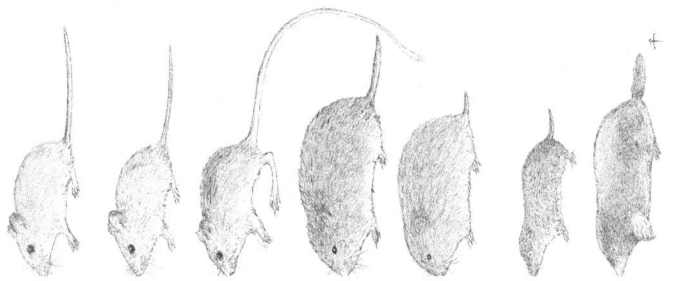

Animal Pests

(**Image above starting at left:** All are rodents. Deer Mouse, House Mouse, Woodland Jumping Mouse, Meadow Vole, Southern Bog Lemming, Northern Short-Tailed Shrew, and Hairy-Tailed Mole.)

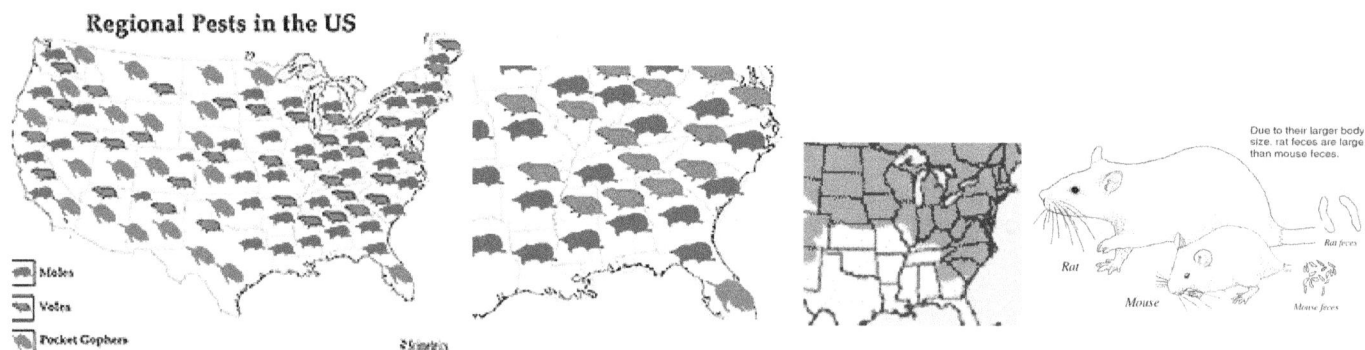

(**Image above on left:** Moles, voles and pocket gophers. **Image in middle:** Dark area is meadow vole population.)

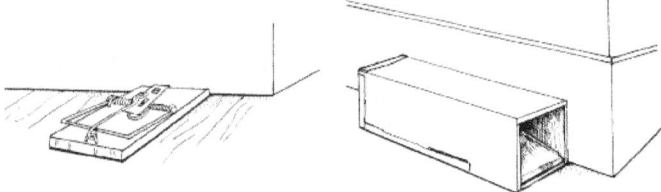

General Prevention for Chipmunks, Mice, Moles, Rats, Voles and Other Rodents

To reduce the number of rodents, **keep your yard and farm clear of debris.** This includes rock piles, brush piles, thick bushes, stacks of firewood, and other places where they might find shelter. Keep grasses near the edges of buildings well trimmed. Voles like to hide in mulch.

Make sure all animal food is in metal or other rodent proof containers. Make sure your food garbage is in metal cans. If composting, cover food with soil or bury in pile. Do not throw on top of it.

Exclusion: Use 1/4 inch hardware cloth to cover holes that rodents use to get into buildings.

General Animal Control

Poison baits (poisoned food) work but can end up killing cats, dogs, other pets, or young children if they eat the poisoned rodent or eat the bait.

Have a few **outdoor cats**. Unfortunately, they also kill wanted birds.

Place many **traps**. (See above images for trap placement against wall.) (For **how to build a trap box**, see September Farm Animals.)

Some nuisance animals can be shot. The **North Carolina Wildlife Resources Commission** and other states require hunting licenses or permits to kill certain wild animals. Check with your license office for details.

 **Birds**

Birds eat garden pests but also eat some fruits and vegetables. Cover trees and shrubs that have fruit with plastic netting to keep the birds out. Try moving a scarecrow every few days. Tie pie pans, pinwheels, foil, strips of plastic, or cds/dvds to its arms or to tree branches.

Plant Health p. 10

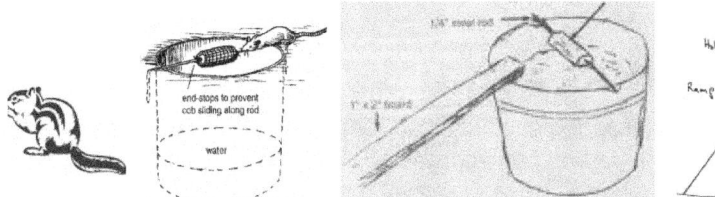 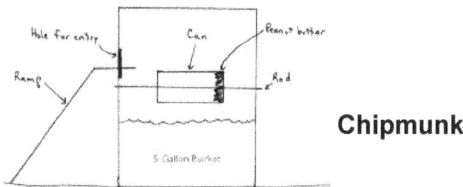

Chipmunk

Eastern chipmunks have 3 dark stripes and are 8-10 inches long. They are active during the day. They eat slugs, snails, insects, small birds, eggs, mice, seeds, fruit, vegetables, seeds, flower bulbs, and pet food.

They are not attracted to poison baits. They are repelled by coyote urine.

The most practical and effective method for chipmunk control is live trapping. Put **snap traps** in a box with 2 small chipmunk-size holes in it. (Put traps in boxes so cats, dogs and children do not get hurt by the snap traps.) Place the trap by their burrows or under plants, trees or bird feeders where you have seen them. Use any of the food you know they are eating as bait. Bird seed/sunflower seeds with peanut butter is a good bait. Put a little dirt, leaves, etc in the box to make it look like the ground. Do not set the trigger on your traps for 2-3 days, allowing chipmunks to get used to going inside a strange box. (For **how to build a trap box**, see September Farm Animals.)

Or fill a **5 gallon bucket** half full of water with the surface covered with some sunflower or safflower seeds (they float). Some people like a 3 gallon bucket better. Some fill it up all the way with water. Place a narrow board (such as 1 inch x 2 inch by 4 feet) on top of the bucket that goes out into the middle of the bucket. The other end is outside the bucket where a chipmunk (or other rodent) can reach. (See above images.) Put some seeds on the board. You can put some peanut butter at the end of the diving board.

The chipmunk eats the seeds on the board, gets to the end and dives or falls into the water. Or make the ramp and diving board 2 different pieces. The diving board is not attached so falls in the water with the chipmunk. Or use a 2 x 4 inch board across the entire top of the bucket. This also works for **rats**. Or build it with a **corn cob** (see above image).

If **squirrels** are eating the seeds, place a board that covers the bucket, with an opening just large enough for a chipmunk, but too small to allow a squirrel in.

Or try an accurate pellet rifle with scope to kill them. Or put used kitty litter around their burrows. Or get cats.

 Deer

Deer can leap 8 feet or more over a fence. An **electric fence** is the best way to stop deer eating plants. Ideally it would be 10 feet tall. Or put up an **8-10 foot tall woven wire fence**. However, **with double fencing you can have a much shorter fence.** Deer can not leap both tall and wide at the same time. It also works because deer are near sighted and do not have good depth or distance perception. You can also use solid fencing since deer will not jump if they can not see the other side.

Various double fence methods: (The outside fence is shorter or the same height as the inside fence.)
1. Use an 18-gauge, high-tensile wire at 18 inches above the ground. Then 3 feet inside the enclosed area, put 2 more wires. The first is 10 inches and the second is 24 inches above the ground.
2. Put up 2 electrified fences that are both 4-5 feet tall placed 3-4 feet apart. Each fence consists of 2 wires, one at 15 inches and the other at 45 inches.
3. Put up a woven wire 4 foot high fence then add 3 strands of wire above it. Then put up another fence 3 feet high about 2-3 feet away on the outside that is a single strand of wire or plastic trimmer string. Or use chicken wire or poultry netting.

Spraying a lot of **bloodmeal on trees and bushes** where deer can reach helps. Hot pepper spray helps. (See **How to Make Organic Insect or Pest Sprays** below.)

 **Gophers** (pocket gophers) spend almost all their time underground in the burrow. Their front teeth are always outside their mouth. Mounds are formed as the gopher digs its tunnel and pushes the loose dirt to the surface. They create fan shaped mounds clustered in an area. The hole, which is off to one side of the mound is usually plugged (filled with dirt). According to the above chart, gophers are **not found in North Carolina**.

Body grip or claw-type traps set below ground in tunnels kills gophers.

 **Mice**

The adult **house mouse** is small, about 3-4 inches long excluding tail. The house mouse has large ears, pointed nose and small eyes. The tail is as long as the head and body combined. They eat almost anything but prefer grains, seeds or sweet food. They are usually active at night. They rarely travel more than 10-30 feet from their home. They can squeeze through any opening larger than 1/4 inch.

Deer and white-footed mice are 6 inches long. Their upper body is brown, the lower body is white. They live in burrows.

Place **mice snap traps** up against walls, behind objects, and in secluded areas where mouse droppings, gnawing and damage are evident. Snap traps should be oriented perpendicular to the wall, with the trigger end against the wall. (See image above at start of Animals Pests.) Place 6-10 feet apart. Expanded trigger snaps work better.

To prevent cats, dogs, other pets, and children from getting hurt by snap traps, put traps in boxes. (For **how to make a trap box**, see September Farm Animal.)

For successful **baiting** of mice, have many bait placements containing a small amount of bait, rather than a large amount of bait in fewer locations. Replace baits periodically. **Mice are curious and not afraid of a new object.**

The **best bait** is one that is accepted by the rodents and not easily removed from the trigger. In sites where food is abundant but nesting material is scarce, soft string, cotton balls, or strips of cloth are attractive to female mice and rats. To enhance the material, apply one or two drops of vanilla extract.

Tie the bait down to the trap or use a sticky bait, like peanut butter, that can not be carried away. When using a sticky bait, smear a small amount on the top and bottom of the expanded trigger. Some solid baits, like cheese, marshmallows or chocolate, can be melted onto the trigger with a match. Use a piece of thread or dental floss to tie down solid baits such as bacon, gum drops or raisins.

After a trap has been used for a while, you may need to apply oil. Use a small amount of vegetable oil or bacon grease that attracts rodents. Do not use machine oil. (See **bucket traps** in Chipmunk above.)

 **Moles**

Moles eat grubs, insects and earthworms, not plants. However, they damage plants and their tunnels make it easy for mice, gophers and voles to eat plant roots. They are almost blind. They have big front feet.

Mole mounds are circular with a plug (filled with dirt) in the middle. Moles rarely go out of their burrows.

Get rid of moles by getting rid of beetle grubs with milky disease spores. But this can take several years to work.

Moles can be trapped with **mole traps** placed at tunnel entrances or in tunnels (see above image). This is the best method but has to be done all the time.

Voles will travel in mole tunnels and eat the roots of plants. Wolves, owls, hawks, coyotes, foxes, weasels, cats and some dogs (especially terrier breeds and ratting dogs) are predators of moles and voles.

 **Rabbits**

Put up a **3 foot tall chicken wire or poultry netting fence** with mesh 1 inch or smaller. You may need to dig the fence 6 inches into the ground. Putting rocks along the fence helps so they do not dig under.

Rabbits **chew on young trees**. Their gnawing begins several inches above the soil line. Rabbits have much larger teeth than voles. To protect trees from rabbit chewing in the winter, put 1/4 inch hardware cloth or wire mesh (1 1/2 to 2 feet tall, 2-3 inches into the ground) around the tree. This also keeps mice and voles away.

Some say spraying a lot of bloodmeal on trees and bushes where rabbits can reach helps. Hot pepper spray helps. (See **How to Make Organic Insect or Pest Sprays** below.)

 **Raccoons**

It is a medium-sized mammal native to North America. It weighs 8-20 pounds. It **usually only moves around at night.** It will eat almost anything. It is very good at manipulating things with its paws. They are intelligent. They can get rabies and distemper.

They kill poultry. They eat farm crops and garden plants. They eat food out of garbage so keep in metal trash can. You may have to tie or wire the lids on. They can be caught in Havahart live traps that are at least 12 x 12 x 32 inches.

See General Prevention and Control at start of Animals Pests. **Keep all food in rodent-proof containers.**

 Rats

A **Norway or Roof rat** moves within a diameter of 98 to 164 feet from its home. Both species have droppings about the size of a black or red bean. **They are creatures of habit.** Their favorite food is pet food.

Rats are very cautious of any new object so place unset traps in a new location for a week or two. Then set enough **rat snap traps** to kill a large percentage of the population (**mass trapping**) before the rodents become trap shy. You may need as many as 2-3 dozen rat traps set. The spring on a rat snap trap is very strong. It can break a finger.

To prevent cats, dogs, other pets, and children from getting hurt by snap traps, put traps in boxes. (For **how to make a trap box**, see September Farm Animal.)

Use a snap trap with an expanded trigger. Place them along the wall where the rodent is foraging. Put the trigger closest to the wall. They do not need bait but if you want to use bait try pieces of apple, potato, raw bacon, or peanut butter spread on a cotton ball. Attach the bait to the trap with thread. You can attach the trap to the ground with a nail or something else so the rat does not carry it away. (See a **bucket trap** above in Chipmunk.)

The **best bait** is one that is accepted by the rodents and not easily removed from the trigger. In sites where food is abundant but nesting material is scarce, soft string, cotton balls, or strips of cloth are attractive to female mice and rats. To enhance the material, apply one or two drops of vanilla extract.

 Squirrels

A squirrel is a small to medium sized rodent. They eat nuts, seeds, conifer cones, fruits, fungi and green vegetation. They will eat meat if very hungry. **Usually about half the squirrels in a population die each year.** In the wild, squirrels rarely live over 4 years old but in captivity they may live 10 years or more.

Damage: In nut orchards, squirrels can eat nuts prematurely and carry away mature nuts. They gnaw on wires, enter buildings, and build nests in attics. In gardens, squirrels may eat seeds, mature fruit, or grains such as corn.

Close openings to attics and other areas of buildings so they can not get inside but make sure not to lock them inside. They cause a great deal of damage trying to get out. A squirrel excluder can be made by mounting an 18-inch section of 4-inch plastic pipe over an opening. The pipe should point down at a 45 degree angle.

Rat snap traps can catch small squirrels. **Wire cage traps** (Havahart) with minimum size 5 x 5 x 20 inches, single door, can be used to capture them alive. Good baits are slices of orange/apple, shelled walnuts/pecans, peanut butter, corn and sunflower seeds.

 Voles (sometimes called meadow mice or field mice)

It is a small rodent that looks like a mouse but has a stouter body, shorter hairy tail, and smaller ears/eyes. They are usually brown or gray. They are 5-8 inches long including the tail. There are 23 species in the United States.

They are a big pest in gardens. They eat plants, roots, bulbs, tubers, seeds, and bark. They are active day and night, all year. If planting bulbs or tubers, put some small gravel in the hole to help protect them. Hardware cloth with a mesh of 1/4 inch or less can be put around trees to protect them. Bury 6 inches deep.

Wolves, owls, hawks, coyotes, foxes, weasels, cats and some dogs (especially terrier breeds and ratting dogs) are predators of voles.

(See moles above. For **more about voles and how to make trap boxes**, see September Farm Animals.)

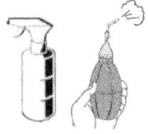

Sprays and Dusts to Control Plant Diseases and Pests

(For **dormant oil spraying and copper/sulfur spraying,** see February Garden-Maintenance.)

Sprays and dusts for plant diseases do not cure the plant, instead they **provide protection against further infection.** So it is important to apply them early in the disease. Badly infested plants are probably too far gone to have them do any good. **Never spray insecticides when fruit trees are in bloom.** It kills the bees that pollinate the blossoms. Always read and follow label directions when using any pesticide. Contact your County Extension Office.

Non-Oil Sprays (as primary ingredient)

1. Sulfur, lime-sulfur and bordeaux mix are fungicides acceptable for use in organic gardens. Sulfur can be used as a dust, wettable powder, paste or liquid. Lime-sulfur is more toxic than just sulfur. However, they do kill beneficial organisms too. Use only as a last resort. Contact your County Extension Office.

a. Sulfur sprays have been used for hundreds of years. They do damage plants in hot weather especially grapes, melons and tomatoes. Use when temperatures are less than 90 degrees. **Do not use sulfur on plants within 3-4 weeks of applying oil sprays** such as horticultural oil.

Use wettable sulfur powder. Add 1 ounce of powder per gallon of water for a general application. Amounts vary for different uses. Cover sprayer and mix. Shake periodically while apply to plants. Use on spider mites, psyllids, thrips and many fungi and mildews.

b. Lime-sulfur is good against anthracnose, apple scab, blight, brown rot on peaches, leaf spot, powdery mildew, scab, and scales. It is better used as a dormant spray (in the winter or early spring when plants are dormant). Best used early spring. Toxic to mammals.

It is made by cooking sulfur in water with lime or calcium hydroxide. **Do not spray within 3-4 weeks of using horticultural oil.** Do not spray if above 85 degrees. Spray raspberries infected with anthracnose or blight when buds first show silver. Spray currants and gooseberries infected with anthracnose at bud break and again 10-15 days later.

c. Copper-based fungicides control many fungi and are also effective against some bacteria. Good against anthracnose, bacterial leaf spot, black rot, blights, downy mildew, peach leaf curl, and Septoria leaf spot. Copper is toxic to fish. Spray early in the morning in dry, sunny weather so plants dry. Spray entire plant including under leaves.

d. Bordeaux mix is copper sulfate with lime and water. It is a fungicide that controls bacterial leaf spots, blights, anthracoses, downy mildews and cankers. It also repels many insects. (For bordeaux mix see below "**How to Make Organic Fungicide Sprays and Solutions**".)

2. Caustic Soda (Lye) orchard spray controls sucking mites, aphids, Rutherglen Bug and thrips but does not seem to affect spiders. It can be made as follows: Grate 2/3 of a cake of soap into 1 gallon of hot water. Optionally, add 10 ounces cooking oil. Add 2 teaspoons caustic soda (sodium hydroxide, NaOH, lye) to the mix while still hot. Let cool to room temperature. Only spray on affected parts of trees. Do not spray during hot part of day. Contact your County Extension Office.

3. Pyrethrin (See above **Products to Control Insects**.)

Pyrethrin daisies (chrysanthemum cinerariifolium or coccineum) kill insects on contact. Dried flowers are called pyrethrum. Extracts are called pyrethrin. Use against aphids, cabbage loopers, celery leaftiers, codling moths, Colorado potato beetles, leafhoppers, Mexican bean beetles, spider mites, stink bug, thrips, tomato pinworms, and whiteflies. Not as effective against diamondback moths, flea beetles, cabbageworms, pear psyllas, or tarnished plant bugs.

Moderately toxic to mammals. Does kill ladybugs.

You can grow pyrethrum (chrysanthemum) daisies. Pick blossoms in full bloom and hang in dry, sheltered, dark place. When dry, store in tightly sealed jar in dark, cool location. To use, grind the flowers with mortar and pestle, then add a little liquid dish soap and enough water to make a spray. Experiment with different concentrations such as soak 1 to 4 teaspoons pyrethrin in 1 gallon of warm water for three hours. (For **sow/grow chrysanthemum**, see March Greenhouse.)

Oil Sprays

1. Horticultural or Botanical Oils are lightweight oils, either petroleum or vegetable based. Most horticultural oil sprays are light enough to be used all year. They are called dormant, superior, summer or supreme oil. They do not harm beneficial insects a lot. Contact your County Extension Office.

    a. Dormant oil is used on woody plants such as fruit trees during the dormant season. It originally referred to heavier weight oils that were unsafe to use on plants after they broke dormancy. But now they have been replaced with lightweight oils that can be applied to plant foliage (leaves) too. So now dormant oil refers to the time of application rather than to the type of oil.

    b. Summer oil is used on plants when foliage (leaves) are present (foliar oils). It now refers to the time an application is made rather than to the properties of the oil.

    c. Supreme oil is similar to superior oil.

    d. Superior oil is a summer use oil that allows year-round use without phytotoxicity (leaf damage by the sun). Also called narrow-range oil.

Horticultural oils are applied as a dilute spray on plant surfaces to control aphids, mealybugs, scale, mites and other insects. **They smother pests.** They are only effective if applied directly to the pest and do not provide residual control.

A little toxic to mammals. Do not use if temperature is above 85 degrees or below freezing. **Do not use within 1 month of using a sulfur spray** (sulfur, lime-sulfur or Bordeaux mix). Do not spray water-stressed (drought) plants.

Use 3% solution for early spring dormant applications. Use 2% solution against insects/mites when plants have leaves. Use 1% solution on sensitive plants such as cucumber, rutabaga and tomatoes. To make 2% solution, use 1/3 cup oil into 1 gallon of water. For 3% solution use 1/2 cup of oil per gallon. For 1% solution, use 2 1/2 tablespoons oil per gallon.

Spray oils on trees at least 6 weeks apart. Spray oils on plants weekly. **Do not spray within 24 hours of any other sprays or dusts.** Spray early morning or evening.

2. White oil (mineral oil) spray: Mix 1 cup vegetable or white mineral oil, 1 1/2 cups water, 2 teaspoons dish soap or Murphy's soap. The soap helps the oil stick to the insect. This **suffocates most insects**. It is effective in the control of soft body insects such as aphids, scale, mealybug, mites, citrus leaf miner and smooth skinned caterpillars.

3. Neem oil spray (azadirachtin)

Neem oil is extracted from the seeds of the neem tree. It is a **broad spectrum insect repellant and poison.** Plants can take it up through their roots and spread it throughout the plant. Good for Colorado potato beetles, corn earworms, cucumber beetles, flea beetles, leafminers, Mexican bean beetles, mites. Use against aphids, gypsy moth, loopers, mealybugs, thrips, and whiteflies.

Not toxic to mammals. Does not usually harm honey bees, spiders and other beneficial insects. But it can harm parasites that kill unwanted insects. Repeated use can lead to **pest resistance** so alternate with insecticidal soap or other treatment. Neem stops working in 5-7 days. It may take some time for insects to be killed.

To make neem oil pesticide: Mix 2 quarts warm water with 1/2 teaspoon of mild liquid soap like Castile. Mix and then slowly add 1/2 ounce of neem oil.

4. Insecticidal Soap Sprays

It is any potassium fatty acid soap such as **liquid soap** but not detergent. Do not use antibacterial soap since it may cause damage to plants. The best to use are **Ivory Liquid or Shaklee Basic H.**

Use insecticidal soap sprays against soft-bodied insects such as aphids, mealybugs, and whiteflies. Also good against chiggers, earwigs, fleas, mites, scales, thirps, and ticks. Not as effective against beetles and caterpillars. Many pollinators and predatory insects such as Lady beetle adults, bumble bees, and syrphid flies are relatively unaffected.

It can cause phototoxic harm (sun damage) to some leaves so test on a small area first. Do not use on plants with thin cuticles (protective waxy covering on leaves) such as beans, Chinese cabbage, cucumbers, ferns, gardenias, Japanese maples, nasturtiums, and young peas. Tomatoes and potatoes are less susceptible to damage. Thick leaved plants such as cabbage are not damaged except it may reduce yields.

Spray on undersides of leaves too. Do not use more than 3 times on any annual plant. **It only works on direct contact with pests.**

To make your own insecticidal soap, add 1 to 2 tablespoons liquid soap to 1 quart water. Mix and spray.

Dusts

1. Diatomaceous Earth (DE, diatoms)

It is a nonselective (kills good and bad insects), abrasive dust (powder) made from fossilized silica shells of algae called diatoms. It kills by projecting little needles into insects and by dehydration.

Dust plants when they are wet. Dust plants and soil to control crawling pests such as slugs and snails. (For slugs and snails, see **How to Make Traps and Deterrents for Slugs and Snails** below.) Dust around cabbage, onion and other seedlings to kill root maggots and other pests. On plants it kills soft bodied pests such as aphids, caterpillars, leafhoppers and thirps.

Mix with liquid dish soap and water to make a thick slurry and paint on tree trunks. Do not use antibacterial soap since it may cause damage to plants. The best to use are **Ivory Liquid or Shaklee Basic H.** (For **whitewashing trunks**, see June Garden-Maintenance.)

Can also **put in stored grain to kill insects**. Use food grade not pool grade diatoms. Fifty pounds of diatoms protects 7 tons of grain from insects. That is 1 cup to 50 pounds of grain, or 1 pound for 300 pounds of grain.

And add food grade (not pool grade) diatoms to animal feed to **control internal parasites**. Use 5% diatoms in feed once every 3 months or when you buy a new animal. (For **where to buy**, see fertilizers in "Garden Tips".)

2. Rotenone Dust

It occurs naturally in the roots and stems of several plants such as the jicama vine plant. It is a broad spectrum insecticide so it also kills beneficial insects. It kills potato beetles, cucumber beetles, flea beetles, cabbage worms, raspberry bugs, and asparagus bugs, as well as most other arthropods (invertebrate having jointed limbs and a segmented body with an exoskeleton). Contact your County Extension Office.

It is moderately toxic to mammals and some people are allergic to it. It is very toxic to birds, fish and swine. It is no longer recommended for organic gardens. It is used as a dust-on powder to reduce mites on chickens and other poultry. (For **mites and lice**, see March Farm Animals.)

3. Sabadilla Dust

It is a botanical insecticide made from the seeds of a small perennial bulb in the Lily family. It is moderately toxic to mammals and is very toxic to some people. It kills honey bees. It is not recommended for an organic garden. It's dust can be highly irritating to the eyes. Sold under the trade names Red Devil or Natural Guard. It is effective against caterpillars, leaf hoppers, thrips, stink bugs and squash bugs.

4. Sulfur Dust (see **non-oil sprays** above for information about sulfur.)

How to Make Traps for Flying Insects

1. Board or Paper #1- Choose a bright colored piece of construction paper such as orange, yellow or bright pink. Most bugs are attracted to bright colors. Spread the sticky paste (see formula below) on the paper with a paintbrush. Punch a hole at the top of the paper and attach a string. Hang it where flying insects will find it. **Catches good and bad insects.**

2. Board or Paper #2- For longer lasting traps, use 1/4 inch plywood, masonite or similar material attached to garden stakes or wires for hanging. Make trap 3 inches x 5 inches or any larger size. Paint boards bright yellow. The best yellows are 'Federal Safety Yellow #659' from Rustoleum or 'Saturn Yellow' from Day Glo. Traps should be at the height of the plant but not so close that the plants get stuck on the traps. If needed, put chicken wire or hardware cloth around the traps to keep plants off. It is better if they face away from the sun. Clean with a paint scraper or wipe with cloth soaked in oil.

#1 Sticky Formula- Mix 1 quart corn syrup and 1 quart water in pan. Bring to boil. A non-drying paste.
#2 Sticky Formula- Mix equal parts petroleum jelly (Vaseline) or mineral oil with liquid dish soap.
#3 Sticky Formula- Buy a commercial product such as Bug Gum, Stickem, Tanglefoot or Tangletrap. Some are very thick and may need some paint thinner added to capture small, lightweight pests.
#4 Sticky Formula- Mix 2 parts Vaseline (petroleum jelly) with 1 part detergent or insecticidal soap.

3. Red Apple Maggot Trap- Buy reusable red spheres from a garden center, or use any red, apple-sized sphere. Install an eye screw on it, and put a wire or string on it to hang in the apple or other fruit tree. Use one of the sticky formulas above and coat on the sphere. Re-apply the sticky coating every 2 weeks.

Set the traps out 3-4 weeks **after petal fall (mid-June)** and leave until after harvest. Hang 1 trap for every dwarf tree, 2-3 traps for every semi-dwarf tree, and 5-6 traps for every standard tree. Apple maggots riddle fruit with holes; codling moths make an entry and exit hole only. (For **apple maggot traps**, see June Garden-Maintenance.)

4. Japanese Beetle Trap- Cut a wide opening in the top of a 1-gallon plastic jug (leave the handle on). Fill it 1/3 with wine, sugar water, and mashed fruit. Add baking yeast to increase fermentation. Let sit in a warm place to ferment. One formula is 1 pint water, 1 banana or non-citrus fruit, 1/2 cup sugar or honey, 1/2 cup wine and 1/4 teaspoon baking yeast. Let sit in a warm place to ferment.

In the **spring** set traps 1-3 feet above the ground in an open, sunny area. Place 20-30 feet downwind from beetle's favorite plants. Use 3 traps for every 1/8 acre. (See **bramble pests** in April Garden-Plant. See **Gramineae family** in "Plant Families".)

5. Ant Trap

Ants are attracted to sugar and/or protein. Use a small plastic container with lid such as a margarine container. Poke 4-5 holes, about the size of a pencil, spaced evenly around the container about an inch from the top. Mix 20 Mule Team Borax Natural Laundry Booster with granulated sugar in a 1 to 3 ratio such as 1 cup Borax and 3 cups sugar to make 6-8 traps. Fill the container so the powder is 1/2 inch or so below the holes. Add enough water to make the mixture slightly soupy or syrupy. Add about 1 teaspoon peanut butter. Stir well. Put the lid on the container.

Any ants that eat it are killed. Some eat it right away, and some is taken back to the colony. It should work in about 2 weeks. Borax is toxic so do not place traps in areas where animals or children can accidentally eat it.

6. Fly Traps

Fly Trap #1- Cut 4-5 small holes around the height where the center of the handle is in a 1 gallon milk container. Mix 1/4 cup sugar/corn syrup and 1/4 cup apple cider vinegar. Pour into container. Fill halfway with water and put on cap. Flies can get in but not out.

Fly Trap #2- Poke holes the size of flies in a jar lid. Put a piece of raw hamburger or other meat inside the jar. Put on lid. Once the flies get inside the jar, they can not get out.

Fly Trap #3- Cut a 2-3 liter plastic bottle into three parts. You do not need the cap. Cut 2/3 of the way down, and again just below the neck. Put water in the large section. Discard the center section. Turn the top part upside down on top of the large section. Put jam or honey on the inside. **To make sure honeybees do not fall for this trap**, put 1/4 cup vinegar in it. Place trap in shade if possible to stop the jam or honey melting. (See below 2 images for how to make and variations.)

7. YellowJacket Trap- It is a **meat bee**. Use an empty half or one gallon plastic milk jug. Cut out 1 or 2 entrance holes somewhere near the top quarter on the side. Fill jug about a quarter with water. Chop up chunks of a hot dog or other meat, and put in water. Put on cap. Hang trap away from where you usually walk. It takes a day or two but the yellowjackets find it. **They can fly in but not out.** Eventually they fall into the water and drown. Also catches flies. (See last 2 images below.)

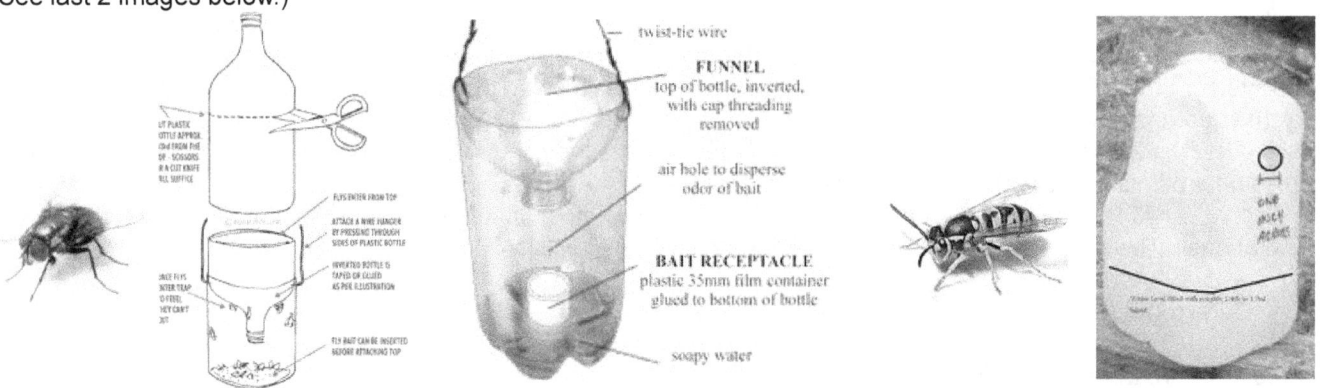

How to Make Traps and Barriers for Crawling Insects

Trap- Mix 1 quart corn syrup and 1 quart water in a pan. Bring to a boil. This creates a non-drying sticky paste. Choose a bright colored piece of construction paper such as orange, yellow or bright pink. (For **other sticky formulas**, see How to Make Traps for Flying Insects above.)

Plant Health p. 17

Cut the paper into 1-inch-wide strips about 1 foot in length. Brush the paste on the top side of the strips and lay them paste side up where crawling insects will find them. Can wrap around plants. Weigh the paper down with something. Catches good and bad insects.

Barriers keep crawling insects such as ants, caterpillars, earwigs, slugs, snails, and sowbugs away from plants.

#1 Barrier- Place a **2 inch wide strip** of wood ashes, diatomaceous earth, sawdust, crushed seashells, or tiny sharp stones around a plant or group of plants. For an individual plant stay within 6 inches of the base of the plant. Diatomaceous earth and ashes do not work well when wet. (For **diatomaceous earth**, see Dusts above.)

#2 Barrier- Make a dehydrating dust paint by mixing 1/4 pound of diatomaceous earth with 1 teaspoon of pure liquid soap such as Ivory. It should be a slurry. Apply to lower **trunk of tree**. It shields bark from sun and deters pests. (For **whitewashing trunks**, see June Garden-Maintenance. For **diatomaceous earth**, see Dusts above.)

#3 Barrier- Place a fence of aluminium foil around beds to keep out **caterpillars** such as armyworms and cutworms. Bury 1 edge of the foil in soil with several inches above soil. Fold top edge away from the plants to form a lip.

#4 Barrier- Cutworms feed at night and hide just under the soil surface during the day. Place barrier around seedlings such as cabbage, eggplant and bean plants. Use cardboard tubes from toilet paper and paper towels. Cut the tubes 2-3 inches long. Place over seedlings when transplanting. Push the tubes a little into the soil. You can also use shallow cans. Remove them once the plants get older. (See Brassicaceae family in "Plant Families". For image see tomatoes in March Greenhouse.)

#5 Barrier- Root maggot flies lay eggs at the base of young vegetable plants especially broccoli, cabbage, cauliflower and onion. When the maggots hatch, they eat the plants. Use tar paper or other heavy, flexible paper. Cut squares or circles 6-8 inches long. Make 1 cut from an edge to the center and then cut a small hole. When planting, put the seed or seedling in the circle and the paper flat on the ground. (See Brassicaceae family in "Plant Families".)

How to Make Traps and Deterrents for Slugs and Snails

Slugs and snails feed at night. Look for shiny trails of slime near chewed holes in leaves. Contact your County Extension Office.

Traps
#1 For a trap use an aluminium pie plate, tuna can or other shallow container. Have top of container level with the soil. Or use a taller container and cut a 1 inch hole in the side that is at soil level when partially buried. Or try a container that is 5-6 inches deep. Bait the traps with beer or a mixture of baking yeast/molasses/water. Put 1 trap every 10 square feet. Experiment.
#2 During the day snails and slugs rest in moist, shady spots. So put out overturned pots, boards, or shingles near plants. Check periodically and remove them.
#3 Put rotting fruit, damp dog/cat food or a cut potato (cut surface down) in a container or on the ground at dusk. The snails/slugs stay and eat. Collect in morning to kill them.

Deterrents (Experiment.)
#1 Snails and slugs are sensitive to getting dried out. A barrier of wood ash sprinkled on the ground absorbs moisture from their bodies.
#2 Other substances snails and slugs may not cross include crushed oyster shells, pea or chat gravel (horticultural grit 4-8 mm), coarse/sharp sand (not play sand), sawdust, garden lime, vermiculite, and diatomaceous earth.

 ## How to Use Floating Row Covers and Other Physical Crop Protectors

Row covers (Reemay) rest lightly on plants or they can be supported with heavy #9 wire or pvc pipe. They come in different weights. They let water and light in but keep pests out if all edges are sealed with soil, rocks, boards, bricks or row cover pins. They also **extend growing season** by 5-10 days. However, in very hot weather, they can get up to 30 degrees hotter.

They **keep out insects** that fly or crawl, rabbits, and birds but they do not keep out pests that come up through the soil. Good to keep out aphids, armyworms, asparagus beetles, cabbage root maggots, carrot rust flies, Colorado potato beetles, cutworms, flea beetles, leafhoppers, and Mexican bean beetles.

Place covers loosely over plants immediately after planting seeds or transplants. May need wire or PVC frame. **Remove covers** permanently when plants are blooming so pollination can take place. Remove when plants get larger or particular pest season is over.

After using wash with soapy water and bleach if there were diseases. Patch holes with duct tape on both sides.

A heavy weight row cover called a **frost blanket** is used in winter to protect plants. Sheets can also be used.
Hot caps are used in **early spring** to protect plants from cold temperatures. They can be plastic milk containers, a cloche (bell jar), or a water filled teepee (Wall-O-Water). They are put on individual plants in the evening and removed in the morning if it is going to be warm and sunny.

 ## How to Make Organic Insect or Pest Sprays

1. All Purpose Insect Spray for Leaf Eating Insects
It kills good and bad insects. It controls aphids, cabbageworms, leafhoppers, squash bugs, and whiteflies. Does not work against Colorado potato beetles, grapeleaf skeletonizers, grasshoppers, red ants, or sowbugs.

#1: Blend an entire **garlic** bulb with 2 cups of water in a blender. You can also add 1 small onion and/or 1 teaspoon powdered cayenne pepper. Mix at high speed for 1-2 minutes. Pour into a container and let sit for up to one day. Strain liquid through cheesecloth. Add 1 gallon of water. Add 1 tablespoon liquid dish soap and mix. Spray on top and bottom of leaves.

#2: Soak 3 ounces of finely minced **garlic** in 2 teaspoons mineral oil for at least 24 hours. Add 1 pint water and 2-3 teaspoons liquid dish soap. Stir well and strain into glass jar for storage. To use, combine 1-2 tablespoons of concentrate with 1 pint water to make a spray.

2. Herbal Sprays
a. Alcohol extracts (tinctures) of hyssop, rosemary, sage, thyme, and white clover help control insects. Tansy repels cabbageworms on cabbage. To make an **alcohol extract**, soak overnight 1 cup packed fresh herb in 1/8 cup of 70% isopropyl alcohol. Strain through cheesecloth, then store in tightly sealed, glass container. To use, add 3 quarts water to the extract. Do not eat if made with isopropyl alcohol since it is not edible. (For **how to make a tincture, decoction or infusion**, see April Garden-Harvest.)

b. Essential oils such as sage and thyme are diluted to a few drops per cup of water.

c. Herbal non-alcohol solutions are made by mashing or blending 1-2 cups packed, fresh leaves with 2-4 cups water. Let soak overnight. Strain through cheesecloth and dilute with another 2-4 cups water. Add a few drops liquid dish soap to help spray stick.

d. Teas made from wormwood and nasturtium (see below) repel aphids. Also try catnip, chives, feverfew, marigold (see below), and rue against leaf eating pests.

3. Hot Pepper Spray
This is used to repel deer, rabbits and other pests from flowers and some vegetables. Put hot peppers and 2 cups of water into a blender. Mix at high speed for 1-2 minutes. Pour into container and let sit for up to one day. Strain liquid through cheesecloth. Add liquid to 1 quart of water. Spray on plants. Re-apply after rain or in one to two weeks.

4. Marigold or Nasturtium Insect Spray
Marigolds and nasturtiums are grown as companion plants to keep pests away. **Marigolds** are good for keeping away tomato hornworms, asparagus beetles, and other leaf-cutting insects. **Nasturtiums** are good for keeping

Plant Health p. 19

away n- aphids, cucumber beetles, leafminers and white flies. Put leaves, flowers and stems in a blender with hot water. Let it soak for a day, then drain. Add more water and apply liberally. (For **growing marigold** see April Greenhouse. For **growing nasturtium** see June Garden-Plant. See **companion planting** in "Garden Tips".)

5. Soap and Soap/Oil Insect Repellent Sprays

#1 Use a pure soap such as **Ivory Snow, Ivory Liquid or Shaklee's Basic H**. Perfumes, whiteners, etc in unpure soaps can damage plants. Or use dishwashing liquid. Do not use anti-bacterial soap because it can damage plants. Put one tablespoon of soap per gallon of water into a sprayer. Spray on top and bottom of leaves.

#2 Mix 1 tablespoon dishwashing liquid (such as Ivory) or Shaklee's Basic H with 1 cup cooking oil. Mix 1 to 2 1/2 teaspoons with 1 cup water. (Do not use antibacterial soap since it may damage plants.)

For #1 or #2 you can mix soaps with **other ingredients** such as BT, cooking oil, horticultural oil or pyrethrin. Re-apply after rain or in one to two weeks. Contact your County Extension Office.

To control **aphids**, spray early spring and again when winged females arrive. Use yellow sticky traps to monitor populations. If still a problem, apply every 2 weeks. Spray during cool, humid, foggy weather. (For **sticky traps** see How to Make Traps for Flying Insects above.)

To control **other insects**, spray as soon as nymphs (they hatch out of eggs) begin feeding.

To control **mites**, use a high-pressure sprayer that washes away pests. Spray again in 7-10 days.

Use to **control pests** on carrots, celery, cucumber, eggplant, lettuce, and peppers.

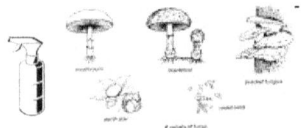

How to Make Organic Fungicide Sprays and Solutions

Remove leaves that are the most badly infected. Burn them or bury someplace else far away.

1. Baking Soda (sodium bicarbonate) Fungicide Spray #1

Dissolve 1 teaspoon baking soda in 1 quart warm water. Add 1 teaspoon liquid dish soap (such as Ivory) or Shaklee's Basic H or **insecticidal soap**. Spray on tops and bottoms of leaves. Helps with black spot and powdery mildew.

2. Baking Soda (sodium bicarbonate) Fungicide Spray #2

Mix 1 tablespoon baking soda with 1-2 teaspoons each of liquid dish soap (such as Ivory or use Shaklee's Basic H) and **cooking or horticultural oil**.

3. Baking Soda (sodium bicarbonate) Fungicide Spray #3

Mix 3 tablespoons baking soda, 1 tablespoon bleach, 1 teaspoon dishwashing liquid with 1 gallon of water. Spray on top and bottom of leaves. Apply a heavier dose on infected leaves and only lightly on unaffected leaves since bleach can harm leaves. Do not use more bleach than the amount above. Helps with powdery mildew and other fungus.

3. Bordeaux Mix- a Fungicide with Insecticidal and Anti-Bacterial Properties

Good for anthracnose, bacterial leaf spots, black spot on roses, black rot on grapes, fire blight on apples and pears, peach leaf curl, powdery mildew, rust, and wilts. Contact your County Extension Office.

It can burn foliage. It can damage apples, tomatoes and roses especially in cool, overcast weather. Best to apply just **before plants leaf out in spring**. Spray only once during dormant season. Bordeaux is **copper sulfate** (bluestone) and **hydrated lime** that is made into a wettable powder dusted on plants or mixed with water and sprayed.

Mix hydrated lime or any powdered lime with powdered copper sulfate. For a **dormant season spray**, mix 6 1/2 teaspoons of copper sulfate in 1 pint water (this may take a few minutes). Then in another container mix 3 tablespoons hydrated lime with 1 pint water. Then filter each through cheesecloth or similar material into the same 1 gallon container. Add 3 quarts water or enough to fill the container.

(For more about **copper sulfate**, see above "Sprays and Dusts". For **spraying fruit trees with dormant oil**, see February Garden-Maintenance.)

Bleach (sodium and calcium hyprochlorite) Solution to Disinfect Tools and Cuttings

Use bleach solution to disinfect greenhouse and garden tools, potting benches, potting flats, seed starting equipment, and cuttings. Dip plant cuttings before rooting in a 10% solution (1 part bleach to 9 parts water). Disinfect tools, traps, cages, and row covers with a 10% solution. Drench seedling soil with a 2% solution (2 1/2 tablespoons bleach to 1 gallon water). (For **seedlings and damping off**, see "Garden Tips".)

INDEX

Sections in the Index:
1. *Alphabetical Index: Fruits, Grain, Herbs, Nuts, Seeds and Vegetables*
2. *Plants Indexed by How Used*
3. *General Index*

1. Index of Fruits, Grain, Herbs, Nuts, Seeds and Vegetables by Common Name
(Months in bold have more information.)

Alfalfa- sprouts	**December Greenhouse**
Amaranth	**June Garden-Plant, September Garden-Harvest**
Anise Hyssop (mint)	March Garden-Plant, **May Garden-Plant**
Apple	**February**/June/July Garden-Maintenance, **April Garden-Plant, August**/Sept/Oct Garden-Harvest
Apricot	**February**/June Garden-Maintenance, **April Garden-Plant, August Garden-Harvest**
Arnica	**March Greenhouse**, May Garden-Plant
Arugula	**February**/September/October/November Greenhouse
Asian Pear	**February Garden-Maintenance, April Garden-Plant, August Garden-Harvest**
Asparagus	March/July/**Sept**/Oct Garden-Maint., March Greenhouse, **April Garden-Plant, April Garden-Harvest**
Bamboo	**April Garden-Plant, April Garden-Harvest**
Barley	**October Garden-Plant**
Basil	**April Greenhouse**, June Garden-Plant, July/September Garden-Harvest
Beans	**May Garden-Plant, August/September Garden-Harvest**
Beans- Fava	**March Garden-Plant**
Beans- Lima	**June Garden-Plant**
Beetberry	**May Garden-Plant**
Beets	**March**/August Garden-Plant, July/**October Garden-Harvest**, September Greenhouse
Blackberry	**April Garden-Plant, July Garden-Harvest**, February/August/**October Garden-Maintenance**
Blueberry	**February**/March/June/Dec Garden-Maintenance, **April Garden-Plant, July Garden-Harvest**
Broccoli	**March Greenhouse**, June/April Garden-Plant, **October Garden-Harvest**
Broccoli Raab	**March Greenhouse**, April Garden-Plant
Brussels Sprouts	**July Garden-Plant, October Garden-Harvest**
Buckwheat	**June**/July Garden-Plant
Burdock	**March Garden-Plant**, March/April/**June Garden-Harvest**
Burnet- Salad	**May Garden-Plant**, June Garden-Maintenance
Cabbage	March Greenhouse, **April**/June Garden-Plant, **October Garden-Harvest**
Catnip (a mint)	**May Garden-Plant**
Calendula	**May Garden-Plant**
Canola/Rape (kale)	**March**/August Garden-Plant, August/**December Greenhouse**
Carrot	March/**June Garden-Plant**, May/**September**/October Garden-Harvest, August/Dec Greenhouse
Cauliflower	**April Greenhouse**, May/June Garden-Plant, **September Garden-Harvest**
Celeriac	**May Garden-Plant, October Garden-Harvest**
Celery	**March Greenhouse**, May Garden-Plant, **August Garden-Harvest**
Chamomile- germ.	**May Garden-Plant**
Chamomile-roman	**May Garden-Plant**
Cherry (sand)	**April Garden-Plant, July Garden-Harvest**
Chicory	**May Garden-Plant**
Chickweed	**April Garden-Plant**

Index of Plants by Common Name p. 1

Chives	March/April Garden-Maintenance, **May Garden-Plant**
Chrysanthemum	**March Greenhouse**
Cilantro	**March Garden-Plant**
Claytonia	**February**/September/October/November Greenhouse
Clover	**March Garden-Plant**
Collard Greens	**March**/August Garden-Plant, August Greenhouse
Comfrey	**March Garden-Plant**, April Garden-Maintenance, **June**/July/August/September Garden-Harvest
Coriander	**March Garden-Plant** (in cilantro)
Corn	**May Garden-Plant, August/September**/October Garden-Harvest
Corn- field	**May Garden-Plant, September**/October Garden-Harvest
Cotton	**May Garden-Plant,** September Garden-Harvest
Creasy Green	**March**/August Garden-Plant, September/October Greenhouse
Cress- garden	**August Garden-Plant**
Cucumber	**April**/August Greenhouse, May Garden-Plant
Currant	**February** Garden-Maintenance, **April Garden-Plant, June Garden-Harvest**
Dandelion	**March Garden-Plant**, July/August Greenhouse
Daylily	**June**/July Garden-Harvest
Dill	**March Garden-Plant**
Dock- yellow	**April Garden-Harvest**
Echinacea	**April Garden-Plant**, July/**October Garden-Harvest**
Eggplant	**April Greenhouse**, June Garden-Plant
Elderberry	Feb/March/July/Oct Garden-Maintenance, **April Garden-Plant**, June/**Aug**/Sept Garden-Harvest
Endive	**August Garden-Plant**, September Greenhouse
Escarole	**August Garden-Plant**, September Greenhouse
Fava Beans	**March Garden-Plant**
Fennel- Bronze	**April Greenhouse**
Fennel- Florence	**April Greenhouse**, July Garden-Plant
Fenugreek	**May Garden-Plant**
Fern- Fiddlehead	**April Garden-Harvest**
Fescue	**March Garden-Plant**
Feverfew	**May Garden-Plant**
Fiddlehead Fern	**April Garden-Harvest**
Fig	**February**/July/November Garden-Maintenance, **April Garden-Plant, August Garden-Harvest**
Filbert (Hazelnut)	**April Garden-Plant, September Garden-Harvest**
Flax	**March Garden-Plant, July Garden-Harvest**
Forsythia	**April Garden-Harvest**
Garden Cress	**August Garden-Plant**
Garlic	April Garden-Maintenance, **July**/August Garden-Harvest, March/April/**September Garden-Plant**
Ginseng	**September Garden-Harvest**
Globe Artichoke	**February**/April Greenhouse, June Garden-Plant, **August**/September Garden-Harvest
Good King Henry	**May Garden-Plant**
Gooseberry	February Garden-Maintenance, **April Garden-Plant, June Garden-Harvest**
Gourd	**May Garden-Plant, September Garden-Harvest**
Grape	**February**/March/Dec Garden-Maintenance, **April Garden-Plant**, September Garden-Harvest
Ground Cherry	**March Greenhouse**, May Garden-Plant, **July Garden-Harvest**
Hairy Vetch	September Garden-Plant
Hardy Kiwi	**February**/June/July/Oct Garden-Maintenance, **April Garden-Plant, September Garden-Harvest**
Hay	June Garden-Harvest
Hazelnut (Filbert)	April Garden-Plant, September Garden-Harvest
Hickory Nuts	September Garden-Harvest
Hops	April Garden-Plant, August Garden-Harvest
Horehound	May Garden-Plant

Horseradish	**March Garden-Plant, October Garden-Harvest**
Kale	**March**/August Garden-Plant, August/December Greenhouse
Kohlrabi	March Greenhouse, **August Garden-Plant, October Garden-Harvest**
Jerusal. Artichoke	March/April/**October Garden-Harvest, May Garden-Plant**, July Garden-Maintenance
Lambs Quarters	**April Garden-Harvest**
Leek	**February Greenhouse**, May Garden-Plant, **September**/October Garden-Harvest
Lemon Balm (mint)	**May Garden-Plant**
Lettuce	**March**/August Garden-Plant, August/September Greenhouse
Lima Beans	**June Garden-Plant**
Lovage	**September Garden-Plant**, October Garden-Maintenance
Mache	**February**/September/October/November Greenhouse
Marigold	**April Greenhouse**, May Garden-Plant
Marjoram	**May Garden-Plant**
Marsh Mallow	**June Garden-Plant**
Melon	**April Greenhouse**, May Garden-Plant, **August Garden-Harvest**
Mizuna	**February**/September/October/November Greenhouse
Mint	**May Garden-Plant**
Minutina	**February**/September/October/November Greenhouse
Morel Mushroom	**April Garden-Harvest**
Mugwort (wormwood)	**May Garden-Plant**
Mullein	April/**August Garden-Plant**
Mushroom- S Mane	**June**/September Garden-Harvest
Mushroom-Shiitake	**February Garden Plant**
Mustard Greens	**February**/September Greenhouse, March/August Garden-Plant
Nasturtium	**June Garden-Plant**
Nectarine	**February Garden-Maintenance, April Garden-Plant, August Garden-Harvest**
Nettle- Stinging	**March Garden-Harvest, April Garden-Plant**, May Garden-Maintenance
Oak (acorn)	**September Garden-Harvest**
Oats	**March**/September Garden-Plant, **July Garden-Harvest**
Okra	April Greenhouse, **May Garden-Plant**
Onion- annual	**February Greenhouse**, April Garden-Maintenance, April Garden-Plant, **July**/Aug Garden-Harvest
Onion- perennial	April Garden-Maintenance, **September Garden-Plant, July**/August Garden-Harvest
Oregano	**May Garden-Plant**
Parsley	**April**/October Garden-Plant, August/September/October Greenhouse
Parsleyroot	July/October Garden-Plant, **October Garden-Harvest**
Parsnip	**April Garden-Plant**, April/**September**/December Garden-Harvest
Pea	**March Garden-Plant**
Peach	**February Garden-Maintenance, April Garden-Plant, August Garden-Harvest**
Peanut	**June Garden-Plant**
Pear	**February Garden-Maintenance, April Garden-Plant, August Garden-Harvest**
Pepper	**March Greenhouse**, May Garden-Plant
Petasite	**April Garden-Harvest**
Plantain	**April Garden-Plant**
Plum	**February Garden-Maintenance, April Garden-Plant, August Garden-Harvest**
Pokeweed	**April Garden-Harvest**
Potato	**March Greenhouse, April**/May/June Garden-Plant, June/July/Aug/Sept/**Oct Garden-Harvest**
Pumpkin	**May Garden-Plant, October Garden-Harvest**
Purslane	**May Garden-Plant**
Radicchio	**February Greenhouse**, August/September Garden-Plant

Index of Plants by Common Name p. 3

Radish- summer	**April**/August/September Garden-Plant, August/September Greenhouse
Radish- winter	**April**/**August**/September Garden-Plant, August/September Greenhouse
Ramp	**March Garden-Harvest**
Rape (in kale)	**March**/August Garden-Plant, August/December Greenhouse
Raspberry	**April Garden-Plant, July Garden-Harvest**, February/July/August/**October Garden-Maintenance**
Rhubarb	**March Garden-Plant**, April Garden-Maintenance, **April**/May Garden-Harvest
Rosa Rugosa	February Garden-Maintenance, **April Garden-Plant**, April/**September Garden-Harvest**
Rosemary	**May Garden-Plant**
Rutabaga	**June**/July Garden-Plant, **October Garden-Harvest**
Rye- Winter	**May Garden-Harvest, October Garden-Plant**
Ryegrass	**September Garden-Plant**
Sage	**May Garden-Plant**
Saint Johns Wort	**May Garden-Plant**
Salad Burnet	**May Garden-Plant**, June Garden-Maintenance
Salsify	**May Garden-Plant**, April/**September Garden-Harvest**
Savory- Winter	**May Garden-Plant**
Scorzonera	**May Garden-Plant** (in salsify)
Sea Kale	March/**May Garden-Plant**, November Garden-Maintenance
Shaggy Mane	**June/September Garden-Harvest**
Shiitake Mushroom	February Garden Plant
Shallot	**September Garden-Plant**
Sorghum	**May Garden-Plant, October Garden-Harvest**
Sorrel	**May Garden-Plant**, August/September/October Greenhouse
Soybeans	**May Garden-Plant**
Spinach	March/August Garden-Plant, May/**October Garden-Harvest**, August/September/Oct Greenhouse
Squash- summer	**May Garden-Plant**
Squash- winter	April Greenhouse, **May Garden-Plant**, October Garden-Harvest
Stinging Nettle	**March Garden-Harvest, April Garden-Plant**, May Garden-Maintenance
Strawberry	March/**April Garden-Plant, May**/June/July Garden-Harvest, May/**June**/July Garden-Maintenance
Sunchoke	March/April/**October Garden-Harvest**, **May Garden-Plant**, July Garden-Maintenance
Sunflower	**June Garden-Plant, September Garden-Harvest**
Sweet Potato	March/April **Greenhouse, June Garden-Plant, October Garden-Harvest**
Sweet Violet	**May Garden-Plant**
Swiss Chard	**May**/August Garden-Plant, August/September Greenhouse
Tarragon	**April Greenhouse**, May Garden-Plant
Thyme	**May Garden-Plant**
Tobacco	**April Greenhouse**, June Garden-Plant, **August Garden-Harvest**
Tomatillo	**March**/May Greenhouse, June Garden-Plant, **August Garden-Harvest**
Tomato	**March**/May Greenhouse, May/June Garden-Plant, **August**/October Garden-Harvest
Turnip	March/**August Garden-Plant, October Garden-Harvest**
Valerian	**May Garden-Plant**, June/August/**October Garden-Harvest**, August Garden-Maintenance
Vetch-hairy	**September**/October Garden-Plant
Violet- Sweet	**May Garden-Plant**
Walnut	February Garden-Maintenance, **April Garden-Plant, September Garden-Harvest**
Watermelon	April Greenhouse, May Garden-Plant, **August Garden-Harvest**
Watercress	**April**/September Greenhouse
Wheat- Winter	**May Garden-Harvest, October Garden-Plant**
Winter Savory	**May Garden-Plant**
Winter Squash	April Greenhouse, **May Garden-Plant, October Garden-Harvest**
Wormwood	**May Garden-Plant**
Yarrow	**May Garden-Plant**
Yellow Dock	**April Garden-Harvest**

2. Index of Plants Categorized by How Used (Months in bold have more information.)

Fruit- Annual or Biennial (no tomatoes)
Fruits & Nuts- Perennial
Grain / Grass / Seeds / Fiber and Cover Crops
Herbs & Flowers- Annual or Biennial,
Herbs & Flowers- Perennial
Leafy Greens- Annual or Biennial (no herbs)
Leafy Greens- Perennial (no herbs)
Root Crops (all)
Tomato and Related
Vegetables- Annual or Biennial (no root crops, no leafy greens)
Vegetables- Perennial (no root crops, no leafy greens)

 ### Fruit- Annual or Biennial (no tomatoes)

Beetberry	**May Garden-Plant**
Melon	**April Greenhouse**, May Garden-Plant, **August Garden-Harvest**
Watermelon	**April Greenhouse**, May Garden-Plant, **August Garden-Harvest**

 ### Fruit & Nuts- Perennial

Apple	**February**/June/July Garden-Maintenance, **April Garden-Plant**, **August**/Sept/Oct Garden-Harvest
Apricot	**February**/June Garden-Maintenance, **April Garden-Plant**, **August Garden-Harvest**
Asian Pear	**February Garden-Maintenance**, **April Garden-Plant**, **August Garden-Harvest**
Blackberry	**April Garden-Plant**, **July Garden-Harvest**, February/August/**October Garden-Maintenance**
Blueberry	**February**/March/June/Dec Garden-Maintenance, **April Garden-Plant**, **July Garden-Harvest**
Cherry (sand)	**April Garden-Plant**, **July Garden-Harvest**
Currant	**February** Garden-Maintenance, **April Garden-Plant**, **June Garden-Harvest**
Elderberry	Feb/March/July/Oct Garden-Maintenance, **April Garden-Plant**, June/**Aug**/Sept Garden-Harvest
Fig	**February**/July/November Garden-Maintenance, **April Garden-Plant**, **August Garden-Harvest**
Filbert (Hazelnut)	**April Garden-Plant, September Garden-Harvest**
Gooseberry	**February Garden-Maintenance**, **April Garden-Plant**, **June Garden-Harvest**
Grape	**February**/March/Dec Garden-Maintenance, **April Garden-Plant**, September Garden-Harvest
Hardy Kiwi	**February**/June/July/Oct Garden-Maintenance, **April Garden-Plant**, **September Garden-Harvest**
Hazelnut (Filbert)	**April Garden-Plant, September Garden-Harvest**
Hickory Nuts	September Garden-Harvest
Nectarine	**February Garden-Maintenance**, **April Garden-Plant**, **August Garden-Harvest**
Oak (acorn)	September Garden-Harvest
Peach	**February Garden-Maintenance**, **April Garden-Plant**, **August Garden-Harvest**
Pear	**February Garden-Maintenance**, **April Garden-Plant**, **August Garden-Harvest**
Plum	**February Garden-Maintenance**, **April Garden-Plant**, **August Garden-Harvest**
Raspberry	**April Garden-Plant**, **July Garden-Harvest**, February/July/August/**October Garden-Maintenance**
Rosa Rugosa	**February Garden-Maintenance**, **April Garden-Plant**, April/**September Garden-Harvest**
Strawberry	March/**April Garden-Plant**, **May**/June/July Garden-Harvest, May/**June**/July Garden-Maintenance
Walnut	**February Garden-Maintenance**, **April Garden-Plant**, **September Garden-Harvest**

 ### Grain, Grass, Seeds, Fiber and/or Cover Crops

Amaranth	**June Garden-Plant**, **September Garden-Harvest**
Bamboo	**April Garden-Plant**, **April Garden-Harvest**

Barley	**October Garden-Plant**
Buckwheat	**June**/July Garden-Plant
Clover	**March Garden-Plant**
Coriander	**March Garden-Plant** (in cilantro)
Corn- field	**May Garden-Plant, September**/October Garden-Harvest
Cotton	**May Garden-Plant,** September Garden-Harvest
Fenugreek	**May Garden-Plant**
Fescue	**March Garden-Plant**
Flax	**March Garden-Plant, July Garden-Harvest**
Hairy Vetch	**September Garden-Plant**
Hay	**June Garden-Harvest**
Oats	**March**/September Garden-Plant, **July Garden-Harvest**
Ryegrass	**September Garden-Plant**
Rye- Winter	**May Garden-Harvest, October Garden-Plant**
Sorghum	**May Garden-Plant, October Garden-Harvest**
Sunflower	**June Garden-Plant, September Garden-Harvest**
Vetch-hairy	**September**/October Garden-Plant
Wheat- Winter	**May Garden-Harvest, October Garden-Plant**

Herbs & Flowers- Annual or Biennial

Basil	**April Greenhouse**, June Garden-Plant, July/September Garden-Harvest
Calendula	**May Garden-Plant**
Chamomile- germ.	**May Garden-Plant**
Cilantro	**March Garden-Plant**
Chickweed	**April Garden-Plant**
Dill	**March Garden-Plant**
Fennel- Florence	**April Greenhouse**, July Garden-Plant
Fenugreek	**May Garden-Plant**
Marigold	**April Greenhouse**, May Garden-Plant
Marjoram	**May Garden-Plant**
Mullein	April/**August Garden-Plant**
Nasturtium	**June Garden-Plant**
Parsley	**April**/October Garden-Plant, August/September/October Greenhouse
Purslane	**May Garden-Plant**

Herbs & Flowers- Perennial

Anise Hyssop (mint)	March Garden-Plant, **May Garden-Plant**
Arnica	**March Greenhouse**, May Garden-Plant
Catnip (a mint)	**May Garden-Plant**
Chamomile-roman	**May Garden-Plant**
Chrysanthemum	**March Greenhouse**
Chives	March/April Garden-Maintenance, **May Garden-Plant**
Daylily	**June**/July Garden-Harvest
Dock- yellow	**April Garden-Harvest**
Echinacea	**April Garden-Plant**, July/**October Garden-Harvest**
Forsythia	**April Garden-Harvest**
Ginseng	**September Garden-Harvest**
Hops	**April Garden-Plant, August Garden-Harvest**
Fennel- Bronze	**April Greenhouse**
Feverfew	**May Garden-Plant**
Forsythia	**April Garden-Harvest**

Horehound	May Garden-Plant
Lemon Balm (mint)	May Garden-Plant
Marsh Mallow	June Garden-Plant
Mint	May Garden-Plant
Mugwort (wormwood)	May Garden-Plant
Oregano	May Garden-Plant
Plantain	April Garden-Plant
Rosemary	May Garden-Plant
Sage	May Garden-Plant
Saint Johns Wort	May Garden-Plant
Savory- Winter	May Garden-Plant
Sweet Violet	May Garden-Plant
Tarragon	**April Greenhouse**, May Garden-Plant
Thyme	May Garden-Plant
Valerian	**May Garden-Plant**, June/August/**October Garden-Harvest**, August Garden-Maintenance
Violet- sweet	May Garden-Plant
Winter Savory	May Garden-Plant
Wormwood	May Garden-Plant
Yarrow	May Garden-Plant

Leafy Greens- Annual or Biennial (no herbs, no perennials)

Arugula	**February**/September/October/November Greenhouse
Cabbage	March Greenhouse, **April**/June Garden-Plant, **October Garden-Harvest**
Chicory	**May Garden-Plant**
Collard Greens	**March**/August Garden-Plant, August Greenhouse
Claytonia	**February**/September/October/November Greenhouse
Creasy Green	**March**/August Garden-Plant, September/October Greenhouse
Cress- garden	**August Garden-Plant**
Endive	**August Garden-Plant**, September Greenhouse
Escarole	**August Garden-Plant**, September Greenhouse
Garden Cress	**August Garden-Plant**
Kale	**March**/August Garden-Plant, August/December Greenhouse
Lambs Quarters	**April Garden-Harvest**
Lettuce	**March**/August Garden-Plant, August/September Greenhouse
Mache	**February**/September/October/November Greenhouse
Minutina	**February**/September/October/November Greenhouse
Mizuna	**February**/September/October/November Greenhouse
Mustard Green	**February**/September Greenhouse, March/August Garden-Plant
Radicchio	**February Greenhouse**, August/September Garden-Plant
Spinach	**March**/August Garden-Plant, May/**October Garden-Harvest**, August/September/Oct Greenhouse
Swiss Chard	**May**/August Garden-Plant, August/September Greenhouse

Leafy Greens- Perennials (no herbs)

Burnet- Salad	**May Garden-Plant**, June Garden-Maintenance
Chicory	**May Garden-Plant**
Comfrey	**March Garden-Plant**, April Garden-Maintenance, **June**/July/August/September Garden-Harvest
Dandelion	**March Garden-Plant**, July/August Greenhouse
Good King Henry	**May Garden-Plant**
Lovage	**September Garden-Plant**, October Garden-Maintenance
Nettle- Stinging	**March Garden-Harvest, April Garden-Plant**, May Garden-Maintenance

Salad Burnet	**May Garden-Plant**, June Garden-Maintenance
Sea Kale	March/**May Garden-Plant**, November Garden-Maintenance
Sorrel	**May Garden-Plant**, August/September/October Greenhouse
Stinging Nettle	**March Garden-Harvest, April Garden-Plant**, May Garden-Maintenance
Watercress	**April**/September Greenhouse
Yellow Dock	**April Garden-Harvest**

Root Crops (annual, biennial and perennial)

Beets	**March**/August Garden-Plant, July/**October Garden-Harvest**, September Greenhouse
Burdock	**March Garden-Plant**, March/April/**June Garden-Harvest**
Carrot	March/**June Garden-Plant**, May/**September**/October Garden-Harvest, August/Dec Greenhouse
Celeriac	**May Garden-Plant, October Garden-Harvest**
Garlic	April Garden-Maintenance, **July**/August Garden-Harvest, March/April/**September Garden-Plant**
Horseradish	**March Garden-Plant, October Garden-Harvest**
Jerusal. Artichoke	March/April/**October Garden-Harvest, May Garden-Plant**, July Garden-Maintenance
Kohlrabi	March Greenhouse, **August Garden-Plant, October Garden-Harvest**
Leek	**February Greenhouse**, May Garden-Plant, **September**/October Garden-Harvest
Onion- annual	**February Greenhouse**, April Garden-Maintenance, April Garden-Plant, **July**/Aug Garden-Harvest
Onion- perennial	April Garden-Maintenance, **September Garden-Plant, July**/August Garden-Harvest
Parsleyroot	**July**/October Garden-Plant, **October Garden-Harvest**
Parsnip	**April Garden-Plant**, April/**September**/December Garden-Harvest
Potato	**March Greenhouse, April**/May/June Garden-Plant, June/July/Aug/Sept/**Oct Garden-Harvest**
Radish- summer	**April**/August/September Garden-Plant, August/September Greenhouse
Radish- winter	**April**/**August**/September Garden-Plant, August/September Greenhouse
Ramp	**March Garden-Harvest**
Rutabaga	**June**/July Garden-Plant, **October Garden-Harvest**
Salsify	**May Garden-Plant**, April/**September Garden-Harvest**
Scorzonera	**May Garden-Plant** (in salsify)
Shallot	**September Garden-Plant**
Sunchoke	March/April/**October Garden-Harvest, May Garden-Plant**, July Garden-Maintenance
Sweet Potato	**March**/April Greenhouse, June Garden-Plant, **October Garden-Harvest**
Turnip	March/**August Garden-Plant, October Garden-Harvest**

● Tomato and Related Fruit

Ground Cherry	March Greenhouse, May Garden-Plant, **July Garden-Harvest**
Tomatillo	**March**/May Greenhouse, June Garden-Plant, **August Garden-Harvest**
Tomato	**March**/May Greenhouse, May/June Garden-Plant, **August**/October Garden-Harvest

Vegetables- Annual & Biennial (no root crops, no leafy greens)

Alfalfa- sprouts	December Greenhouse
Beans	**May Garden-Plant, August/September Garden-Harvest**
Beans- Fava	**March Garden-Plant**
Beans- Lima	June Garden-Plant
Broccoli	**March Greenhouse**, June/April Garden-Plant, **October Garden-Harvest**
Broccoli Raab	**March Greenhouse**, April Garden-Plant
Brussels Sprouts	**July Garden-Plant, October Garden-Harvest**
Cauliflower	**April Greenhouse**, May/June Garden-Plant, **September Garden-Harvest**

Celery	**March Greenhouse**, May Garden-Plant, **August Garden-Harvest**
Corn	**May Garden-Plant**, **August/September**/October Garden-Harvest
Cucumber	**April**/August Greenhouse, May Garden-Plant
Eggplant	**April Greenhouse**, June Garden-Plant
Fava Beans	**March Garden-Plant**
Globe Artichoke	February/April Greenhouse, June Garden-Plant, **August**/September Garden-Harvest
Gourd	**May Garden-Plant**, **September Garden-Harvest**
Lima Beans	**June Garden-Plant**
Morel Mushroom	**April Garden-Harvest**
Mushroom- S Mane	**June/September Garden-Harvest**
Mushroom-Shiitake	**February Garden Plant**
Okra	April Greenhouse, **May Garden-Plant**
Pea	**March Garden-Plant**
Peanut	**June Garden-Plant**
Pepper	**March Greenhouse**, May Garden-Plant
Pumpkin	**May Garden-Plant**, **October Garden-Harvest**
Shaggy Mane	**June/September Garden-Harvest**
Soybeans	**May Garden-Plant**
Squash- summer	**May Garden-Plant**
Squash- winter	April Greenhouse, **May Garden-Plant**, October Garden-Harvest
Tobacco	**April Greenhouse**, June Garden-Plant, **August Garden-Harvest**
Winter Squash	April Greenhouse, **May Garden-Plant**, **October Garden-Harvest**

Vegetables- Perennial (no root crops, no leafy greens)

Asparagus	March/July/**Sept**/Oct Garden-Maint., March Greenhouse, **April Garden-Plant**, **April Garden-Harvest**
Bamboo	**April Garden-Plant, April Garden-Harvest**
Fiddlehead Fern	**April Garden-Harvest**
Fern- Fiddlehead	**April Garden-Harvest**
Petasite	**April Garden-Harvest**
Pokeweed	**April Garden-Harvest**
Rhubarb	**March Garden-Plant**, April Garden-Maintenance, **April**/May Garden-Harvest

3. General Index (no plant names)

Acres Needed to Grow Food: Preface
Amaranthaceae Family (amaranth): Plant Families
Amaryllidaceae Family (onion): Plant Families
Amaranth Family (Amaranthaceae): Plant Families
Animal Bedding: January Farm Animals
Animal Feed Supplements: April Farm Animals (minerals), November Farm Animals
Animal (Nuisance) Control: Ways to Improve Plant Health
Animal Pests (Large): Ways to Improve Plant Health
Annual/Biennial/Perennial: How to Save Seeds
Ant Traps: Ways to Improve Plant Health
Apiaceae or Umbelliferae Family (parsley, carrot, fennel, parsnip): Plant Families
Appalachian Folklore: September Garden-Harvest, November Garden-Maintenance, Garden Tips
Apple Maggot Fly Traps for Fruit Trees: June Garden-Maintenance
Ass or Donkey Care: February Farm Animals
Astrological Signs for Planting: Garden Tips

Bacterial Disease (Plants): Ways to Improve Plant Health
Bagging Fruit in Trees: June Garden-Maintenance
Bantam (Banty, Bantie) Chickens: April Farm Animals
Bedding (Animal): January Farm Animals
Bee (Mason) Homes: December Farm Animals
Beekeeping (Honey): January Farm Animals plus in most months
Beet Pulp: November Farm Animals, Garden Tips (pasture)
Beneficial Insects including Pollinators: Ways to Improve Plant Health under "Insects and Nematodes"
Biennial/Annual/Perennial: How to Save Seeds
Bird Pests: Ways to Improve Plant Health
Blooming and Pollination- Fruit and Nut Trees: April Garden-Plant
Books- Good Gardening: March Garden-Plant (books are listed in all months)
Borers- Apple and Fruit Trees: May Garden-Maintenance
Brambles (Raspberries and Blackberries): April Garden-Plant, October Garden-Maintenance
Brassica Rapa: Includes field mustard, mizuna, turnips, turnip rape, and turnip mustard. See mizuna and turnip listings.
Brassicaceae or Cruciferae Family (cabbage+): Plant Families
Breeding- Sheep and Goats: September Farm Animals
Broadcast Spreaders and Garden Seeders: Garden Tips
Brooding Baby Birds and Incubating: March Farm Animals
Broody Hens (Poultry): April Farm Animals
Bushel- How Much Is It: Garden Tips
Butcher Animals: November Farm Animals

Calories and High Yield Crops: Preface
Carbon Crops: Preface
Cat Care and Breeding: February Farm Animals
Cattle (Cow) Care and Breeds: February Farm Animals
Chenopodiaceae Family (beets, spinach+): Plant Families
Chicken Care and Breeds: February/April Farm Animals
Chicken and Poultry- Broody Hens: April Farm Animals
Chicken and Poultry- Egg Production: April Farm Animals
Chicken and Poultry- Egg Storage: July Farm Animals
Chicken and Poultry- Garlic Water: January Farm Animals
Chicken and Poultry- Incubating Eggs: March Farm Animals

General Index p. 1

Chicken and Poultry- Mites and Lice: March Farm Animals
Chicken and Poultry- Molting: September Farm Animals
Chickens- Bantam (Banty/Bantie) and Standard: April Farm Animals
Chill Hours and Dormancy- Fruit and Nut Trees: April Garden-Plant
Chipmunk Pests: Ways to Improve Plant Health
Cleaning Barn and Coop: March Farm Animals
Clean Farm and Tools: November Garden-Maintenance
Coccidia (Coccidiosis) in Animals: April Farm Animals
Cold Frame and Hoop House: January Garden-Maintenance
Companion Planting: Garden Tips
Compositae or Thistle Family (sunflower+): Plant Families
Convolvulaceae Family (sweet potato): Plant Families
Cool Season Cover Crops: March/September/October Garden Plant, Garden Tips
Cool Season Crops (list): Preface
Cool Season Root Crops Planted: July/August Garden-Plant
County Cooperative Extension Service: March Garden-Plant
Cover Crops, Grains and Grass: March/May/June/September/October Garden-Plant, Garden Tips
Cow (Cattle) Care and Breeds: February Farm Animals
Crop Rotation and Cover Crops: Garden Tips
Crops- Cool/Warm/Hot Season, Main, Leafy, Root, Calorie/Yield, Carbon: Preface
Cruciferae or Brassicaceae Family (cabbage+): Plant Families
Cucurbitaceae Family (melon, squash, gourd): Plant Families

Damping Off of Seedlings: Garden Tips
Decoction with Herbs: April Garden-Harvest
Deer Pests: Ways to Improve Plant Health
Diatomaceous Earth (DE or diatoms): May Garden-Harvest (under Storing Grain), April Farm Animals (worming)
Disbudding (horns): February Farm Animals (under Goats)
Disinfecting Tools: Ways to Improve Plant Health
Dog Care and Breeding: February Farm Animals
Donkey or Ass Care: February Farm Animals
Dormant Oil Spray for Fruit and Nut Trees: February Garden-Maintenance
Draft Horses: February Farm Animals
Drying and Storing Grain: May Garden-Harvest
Duck Care and Breeds: February Farm Animals
Dusts and Sprays (Plants): Ways to Improve Plant Health

Ear Mites: February Farm Animals (under Rabbits but may be useful in other animals)
Egg Production: February Farm Animals
Egg Storage: July Farm Animals
Extension Service (County Cooperative): March Garden-Plant

Fabaceae or Leguminosae Family (legume, bean, pea): Plant Families
Fall Sowing: September Garden-Plant
FAMACHA (Worm testing with eyelid): April Farm Animals
Fecal Testing for Worms: April Farm Animals
Fermentation of Seeds: How to Save Seeds
Fertilizing and Fertilizers: March/April Garden-Maintenance, Garden Tips, Ways to Improve Plant Health
Fire Blight in Fruit Trees: June Garden-Maintenance
Firewood and Logging: October Garden-Maintenance
First Killing Frost in North Carolina: October in beginning section
Flowers- Edible: May Garden-Harvest
Flowers- Edible: May Garden-Harvest
Fly Traps: Ways to Improve Plant Health
Folklore- Appalachian: September Garden-Harvest, November Garden-Maintenance, Garden Tips
Food Growing- Acres Needed: Preface
Food (Solar) Dehydrator: May Garden-Harvest

Food Storage: Preface (and information throughout calendar), May Garden-Harvest (under Storing Grain)
Food Storage and Seed Saving Books: August/September Garden-Harvest
Food Supplements (kelp, beet pulp): November Farm Animals
Forage or Pasture: May/November Farm Animals, Garden Tips
Forage (Pasture) Crops: Garden Tips
Foraging: September Garden-Harvest (nuts), though all types in most months
Foraging in Appalachia DVDs: Preface
Frost (First Killing in Fall) in North Carolina: October Garden-Maintenance
Frost (Last Killing in Spring) in North Carolina: May Garden-Maintenance
Frost Hardy Greens and Vegetables (list): February Greenhouse
Frost Predicting and Protection: March Garden-Plant
Fruit and Nut Trees- Borers: May Garden-Maintenance
Fruit and Nut Trees- Burlap Bags and Insects: May Garden-Maintenance
Fruit and Nut Trees- Chill Hours and Dormancy: April Garden-Plant
Fruit and Nut Trees- Dwarf, Semi-Dwarf and Standard: April Garden-Plant
Fruit and Nut Trees- Fertilizing: March Garden-Maintenance
Fruit and Nut Trees- Fungicide Sprays: February Garden-Maintenance
Fruit and Nut Trees- How to Plant: April Garden-Plant
Fruit and Nut Trees- Oil (Dormant) Sprays: February Garden-Maintenance
Fruit and Nut Trees- Pollination and Blooming: April Garden-Plant
Fruit and Nut Trees- Pruning: February Garden-Maintenance
Fruit and Nut Trees- Thinning: June Garden-Maintenance
Fruit and Nut Trees- Sticky Insect Bands: April Garden-Maintenance
Fruit and Nut Trees- Tree Guards: September Garden-Maintenance
Fruit and Nut Trees- Whitewashing: June Garden-Maintenance
Fruit Bearing Age, Plant Longevity, Chill Hours, and Annual Fruit Yields: April Garden-Plant
Fruit Chilling Hours: April Garden-Plant
Fruit Trees- Bagging Fruit: June Garden-Maintenance
Fruit Trees- Bearing Age, Plant Longevity, Chill Hours and Fruit Yields: April Garden-Plant
Fruit Trees- Fire Blight: June Garden-Maintenance
Fungi (Fungal) Disease (Plants): Ways to Improve Plant Health, Garden Tips
Fungicide Sprays and Solutions: February Garden-Maintenance, Ways to Improve Plant Health

Garden Seeders and Broadcast Spreaders: Garden Tips
Garlic Water for Poultry: January Farm Animals
Genetic Vigor and Inbreeding: How to Save Seeds
Genetically Modified Organisms (GMO): May Garden-Plant (in corn), How to Save Seeds
Goat Care and Breeds: February/April/September Farm Animals
Gopher Pests: Ways to Improve Plant Health
Grain Harvesting, Testing/Drying/Storing Grain: May Garden-Harvest
Grain Moisture Content: May Garden-Harvest
Grains, Grass and Cover Crops: March/May/June/September Garden-Plant, Garden Tips
Gramineae or Poaceae Family (corn, sorghum): Plant Families
Grass, Cover Crops, Grains: March/May/June/September/October Garden-Plant, Garden Tips
Green Manure Crops: June Garden-Plant
Greenhouse- Annuals: September/October Greenhouse
Greenhouse- Cold Temperatures: February Greenhouse
Greenhouse- Frost Hardy Greens Planting: August Greenhouse
Greenhouse- Perennial Herbs: March Greenhouse
Greenhouse- Pests and Disease: Ways to Improve Plant Health
Greens (Leafy Crops- list): Preface

Hardening Seedlings Before Planting: Garden Tips
Hardiness Zones (USDA): January
Hardy Greens: February Greenhouse
Hardy Vegetables (list): February Garden-Maintenance
Harvesting Herbs: April Garden-Harvest

Harvesting Perennials: April/September Garden-Harvest
Harvesting and Preparing Seed: How to Save Seeds
Hay- Cuttings and Types: June Garden-Harvest
Hay Making, Grain Harvesting, Testing/Drying/Storing Grain: May Garden-Harvest
Herb Harvesting and Trimming: April/July Garden-Harvest, July Garden-Maintenance
Herb Planting: May Garden-Plant
Herb Trimming / Cutting: July Garden-Maintenance
Herb Decoctions, Infusions and Tinctures: April Garden-Harvest
Herbs- Perennial: March Greenhouse
Herbal Wormers: April Farm Animals
Hog (Pig) Care and Breeds: February Farm Animals
Homing Pigeons: April/August Farm Animals
Honey Beekeeping: January Farm Animals and most months
Honey- Buying: September Farm Animals
Honey Harvest: September Farm Animals
Hoof Trimming (Goats): February Farm Animals
Hoophouse and Cold Frame: January Greenhouse
Horse Care and Breeds: February Farm Animals
Hot Season Crops (list): Preface
Hybrid Vigor: February Farm Animals, How to Save Seeds

Inbreeding and Genetic Vigor (Plants): How to Save Seeds
Inbreeding and Outbreeding Plants: How to Save Seeds
Inbreeding, Linebreeding and Outbreeding Animals: February Farm Animals
Incubating Eggs: March Farm Animals
Infusion with Herbs: April Garden-Harvest
Inoculating Seeds: Garden Tips
Insects- Beneficial: Ways to Improve Plant Health under "Insects and Nematodes"
Insect Control Products: Ways to Improve Plant Health
Insect Damage Symptoms: Ways to Improve Plant Health
Insect or Pest Sprays: Ways to Improve Plant Health
Insect Traps and Barriers for Crawling Insects: Ways to Improve Plant Health
Insect Traps for Flying Insects: Ways to Improve Plant Health
Insects and Nematodes: Ways to Improve Plant Health
Insects- Beneficial: Ways to Improve Plant Health

Labiatae Family (basil): Plant Families
Last Killing Frost in Spring in North Carolina: May Garden-Maintenance
Leguminosae or Fabaceae Family (legume, bean, pea): Plant Families
Lice and Mites (Poultry): March Farm Animals
Liliaceae Family (asparagus): Plant Families
Linebreeding, Inbreeding and Outbreeding Animals: February Farm Animals
Logging and Firewood: October Garden-Maintenance
Logs and Mushrooms: February Garden-Plant

Malvaceae Family (okra): Plant Families
Maggot Fly Traps for Fruit Trees: June Garden-Maintenance
Mason Bee Homes: December Farm Animals
Mice / Mouse Pests: Ways to Improve Plant Health
Milk- Clabbered: February Farm Animals (under Goats)
Milking Animals: February Farm Animals (under Goats)
Minerals- High Magnesium: April Farm Animals
Mites, Lice and Mange (Animals): February Farm Animals
Mites and Lice (Poultry): March Farm Animals
Mole Pests: Ways to Improve Plant Health
Molting (Poultry): September Farm Animals
Moon: Garden Tips (under Appalachian Folklore)

Mushroom Spawn/Spores: February Garden-Plant

Nematodes and Insects: Ways to Improve Plant Health
Nightshade or Solanaceae Family (tomato, potato, eggplant, pepper+): Plant Families
North Carolina Map of Geology: Preface
Nut Trees: See Fruit and Nut Trees in Index
Nutrient Deficiency Symptoms (Plant): Ways to Improve Plant Health

Oil Spray (Dormant) for Fruit and Nut Trees: February Garden-Maintenance
Onion (Amaryllidaceae Family): Plant Families
Open Pollination (and Hybrid): How to Save Seeds
Outbreeding and Inbreeding Plants: How to Save Seeds
Outbreeding, Linebreeding and Inbreeding Animals: February Farm Animals
Oxen: February Farm Animals (under cattle)

Paddocks and Foraging: February/November Farm Animals, Garden Tips
Parasites: April Farm Animals
Pasture or Forage: May/November Farm Animals, Garden Tips
Pasture, Rotational Grazing, Worms: April Farm Animals
Perennial/Annual/Biennial: How to Save Seeds
Perennial Herb Sowing: May Garden-Plant
Perennial Harvesting: April Garden-Harvest
pH and Soil: Garden Tips
Pig Care and Breeds: February Farm Animals
Plant Disease Symptoms: Ways to Improve Plant Health
Plant Bacterial Disease: Ways to Improve Plant Health
Plant Fungi Disease: Ways to Improve Plant Health
Plant Viral Disease: Ways to Improve Plant Health
Plant Nutrient Deficiency Symptoms: Ways to Improve Plant Health
Planting Trees (how to): April Garden-Plant
Poaceae or Gramineae Family (corn, sorghum): Plant Families
Poisoning in Animals: February Farm Animals
Pollination and Blooming- Fruit and Nut Trees: April Garden-Plant
Pollination- Open and Hybrid: How to Save Seeds
Pollination Methods (air/wind, insect, self, hand): How to Save Seeds
Pollinators- Bee (Mason): December Farm Animals
Pollinators- Beneficial Insects: Ways to Improve Plant Health under "Insects and Nematodes"
Polygonaceae Family (rhubarb, buckwheat): Plant Families
Portulacaceae Family (purslane): Plant Families
Potting Mixes or Soils: Garden Tips
Poultry Breeds and Care: February Farm Animals
Poultry- Broody Hens: April Farm Animals
Poultry Care and Breeds: February Farm Animals
Poultry Egg Production: April Farm Animals
Poultry Egg Storage: July Farm Animals
Poultry Garlic Water: January Farm Animals
Poultry Mites and Lice: March Farm Animals
Poultry Molting: September Farm Animals
Propagation by 4 Types of Stem Cuttings: Garden Tips
Propagation by Tip Layering / Rooting: Garden Tips
Pruning Brambles: October Garden-Maintenance
Pruning Fruit and Nut Trees: February Garden-Maintenance
Pruning- Heading Cuts and Thinning Cuts: February Garden-Maintenance
Pullets (Young Hens): April Farm Animals

Rabbit Care and Breeds: February Farm Animals
Rabbit Pests: Ways to Improve Plant Health

Raccoon Pests: Ways to Improve Plant Health
Rat Pests: Ways to Improve Plant Health
Rock Dusts (Stone Meals) and Fertilizers: Garden Tips
Rodent Pests: September Farm Animals, Ways to Improve Plant Health
Root Crops (Cool Season) Planted: March/July/August Garden-Plant
Root Crops (list): Preface
Rooting Cuttings: Garden Tips
Rooting Hormone- Make Your Own: Garden Tips
Rotational Grazing (Pasture): April Farm Animals
Row Covers and Other Physical Crop Protectors: Ways to Improve Plant Health

Schools that Teach Country Living: March Garden-Plant
Seed Fermentation: How to Save Seeds
Seed- Harvesting and Preparing: How to Save Seeds
Seed Moisture and Silica Gel: How to Save Seeds
Seed Saving and Food Storage Books: August Garden-Harvest
Seed Saving and How Difficult: How to Save Seeds
Seed Sources (Buying): March Garden-Plant
Seed Storing for Planting: How to Save Seeds
Seedling Soil Blocks: Garden Tips
Seedlings and Plants- Thinning: Garden Tips
Seedlings- Damping Off: Garden Tips
Seedlings- Hardening Before Planting: Garden Tips
Seeds- How Long Remain Viable: How to Save Seeds
Seeds- How to Stratify: Garden Tips
Seeds- Inoculating: Garden Tips
Sheep Care and Breeds: February/September Farm Animals
Signs for Planting: Garden Tips
Slug and Snail Traps: Ways to Improve Plant Health
Soil Blocks for Seedlings: Garden Tips
Soil pH: Garden Tips
Soil Sterilization: Garden Tips
Soil Types and Weeds (What They Tell): Garden Tips
Solanaceae or Nightshade Family (tomato, potato, eggplant, pepper+): Plant Families
Solar Food Dehydrator: May Garden-Harvest
Sprays and Dusts (Plants): Ways to Improve Plant Health
Sprays and Solutions (Fungicide): Ways to Improve Plant Health
Sprouts to Eat: December Greenhouse
Squirrel Pests: Ways to Improve Plant Health
Standard and Bantam Chickens: April Farm Animals
Storing Grain: May Garden-Harvest
Stratify Seeds: Garden Tips
Swine / Hog /Pig Care and Breeds: February Farm Animals

Tetragoniaceae Family (New Zealand spinach): Plant Families
Three Sisters- Corn, Beans, Winter Squash: May Garden-Plant
Thinning Fruit and Nut Trees: June Garden-Maintenance
Thinning Seedlings and Plants: Garden Tips
Thistle or Compositae Family (sunflower+): Plant Families
Threshing and Winnowing Grain: May Garden-Harvest
Tilling: Garden Tips (in Soil and Crops)
Tincture with Herbs: April Garden-Harvest
Tool Care: February Garden-Maintenance
Tools- Disinfecting: Ways to Improve Plant Health
Trap Boxes: September Farm Animals
Traps and Barriers for Crawling Insects: Ways to Improve Plant Health

Traps and Deterrents for Slugs and Snails: Ways to Improve Plant Health
Traps for Flying Insects: Ways to Improve Plant Health
Tree Insect Control: April Garden-Maintenance (sticky bands), May Garden-Maintenance (burlap bags and borers), June Garden-Maintenance (maggot traps, bagging fruit and fire blight). See Fruit and Nut Trees.
Tree Guards: September Garden-Maintenance
Trellis Systems: February Garden-Maintenance (under Grapes)
Turkey Care and Breeds: February Farm Animals
TV Shows (Country / Rural): March Garden-Plant

Umbelliferae or Apiaceae Family (parsley, carrot, fennel, parsnip): Plant Families

Vaccinations in Animals: February Farm Animals
Vegetables- Cut Off Tips, Tops, Flowers, Small Fruit: September Garden-Maintenance
Vegetables- Fall Sowing: September Garden-Plant
Vegetables- Hardy: February Garden-Maintenance
Vegetables- Heavy and Light Feeders: Garden Tips (in Soil and Crops)
Vegetables- Sow Cool Season: July/August Garden-Plant
Viral Disease (Plants): Ways to Improve Plant Health
Vole Boxes (Trapping): September Farm Animals
Vole Pests: Ways to Improve Plant Health

Warm Season Crops (list): Preface
Warm Season Sowing: July Garden-Plant
Warm Season (Weather) Cover Crops: May/June Garden-Plant, Garden Tips
Weeds and What They Tell about Soil Type: Garden Tips
Wheat Planting Dates: October Garden-Plant
Whitewashing Fruit and Nut Trees: June Garden-Maintenance
Winnowing and Threshing Grain: May Garden-Harvest
Worming: January/**April**/July Farm Animals

Zones (USDA Hardiness): January

Western North Carolina Farm & Garden Calendar

The Farming and Gardening Survival Book

Good for all eastern states in Hardiness Zones 5, 6, 7.